KT-134-959

Granite-Related Ore Deposits

Geological Society books refereeing procedures

The Society makes every effort to ensure that the scientific and production quality of its books matches that of its journals. Since 1997, all book proposals have been refereed by specialist reviewers as well as by the Society's Books Editorial Committee. If the referees identify weaknesses in the proposal, these must be addressed before the proposal is accepted.

Once the book is accepted, the Society Book Editors ensure that the volume editors follow strict guidelines on refereeing and quality control. We insist that individual papers can only be accepted after satisfactory review by two independent referees. The questions on the review forms are similar to those for *Journal of the Geological Society*. The referees' forms and comments must be available to the Society's Book Editors on request.

Although many of the books result from meetings, the editors are expected to commission papers that were not presented at the meeting to ensure that the book provides a balanced coverage of the subject. Being accepted for presentation at the meeting does not guarantee inclusion in the book.

More information about submitting a proposal and producing a book for the Society can be found on its web site: www.geolsoc.org.uk.

It is recommended that reference to all or part of this book should be made in one of the following ways:

Sial, A. N., Bettencourt, J. S., De Campos, C. P. & Ferreira, V. P. (eds) 2011. *Granite-Related Ore Deposits*. Geological Society, London, Special Publications, **350.**

Bejgarn, T., Årebäck, H., Weihed, P. & Nylander, J. 2011. Geology, petrology and alteration geochemistry of the Palaeoproterozoic intrusive hosted Älgträsk Au deposit, Northern Sweden. *In*: Sial, A. N., Bettencourt, J. S., De Campos, C. P. & Ferreira, V. P. (eds) *Granite-Related Ore Deposits*. Geological Society, London, Special Publications, **350,** 105–132.

GEOLOGICAL SOCIETY SPECIAL PUBLICATION NO. 350

Granite-Related Ore Deposits

EDITED BY

A. N. SIAL
Federal University of Pernambuco, Recife, Brazil

J. S. BETTENCOURT
Universidade de Sao Paulo, Brazil

C. P. DE CAMPOS
Ludwig-Maximilians-Universität, Munich, Germany

and

V. P. FERREIRA
Federal University of Pernambuco, Recife, Brazil

2011
Published by
The Geological Society
London

THE GEOLOGICAL SOCIETY

The Geological Society of London (GSL) was founded in 1807. It is the oldest national geological society in the world and the largest in Europe. It was incorporated under Royal Charter in 1825 and is Registered Charity 210161.

The Society is the UK national learned and professional society for geology with a worldwide Fellowship (FGS) of over 10 000. The Society has the power to confer Chartered status on suitably qualified Fellows, and about 2000 of the Fellowship carry the title (CGeol). Chartered Geologists may also obtain the equivalent European title, European Geologist (EurGeol). One fifth of the Society's fellowship resides outside the UK. To find out more about the Society, log on to www.geolsoc.org.uk.

The Geological Society Publishing House (Bath, UK) produces the Society's international journals and books, and acts as European distributor for selected publications of the American Association of Petroleum Geologists (AAPG), the Indonesian Petroleum Association (IPA), the Geological Society of America (GSA), the Society for Sedimentary Geology (SEPM) and the Geologists' Association (GA). Joint marketing agreements ensure that GSL Fellows may purchase these societies' publications at a discount. The Society's online bookshop (accessible from www.geolsoc.org.uk) offers secure book purchasing with your credit or debit card.

To find out about joining the Society and benefiting from substantial discounts on publications of GSL and other societies worldwide, consult www.geolsoc.org.uk, or contact the Fellowship Department at: The Geological Society, Burlington House, Piccadilly, London W1J 0BG: Tel. +44 (0)20 7434 9944; Fax +44 (0)20 7439 8975; E-mail: enquiries@geolsoc.org.uk.

For information about the Society's meetings, consult *Events* on www.geolsoc.org.uk. To find out more about the Society's Corporate Affiliates Scheme, write to enquiries@geolsoc.org.uk.

Published by The Geological Society from:
The Geological Society Publishing House, Unit 7, Brassmill Enterprise Centre, Brassmill Lane, Bath BA1 3JN, UK

(*Orders*: Tel. +44 (0)1225 445046, Fax +44 (0)1225 442836)
Online bookshop: www.geolsoc.org.uk/bookshop

British Library Cataloguing in Publication Data

A catalogue record for this book is available from the British Library.
ISBN 978-1-86239-321-9

Typeset by Techset Composition Ltd, Salisbury, UK
Printed by MPG Books Ltd, Bodmin, UK

Distributors

North America
For trade and institutional orders:
The Geological Society, c/o AIDC, 82 Winter Sport Lane, Williston, VT 05495, USA
Orders: Tel. +1 800-972-9892
Fax +1 802-864-7626
E-mail: gsl.orders@aidcvt.com

For individual and corporate orders:
AAPG Bookstore, PO Box 979, Tulsa, OK 74101-0979, USA
Orders: Tel. +1 918-584-2555
Fax +1 918-560-2652
E-mail: bookstore@aapg.org
Website: http://bookstore.aapg.org

India
Affiliated East-West Press Private Ltd, Marketing Division, G-1/16 Ansari Road, Darya Ganj, New Delhi 110 002, India
Orders: Tel. +91 11 2327-9113/2326-4180
Fax +91 11 2326-0538
E-mail: affiliat@vsnl.com

Contents

Granite pegmatite systems

Skarn systems

Intrusion-related gold systems

Epithermal gold systems

Iron-oxide–copper–molybdenum systems

Regional geology

Granite-related ore deposits: an introduction

A. N. SIAL[1]*, JORGE S. BETTENCOURT[2], CRISTINA P. DE CAMPOS[3] & VALDEREZ P. FERREIRA[1]

[1]*Federal University of Pernambuco, Dept. of Geology, NEG-LABISE C.P. 7852, Cidade Universitaria, 50670-000 Recife, PE, Brazil*

[2]*Institute of Geosciences, University of São Paulo, Rua do Lago 562, 055089-080, São Paulo, Brazil*

[3]*Department of Earth and Environmental Sciences, LMU Theresienstr. 41 III, Munich 80 333- Germany*

**Corresponding author (e-mail: sial@ufpa.br)*

Abstract: A symposium on Mineralization Associated with Granitic Magmatism was held within the framework of the 33rd IGC in Oslo, Norway, in August 2008. While our initial idea was to bring together field, experimental, and theoretical studies in order to review and summarize the current ideas and recent progress on granite-related mineralization systems, we were caught by surprise realizing that participants were inclined to focus more on ore deposits related to granitic magmatism. This spontaneous shift from granites, the major intended focus of the symposium, to mineralization associated with them, spawned the idea for a special issue on this theme and ultimately to the nine papers assembled here, chosen from about 60 scientific contributions at the symposium. Around twenty oral presentations were given and forty posters were presented at the meeting; the 60 papers were grouped according to the current main granite-related ore systems, as follows; granite-pegmatite, skarn and greisen-veins, porphyry, orogenic gold, intrusion-related, epithermal and porphyry-related gold and base metal, iron oxide–copper–gold (IOCG), and special case studies.

Importance of granite-related mineralization systems: diversity of mineralization styles and related mineral deposits

Granite-related mineral deposits are diverse and complex and include different associations of elements such as Sn, W, U, Th, Mo, Nb, Ta, Be, Sc, Li, Y, Zr, Sb, F, Bi, As, Hg, Fe, Cu, Au, Pb, Zn, Ag, Ga, and other metals. Among these, deposits of rare earth elements (REEs) and other precious and semi-precious metals are vital to current technologies upon which society depends. Granite-related ore systems have been one of the major targets of the mineral exploration industry and have probably received more intensive research study over the last decades than any other type of ore deposits.

Many different authors have attempted to summarize metallogenetic models for granite-related mineral deposits. However, due to the diversity of classes of ore-deposits, styles of mineralization and processes involved in their formation, major reviews focus only on individual classes highlighting the current status of investigation. Only very few papers focus on experiments and modelling processes leading to metal enrichment, although significant physical and chemical studies have been conducted by Candela (1997), Linnen (1998), Piccoli *et al.* (2000), Cline (2003), Ishihara & Chappell (2004), Vigneresse (2007), among others. The main topics of interest discussed by these authors are: magma sources, emplacement mechanisms, diversification processes, diffusion-controlled element distribution, partition coefficients between minerals and melt, solubility and redox conditions.

Modern approaches, new paradigms, mixing and unmixing of magmas and related ore generation

In the last thirty years, growing evidence for the coexistence of acidic and basic magmas has reinforced the importance of basic magmatism in the evolution of granites (e.g. Didier & Barbarin 1991; Bateman 1995). From the contributions of modern fluid dynamics, we know that the mixing process is the interplay between thermal and/or compositional convection and chemical diffusion (Ottino 1989; Fountain *et al.* 2000). This is known as a

From: Sial, A. N., Bettencourt, J. S., De Campos, C. P. & Ferreira, V. P. (eds) *Granite-Related Ore Deposits*. Geological Society, London, Special Publications, **350**, 1–5.
DOI: 10.1144/SP350.1 0305-8719/11/$15.00

largely non-linear process and dependent on the viscosity and density of the end members involved.

Persistent inputs of relatively dense and low viscosity mafic magma into a high viscosity felsic magma chamber enhances convection, diffusion and redistribution of different elements through the different melts, and therefore distribution of rare elements throughout the chamber (Reid *et al.* 1983; Wiebe & Collins 1998; Wiebe *et al.* 2002). This process is known to be non-linear, chaotic and fractal (e.g. Poli & Perugini 2002; Perugini *et al.* 2003; De Campos *et al.* 2008).

To date, only a handful of experimental studies on magma mixing have been targeted on investigations with natural melts or magmas (Kouchi & Sunagawa 1984; Bindeman & Davis 1999; De Campos *et al.* 2004, 2008, 2010). This is partly due to the high temperatures and high viscosities involved. From these experimental results, we know that mixing between basalt and granitic melts may enhance diffusive fractionation of metals and trace elements, such as the rare earth elements, this being a potential additional mechanism for ore concentration (De Campos *et al.* 2008; Perugini *et al.* 2008).

As a counterpart to the mixing process, magmatic differentiation may also lead to liquid immiscibility. This has received limited attention as a major process leading to the formation of large plutons. Its importance, however, has been claimed by Ferreira *et al.* (1994) and Rajesh (2003) to explain the generation of coexisting ultrapotassic syenite and pyroxenite at the Triunfo batholith, Brazil, and an alkali syenite-pyroxenite association near Puttetti, Trivandrum block, South India. In the roof zone of granitic plutons, liquid immiscibility between aluminosilicate and hydrous melts controls the partitioning of B, Na and Fe to the hydrous melts. Veksler & Thomas (2002) and Veksler *et al.* (2002) experimentally confirmed the immiscibility of alumino-silicate and water-rich melts with extreme boron enrichment (5 wt%; Thomas *et al.* 2003). Veksler (2004) noted that more water-rich depolymerized melts in immiscible systems are strongly enriched in B, Na, Fe. Therefore, liquid immiscibility may concentrate the necessary elements for nodule formation in water-rich, highly mobile melt phases, which may percolate through crystal mush and coalesce in discrete bodies (Trumbull *et al.* 2008; Balen & Broska 2011; Ishiyama *et al.* 2011).

Regarding recent models for granite generation, it is important to analyse the new paradigm of discontinuous magma input in the evolution of felsic magmas and the related consequences to ore formation, as proposed by Vigneresse (2004, 2007). This model represents a substantial change in the concept of ore generation in which magma source, emplacement mechanisms and magma mixing processes, together with diffusion/partition coefficients between minerals and melt, solubility and redox conditions, are the main control parameters for element distribution and, therefore, enrichment processes.

New ore genetic models and related exploration models

Despite significant advances, due to new ideas and technologies, in the fields of igneous petrology (e.g. repeated magma intrusion, fluctuating redox through magma-crustal interaction), volcanology, geochemistry (e.g. formation of immiscible sulphide phases, salt melts and vapour-like fluid phases), geophysics, high P–T experiments, and numerical modelling, our present understanding of granite-related mineralization systems and related ore-bearing processes leading to metal concentrations is not yet sufficiently advanced. It is still poorly understood which parameters account for potential prospective targets for a given metallic resource. Despite much fundamental knowledge and new concepts in granite-related ore deposit geology we definitely need better genetic and exploration models. Critically, we need better understanding of the key features informing exploration targeting and discovery. In fact, future global needs for metal resources will require a subsequent surge in mineral exploration programmes, which inevitably rely upon reliable ore deposit models. Such improvement in models of ore formation is only possible through continuing multidisciplinary investigations.

This issue provides a range of studies that are broadly distributed in both space and time, highlighting granite-related ore deposits from Europe (Russia, Sweden, Croatia and Turkey), the Middle East (Iran), Asia (Japan and China) and South America (Brazil and Argentina) spanning from Palaeoproterozoic to Miocene. The nine papers selected for publication in this title fall under the following general themes:

Granite-pegmatite systems

The correlation between grain-size in orogenic granite/pegmatite magma and crystallization age is a topic that has not yet occurred to many petrologists. **Tkachev** (2011) discusses the evolution of orogenic granite-pegmatites through geological time. He focuses on pegmatite bodies both from Russia and other parts of the world. Based on data from the literature, this work quantitatively analyses distinct pegmatite generation intensity and/or evolutionary changes through geological time, bringing a new approach to the driving forces which have not previously been properly addressed.

Pedrosa-Soares *et al.* (2011) examine mineral resources related to the Araçuaí orogen in eastern Brazil. The most remarkable feature of this crustal segment is the huge amount of plutonic rocks of Late Neoproterozoic up to Cambro-Ordovician ages, depicting a long lasting succession of granite production episodes in an area of over 350 000 km^2. Granitic rocks cover one third of the orogenic region, and built up the outstanding eastern Brazilian pegmatite province and the most important dimension stone province of Brazil. This is an example of how granites themselves can represent an economic target of a region.

The role of devolatilization in final stages of granitic melt leading to the formation of tourmaline nodules in Cretaceous peraluminous plutons in Croatia is discussed by **Balen & Broska** (2011).

Skarn systems

Wang *et al.* (2011) analysanalyse the distribution and migration characteristics of Au, Ag, Cu, Pb, and Zn during the ore-forming processes in a skarn deposit near Tongling in the Shizishan area, Anhui Province, China. In this area, ore fields are composed of skarn-type deposits formed around several magmatic plutons, emplaced at about 140 Ma. Self-affine and multi-fractal analyses were used to study the migration and to model changes in the distribution patterns of those ore-forming elements, during the skarn mineralization process.

Cathodoluminescence and fluid inclusions shed some light on the study of mechanisms and timing of generation of skarn mineralizations at contact aureoles in granitic plutons in central Japan (Fe–Cu–Pb and Zn) as demonstrated by **Ishiyama *et al.*** (2011) while **Wang *et al.*** (2011) applied fractal analysis to constrain ore-forming processes in skarns from China.

Iron oxide–copper–molybdenum systems

Delibas *et al.* (2011) focused on Fe–Cu–Mo mineralization in the central Anatolian magmatic complex, in Turkey. In this association volcanic rocks grade from basalt to rhyolite, whilst coeval plutonic rocks range from gabbro to leucogranite. Results of this work highlight the importance of magma mixing and metal unmixing, possibly related to stress relaxation during post-collisional evolution in late Cretaceous times.

Epithermal gold systems

Ebrahimi *et al.* (2011) describe Cenozoic epithermal gold prospects in silica, silica-carbonate and veinlets in felsic to intermediate volcanic and plutonic rocks in Iran. The coexistence of vapour-dominant and liquid-dominant inclusions in the ore stage quartz, hydrothermal breccias, bladed calcite, and adularia suggests that boiling occurred during the evolution of the ore fluids. Mixing and boiling are two principal processes involved in the ore formation in this a low-sulphidation epithermal system.

Intrusion-related gold systems

The petrology and alteration geochemistry of Palaeoproterozoic intrusions that host Au deposits in Sweden is the main theme of the contribution by **Bejgarn *et al.*** (2011) who studied a structurally-controlled mineralization that occurs within zones of proximal phyllic/silicic and distal propylitic alteration. It comprises mainly pyrite, chalcopyrite, sphalerite with accessory Te-minerals, gold alloys, and locally abundant arsenopyrite. During hydrothermal alteration an addition of Si, Fe and K together with an increase in Au, Te, Cu, Zn and As occurred.

Regional geology

Rossi *et al.* (2011) examine the metalliferous fertility of the undeformed Carboniferous San Blas granitic pluton in western Argentina that was emplaced at shallow levels by passive mechanisms. The finding of alluvial cassiterite and wolframite in drainage from this pluton is evidence of the fertile character of this granite. The Sr/Eu ratio and other geochemical features characterize this pluton as fertile, evolved granite with the REE tetrad effect, typical of evolved granites with hydrothermal alteration (greisenization).

The guest editors are thankful to the Geological Society of London for the invitation to organize this Special Publication and, in particular, A. Hills for patience and guidance. A.N. Sial, V.P. Ferreira and J.S. Bettencourt wish to acknowledge the financial support from the Brazilian National Council for Scientific Development (CNPq) and from the 38th IGC Organizers (Geohost Programme) that have helped them to participate in that Meeting in August 2008. The guest editors also would like to express their gratitude to the reviewers who gave their time and effort toward this volume: R. F. Martin (Canada), M. K. Pandit (India), M. Taner (Canada), I. Haapala (Finland), R. S. Xavier (Brazil), A. M. Neiva (Portugal), L. Monteiro (Brazil), D. Atencio (Brazil), R. A. Fuck (Brazil), J. K. Yamamoto (Brazil), R. Hochleitner (Germany), A. Dini (Italy), A. Muller (Germany), C. P. De Campos (Germany) and K. Sekine (Japan).

References

Balen, D. & Broska, I. 2011. Tourmaline nodules: products of devolatilization within the final evolutionary stage of granitic melt? *In*: Sial, A. N., Bettencourt, J. S., De Campos, C. P. & Ferreira, V. P. (eds)

Granite-Related Ore Deposits. Geological Society, London, Special Publications, **350**, 53–68.

BATEMAN, R. 1995. The interplay between crystallization, replenishment and hybridization in large felsic magma chambers. *Earth Science Reviews*, **39**, 91–106.

BEJGARN, T., ÅREBÄCK, H., WEIHED, P. & NYLANDER, J. 2011. Geology, petrology and alteration geochemistry of the Palaeoproterozoic intrusive hosted Älgträsk Au deposit, Northern Sweden. *In*: SIAL, A. N., BETTENCOURT, J. S., DE CAMPOS, C. P. & FERREIRA, V. P. (eds) *Granite-Related Ore Deposits*. Geological Society, London, Special Publications, **350**, 105–132.

BINDEMAN, I. N. & DAVIS, A. M. 1999. Convection and redistribution of alkalis and trace elements during the mingling of basaltic and rhyolitic melts. *Petrology*, **7**, 91–101.

CANDELA, P. 1997. A review of shallow, ore-related granites: textures, volatiles and ore metals. *Journal of Petrology*, **38**, 1619–1633.

CLINE, J. S. 2003. How to concentrate copper? *Science*, **302**, 2075–2076.

DE CAMPOS, C. P., DINGWELL, D. B. & FEHR, K. T. 2004. Decoupled convection cells from mixing experiments with alkaline melts from Phlegrean Fields. *Chemical Geology*, **213**, 227–251.

DE CAMPOS, C. P., PERUGINI, D., DINGWELL, D. B., CIVETTA, L. & FEHR, T. K. 2008. Heterogeneities in Magma Chambers: insights from the behavior of major and minor elements during mixing experiments with natural alkaline melts. *Chemical Geology*, **256**, 131–145.

DE CAMPOS, C. P., ERTEL-INGRISCH, W., PERUGINI, D., DINGWELL, D. B. & POLI, G. 2010. Chaotic mixing in the system Earth: mixing granitic and basaltic liquids. *In*: SKIADAS, C. H. & DIMOTILAKIS, I. (eds) *Chaotic Systems Theory and Applications. Selected Papers from the 2nd Chaotic Modeling and Simulation International Conference (CHAOS2009)*. World Scientific Publishers Co., Singapore. 51–58. doi: 10.1142/9789814299725_0007.

DELIBAŞ, O., DE CAMPOS, C. P. & GENÇ, Y. 2011. Magma mixing and unmixing related mineralization in the Karacaali Magmatic Complex, central Anatolia, Turkey. *In*: SIAL, A. N., BETTENCOURT, J. S., DE CAMPOS, C. P. & FERREIRA, V. P. (eds) *Granite-Related Ore Deposits*. Geological Society, London, Special Publications, **350**, 149–174.

DIDIER, J. & BARBARIN, B. 1991. *Enclaves and Granite Petrology. Developments in Petrology*. Elsevier, Amsterdam.

EBRAHIMI, S., ALIREZAEI, S. & PAN, Y. 2011. Geological setting, alteration, and fluid inclusion characteristics of Zaglic and Safikhanloo epithermal gold prospects, NW Iran. *In*: SIAL, A. N., BETTENCOURT, J. S., DE CAMPOS, C. P. & FERREIRA, V. P. (eds) *Granite-Related Ore Deposits*. Geological Society, London, Special Publications, **350**, 133–148.

FERREIRA, V. P., SIAL, A. N. & WHITNEY, J. A. 1994. Large-scale silicate liquid immiscibility: a possible example from northeastern Brazil. *Lithos*, **33**, 285–302.

FOUNTAIN, G., KHAKHAR, D., MEZIC, V. & OTTINO, J. M. 2000. Chaotic mixing in a bounded three-dimensional flow. *Journal of Fluid Mechanics*, **417**, 265–301.

ISHIHARA, S. & CHAPPELL, B. W. 2004. A special issue of granites and metallogeny: the Ishihara volume. *Resource Geology*, **54**, 213–382.

ISHIYAMA, D., MIYATA, M. *ET AL*. 2011. Geochemical characteristics of Mioce ne Fe–Cu–Pb–Zn granitoids associated mineralization in the Chichibu skarn deposit (central Japan): evidence for magmatic fluids generation coexisting with granitic melt. *In*: SIAL, A. N., BETTENCOURT, J. S., DE CAMPOS, C. P. & FERREIRA, V. P. (eds) *Granite-Related Ore Deposits*. Geological Society, London, Special Publications, **350**, 69–88.

KOUCHI, A. & SUNAGAWA, I. 1984. A model for mixing basaltic and dacitic magmas as deduced from experimental data. *Contributions to Mineralogy and Petrology*, **89**, 17–23.

LINNEN, R. L. 1998. Depth of emplacement, fluid provenance and metallogeny in granitic terranes: a comparison of western Thailand with other Sn-W belts. *Mineralium Deposita*, **33**, 461–476.

OTTINO, J. M. 1989. *The Kinematics of Mixing: Stretching, Chaos and Transport*. Cambridge University Press, Cambridge.

PEDROSA-SOARES, C., DE CAMPOS, C. P. *ET AL*. 2011. Late Neoproterozoic–Cambrian granitic magmatism in the Araçuaí orogen (Brazil), the Eastern Brazilian Pegmatite Province and related mineral resources. *In*: SIAL, A. N., BETTENCOURT, J. S., DE CAMPOS, C. P. & FERREIRA, V. P. (eds) *Granite-Related Ore Deposits*. Geological Society, London, Special Publications, **350**, 25–51.

PERUGINI, D., POLI, G. & MAZZUOLI, R. 2003. Chaotic advection, fractals and diffusion during mixing of magmas: evidence from Lava flows. *Journal of Volcanology Geothermal Research*, **124**, 255–279.

PERUGINI, D., DE CAMPOS, C. P., DINGWELL, D. B., PETRELLI, M. & POLI, G. 2008. Traceelement mobility during magma mixing: preliminary experimental results. *Chemical Geology*, **256**, 146–157.

PICCOLI, P. M., CANDELA, P. A. & RIVERS, M. 2000. Interpreting magmatic processes from accessory phases: titanite, a small-scale recorder of large-scale processes. *Transactions of the Royal Society of Edinburgh, Earth Sciences*, **91**, 257–267.

POLI, G. & PERUGINI, D. 2002. Strange attractors in magmas: evidence from lava flows. *Lithos*, **65**, 287–297.

RAJESH, H. M. 2003. Outcrop-scale silicate liquid immiscibility from an alkali syenite (A-type granitoid)-pyroxenite association near Puttetti, Trivandrum Block, South India. *Contributions to Mineralogy and Petrology*, **145**, 612–627.

REID, J. B., JR., EVANS, O. C. & FATES, D. G. 1983. Magma mixing in granitic rocks of the central Sierra Nevada, California. *Earth and Planetary Science Letters*, **66**, 243–261.

ROSSI, J. N., TOSELLI, A. J., BASEI, M. A., SIAL, A. N. & BAEZ, M. 2011. Geochemical indicators of metalliferous fertility in the Carboniferous San Blas pluton, Sierra de Velasco, Argentina. *In*: SIAL, A. N., BETTENCOURT, J. S., DE CAMPOS, C. P. & FERREIRA, V. P. (eds)

Granite-Related Ore Deposits. Geological Society, London, Special Publications, **350**, 175–186.
THOMAS, R., FOERSTER, H. J. & HEINRICH, W. 2003. The behaviour of boron in a peraluminous granite–pegmatite system and associated hydrothermal solutions: a melt and fluid inclusion study. *Contributions to Mineralogy and Petrology*, **144**, 457–472.
TKACHEV, A. V. 2011. Evolution of metallogeny of granitic pegmatites associated with orogens throughout geological time. *In*: SIAL, A. N., BETTENCOURT, J. S., DE CAMPOS, C. P. & FERREIRA, V. P. (eds) *Granite-Related Ore Deposits*. Geological Society, London, Special Publications, **350**, 7–24.
TRUMBULL, R. B., KRIENITZ, M. S., GOTTESMANN, B. & WIEDENBECK, M. 2008. Chemical and boron-isotope variations in tourmalines from an S-type granite and its source rocks: the Erongo granite and tourmalinites in the Damara Belt, Namibia. *Contributions to Mineralogy and Petrology*, **155**, 1–18.
VEKSLER, I. V. 2004. Liquid immiscibility and its role at the magmatic hydrothermal transition: a summary of experimental studies. *Chemical Geology*, **210**, 7–31.
VEKSLER, I. V. & THOMAS, R. 2002. An experimental study of B-, P- and Frich synthetic granite pegmatite at 0.1 and 0.2 GPa. *Contributions to Mineralogy and Petrology*, **143**, 673–683.
VEKSLER, I. V., THOMAS, R. & SCHMIDT, C. 2002. Experimental evidence of three coexisting immiscible fluids in synthetic granite pegmatite. *American Mineralogist*, **87**, 775–779.
VIGNERESSE, J. L. 2004. Toward a new paradigm for granite generation. *Transactions of the Royal Society of Edinburgh, Earth Sciences*, **95**, 11–22.
VIGNERESSE, J. L. 2007. The role of discontinuous magma inputs in felsic magma and ore generation. *Ore Geology Reviews*, **30**, 181–216.
WANG, Q., DENG, J., LIU, H., WAN, L. & ZHANG, Z. 2011. Fractal analysis of the ore-forming process in a skarn deposit: a case study in the Shizishan area, China. *In*: SIAL, A. N., BETTENCOURT, J. S., DE CAMPOS, C. P. & FERREIRA, V. P. (eds) *Granite-Related Ore Deposits*. Geological Society, London, Special Publications, **350**, 89–104.
WIEBE, R. A. & COLLINS, W. I. 1998. Depositional features and stratigraphic sections in granitic plutons: implications for the emplacement and crystallization of granitic magma chambers. *Journal of Structural Geology*, **20**, 1273–1289.
WIEBE, R. A., BLAIR, K. D., HAWKINS, D. P. & SABINE, C. P. 2002. Mafic injections, in situ hybridization, and crystal accumulation in the Pyramid Peak granite, California. *GSA Bulletin*, **114**, 909–920.

Evolution of metallogeny of granitic pegmatites associated with orogens throughout geological time

A. V. TKACHEV

Vernadsky State Geological Museum, Russian Academy of Sciences, Moscow, Russia (e-mail: tkachev@sgm.ru)

Abstract: Since *c.* 3.1 Ga, pegmatite mineral deposits in orogenic areas have been formed throughout geological time in pulses, alternating with total absence of generating activity. The higher activity peaks of 2.65–2.60, 1.90–1.85, 1.00–0.95, and 0.30–0.25 Ga suggest a quasi-regular periodicity of 0.8 ± 0.1 Ga. This series is dominated by pegmatites of Laurasian blocks. The lower peaks at 2.85–2.80, 2.10–2.05, 1.20–1.15, and the higher one at 0.55–0.50 make up another series represented by pegmatites in Gondwanan blocks only. Each pegmatite class is characterized by a life cycle of its own, from inception to peak through to decline and eventual extinction. The longest cycle is recorded for the rare-metal class deposits, which first appeared in the Mesoarchaean and persisted through all the later eras, deteriorating gradually after the Early Precambrian. Muscovite pegmatites first appeared in the Palaeoproterozoic and reached the end of their life cycle at the Palaeozoic–Mesozoic boundary. The miarolitic class of pegmatite deposits in orogenic setting first came into being in the terminal Mesoproterozoic and dominated the pegmatite metallogeny of many Phanerozoic belts. The evolution of the pegmatite classes was controlled by the general cooling of the Earth and by associated changes in the tectonics of the lithosphere.

Supplementary material: Geochronological data used is available at http://www.geolsoc.org.uk/SUP18435.

Fersman's fundamental works (1931, 1940) and Landes' extensive paper (1935) were the first to present global-scale reviews of the distribution of all types of granitic pegmatites on continents and, most interestingly (for the purposes of this paper), throughout geological time. Schneiderhöhn's book (1961) added little new to geochronological aspects of the topic. The paper by Ovchinnikov *et al.* (1975) was the first to demonstrate entirely new approaches to geochronological synthesis of this sort. Ovchinnikov *et al.* (1975)'s study assessed the intensity of pegmatite generation through the Earth's history not merely in approximate terms (such as 'many', 'few', 'very extensive'), but also presenting rigorous quantifications based on the geochronological data amassed by that time. Shortly after, Ginzburg *et al.* (1979) published one of the most important works analysing global pegmatite metallogeny. In addition to many other aspects, the authors carried out a seamless integration of geochronology and geology for pegmatite provinces worldwide, revealing evolutionary trends of the most important pegmatite classes ('pegmatite formations,' to use the authors' terms). Although the majority of the empirically established evolutionary trends remained unexplained, this was great progress in the field of pegmatite metallogeny. However, over the years, as new information was accumulated, it became clear that not all data used in the book were correct, in terms of present-day geochronological standards.

Since then, there have been no publications of detailed work in the field of the global evolution of granitic pegmatite metallogeny. Some studies addressed global aspects of selected pegmatite classes (Solodov 1985; Makrygina *et al.* 1990; Černý 1991*b*; Zagorsky *et al.* 1997, 1999; Shmakin *et al.* 2007; London 2008). These works analyse all issues pertaining to the intensity of granitic pegmatite generation in the geological record without offering numerical calculations. No previous attempt has been made to unravel the driving forces of the established evolutionary trends. The main purpose of this study is to bridge this gap and propose a new evolutional paradigm.

Data sources

At present, the magmatic origin of granitic pegmatites is a matter of near total consensus. Granitic pegmatites are formed mainly in orogens as a result of crystallization of melts that are produced and variously evolved in thickened continental crust as a result of powerful heat generation (due to mechanical and radiogenic decay processes) and also to slow heat dissipation. In each particular orogen, synkinematic pegmatites are the earliest

From: Sial, A. N., Bettencourt, J. S., De Campos, C. P. & Ferreira, V. P. (eds) *Granite-Related Ore Deposits*. Geological Society, London, Special Publications, **350**, 7–23.
DOI: 10.1144/SP350.2 0305-8719/11/$15.00

type. They are common constituents of migmatitic fields. These pegmatites crystallize from poorly evolved melts. This is why they do not contain any specific minerals that might be used to distinguish them from 'normal' granites. The amount and especially the quality of economic minerals (such as K-feldspar, quartz, and muscovite) in these pegmatites, are not economically attractive. All of these pegmatites should be attributed to a separate class (non-specialized or non-mineralized pegmatites) and present no interest for pegmatite metallogeny studies. They are therefore not discussed in the analysis of pegmatite evolution below.

Crustal granitic melts keep on generating during the post-culmination (extension- or relaxation related) phase of orogen evolution, lasting up to 60 million years (Thompson 1999). These granitoids, not in all cases but quite commonly, are pegmatite-bearing. Some of the pegmatite fields are not accompanied by any reliably identified fertile granitic intrusions. Such relations are most common for deposits of the muscovite and abyssal feldspar–rare-element pegmatite classes (Ginzburg *et al.* 1979; Černý 1991*a*). Miarolitic pegmatites always show clear connection with their related granitoid massifs.

Many of these pegmatites related to the post-culmination orogenic phase are being mined or are of potential interest in the extraction of numerous rare elements, industrial minerals, gems, and specimens for collections. This genetic type of mineral deposits is of particular economic importance as a source of Ta, Li, Rb, Cs, various ceramic and optical raw materials, sheet muscovite (the only natural source), and crushed muscovite. In this paper, all pegmatites that show even the slightest potential for the extraction of these commodities are referred to as 'mineralized pegmatites.'

Mineralization features exhibited in a pegmatite field depend on a number of factors. Amongst others, the crucial factors are fertile magma sources, P–T conditions and duration of melt evolution and crystallization, as well as host rock composition (Ginzburg *et al.* 1979; Kratz 1984; Černý 1991*a*, *b*, *c*). It is the mineralized pegmatites in late-orogenic to post-orogenic settings that are the focus of this study; these are here jointly referred to as 'orogenic pegmatites.'

Pegmatites located in intraplate anorogenic granites (rapakivi, alkaline granites, syenites) may be of economic interest as sources of rare elements, feldspar raw materials, gems, and minerals for collections. However, the number of these deposits are small when compared to orogenic deposits; it was impossible to collect enough representative geochronological data to establish their generation scenario through geological time.

As with the crystallization of any granite, in general the formation of a pegmatite is quite a high-temperature process. Hence, the most reliable results are obtained from the study of U– (Th) –Pb isotope system on zircon, monazite-xenotime, tantaloniobates, and cassiterite, because of the high closure-temperatures and low susceptibility to external thermal and chemical influences (Faure & Mensing 2005). These features of the system are especially important, because pegmatites of some deeply generated fields may have remained in high-temperature conditions for time periods as long as tens of millions of years. Experience shows that Re-Os molybdenite dating results are reliable enough for the purposes of this study.

The K–Ar, Ar–Ar, Rb–Sr or Sm–Nd isotope systems are less resistant to external influences and have lower closure-temperatures. Hence, the results obtained by these methods do not always correctly reflect the time of pegmatite formation or crystallization of other magmas. This inconsistency was statistically confirmed by, for example, Balashov & Glaznev (2006). Nonetheless, some dating results obtained by these methods were used. However, this was only done in the absence of conflicting information and with at least partial support from independent geological and geochronological studies within the same pegmatite field. In most cases, these data are related to Phanerozoic pegmatites.

Unfortunately, there is no representative body of sufficiently accurate dating obtained by modern methods with a special focus on pegmatites. For this reason, in order to create a larger statistical sample, this study relies on the close genetic and temporal relationships between pegmatite fields and granitoid complexes, which differ from area to area. Where possible, age data was collected for those granites that occur within pegmatite fields and for those that are considered to be sources of the pegmatites that are under consideration. For a few provinces, geochronological data alone has been found for granites located outside pegmatite fields. These dates were used in case the researchers of the region were definite in the comagmatic origin of the granites with fertile granites from pegmatite fields. Only zircon and monazite age data were accepted in these cases.

Even the above 'wide span' approach to pegmatite geochronology did not enable the author to collect data precise enough to allow all the pegmatite fields but even some well-known pegmatite belts to be placed within any age range with an accuracy of 25 million years. In some cases, dating measurements from these areas have been made by obsolete methods, and in others, no reliable links between dated granitic intrusions and undated pegmatites from the same area have been revealed. This is the case, in particular, with the Ukrainian Shield, most fields in the Palaeozoides of Central Asia and China, and the Mesozoides of Indo-China and adjacent areas.

In each particular pegmatitic province, only some veins and fertile granitic massifs have been dated, and the number of pegmatitic fields (and hence, the intensity of pegmatite-generating processes) differs essentially from province to province. For this reason, within each particular region (entire tectonic province or part thereof) the known dates have been extrapolated to all pegmatite fields that, according to alternative geological information, may be related to the same stage of generation.

All geochronological data and their sources, the main references to geological information on pegmatite deposits, and extrapolation results are presented in the supplementary material. Depending on the purpose of use, for this data it is distributed along the geochronological scale with steps ranging from 25 to 100 million years.

Data verification: comparison with an independent database on crustal magmatism

The only way to verify the reliability and representativity of the collected database on pegmatite geochronology for the purpose of global interpretation, is to compare it to an independent dataset that to some extent covers the subject of study. For example, this could be a database on the crustal magmatism on the Earth. Conveniently, a recently published analysis of a database on the terrestrial magmatism (Balashov & Glaznev 2006) contains processing results of crustal magmatic dates based on 9808 measurements, mostly U–(Th) –Pb ones. Figure 1 shows comparison between these results and processed dates from the supplementary material. It demonstrates a coincidence for all the main and most minor maxima and minima in both of the independently graphed diagrams. This result shows that the data collected and extrapolated are reliable and representative and can be used in global analysis and synthesis.

Comparison to previous studies

Apart from the above-mentioned publications (Ovchinnikov *et al.* 1975; Ginzburg *et al.* 1979), as far as the author is aware, there is no known published literature worldwide to propose a quantitative analysis of granitic pegmatite, in terms of the intensity of their development or evolutionary changes.

Integrated results in Ovchinnikov *et al.* (1975) are based on a synthesis of 809 dates, including 340 ones from pegmatites of the USSR and the rest from elsewhere. Approximately half of the dates were acquired by the K–Ar method, about a third by the U– (Th) –Pb method (mostly on uraninite), and the rest by the Rb–Sr. Neither this paper nor the extended variant (Ovchinnikov *et al.* 1976) contain a table of measured ages. Only some of the data used in this study were obtained by the authors in their own laboratories. Some of the ages may have been obtained from non-mineralized pegmatites. At present it is impossible to clarify this issue because no list of references for these dates has been published. Unfortunately, the reference list for the dates used in the diagrams illustrating the generation intensity and evolution of granitic pegmatite classes (Ginzburg *et al.* 1979) is incomplete. At the same time, these materials show clearly that the ages used in this work were acquired only from the mineralized pegmatites with a very small number of pegmatites in anorogenic environs.

In both of these works, data was generalized with intervals as large as 100–200 Ma or even more (for some of the Early Precambrian periods). For a better comparison, our data in this case was also generalized with a step of 100 Ma. Then, all three diagrams were plotted in a single chart for comparison (Fig. 2).

Our comparison shows the presence of both similarities and significant differences between geochronological reconstructions. One principal similarity is the distinct pulsation in the intensity of pegmatite formation displayed in all cases through geological time. Another one is the non-ideal, albeit quite apparent, coincidence of peaks in the Neoarchaean ('Kenoran') and Palaeoproterozoic ('Svecofennian'), at the Mesoproterozoic–Neoproterozoic boundary ('Grenvillian'), at the end of the Palaeozoic ('Hercynian'), and in the latest Mesozoic–Cenozoic ('Kimmerian–Alpine'). However, there are quite a number of discrepancies at some principal points, the most important of these are described below.

Geological time 'infilling' in Figure 2a is more complete than in the parts b and c of the chart. The diagram in Figure 2b is the most 'rarefied' among all the diagrams; this may result from the fact that Ginzburg *et al.* (1979) used only the data from significant (in the authors' judgment) pegmatite provinces worldwide.

According to our data, the oldest mineralized (rare-metal) pegmatites originated in the Mesoarchaean *c.* 3.1 Ga ago (Fig. 2c). Judging by the results in Ovchinnikov *et al.* (1975), this event took place *c.* 0.3 Ga earlier (Fig. 2a), whereas, according to Ginzburg *et al.* (1979), it was 0.3 Ga younger

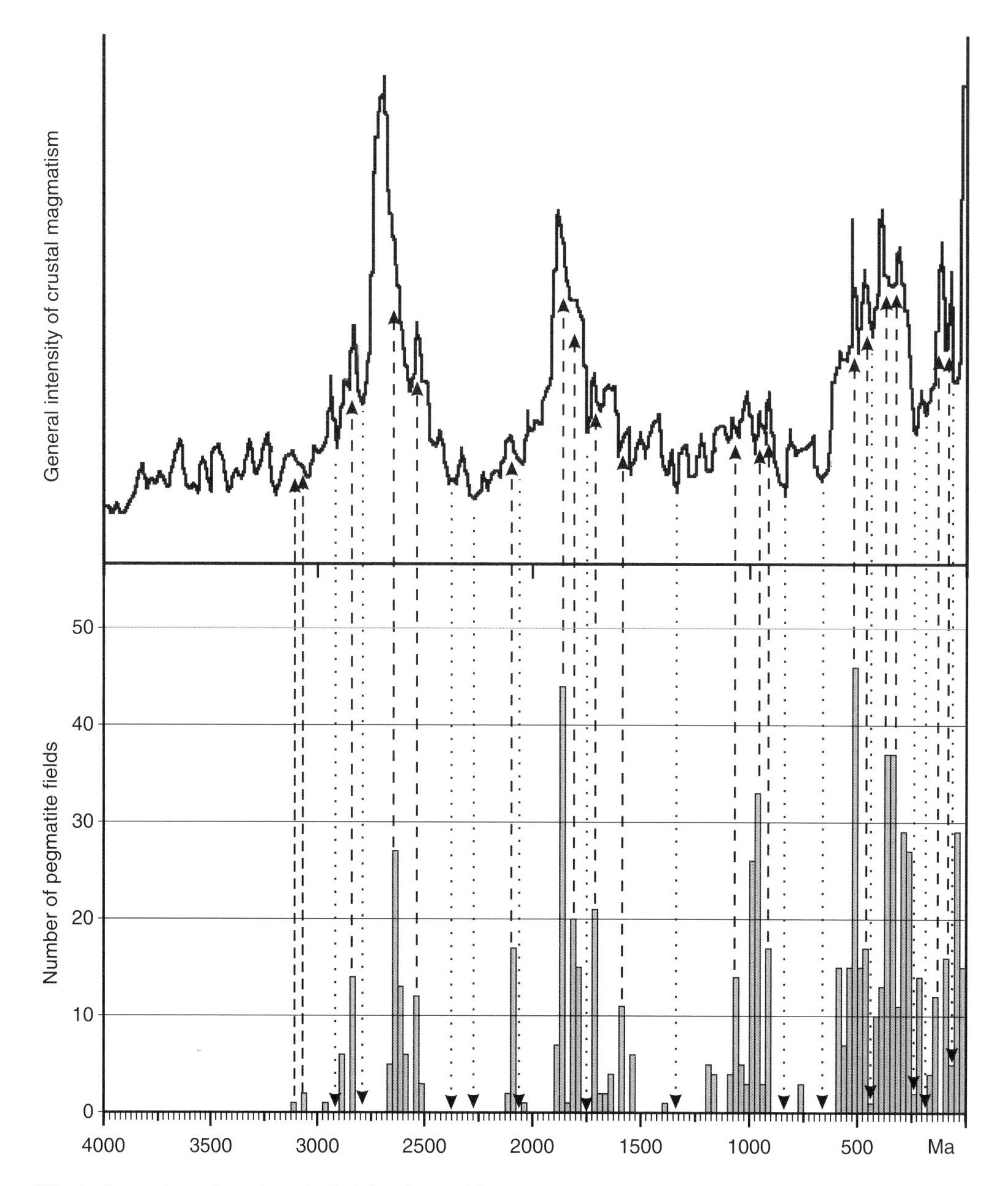

Fig. 1. Comparison of geochronological data for granitic pegmatite fields collected in the course of this study (see supplementary material) and for the continental crust magmatism (modified from Balashov & Glaznev 2006). Correlation lines: dashed for maxima and dotted for minima.

(Fig. 2b). In Černý's well-known review (1991*b*) the first rare-metal pegmatite generation is attributed to the initial phase of the Kenoran orogeny, that is, *c.* 2.75 Ga.

Figure 2a has a notable peak at *c.* 2.3 Ga, which is missing in parts b and c of the figure. The «Pan-Brazilian» pulse (0.5–0.6 Ga) in Figure 2c is stronger than in the other counterparts.

The data in Figure 2c clearly fall into clusters. These clusters have certain signatures of quasi-regular periodicity. This feature is detailed in a special section below. Here it should be pointed out that neither Ovchinnikov *et al.* (1975) nor Ginzburg *et al.* (1979) discuss the issue of cyclicity or periodicity. This might be due to the fact that Figure 2(a, b) gives very little basis for such discussion.

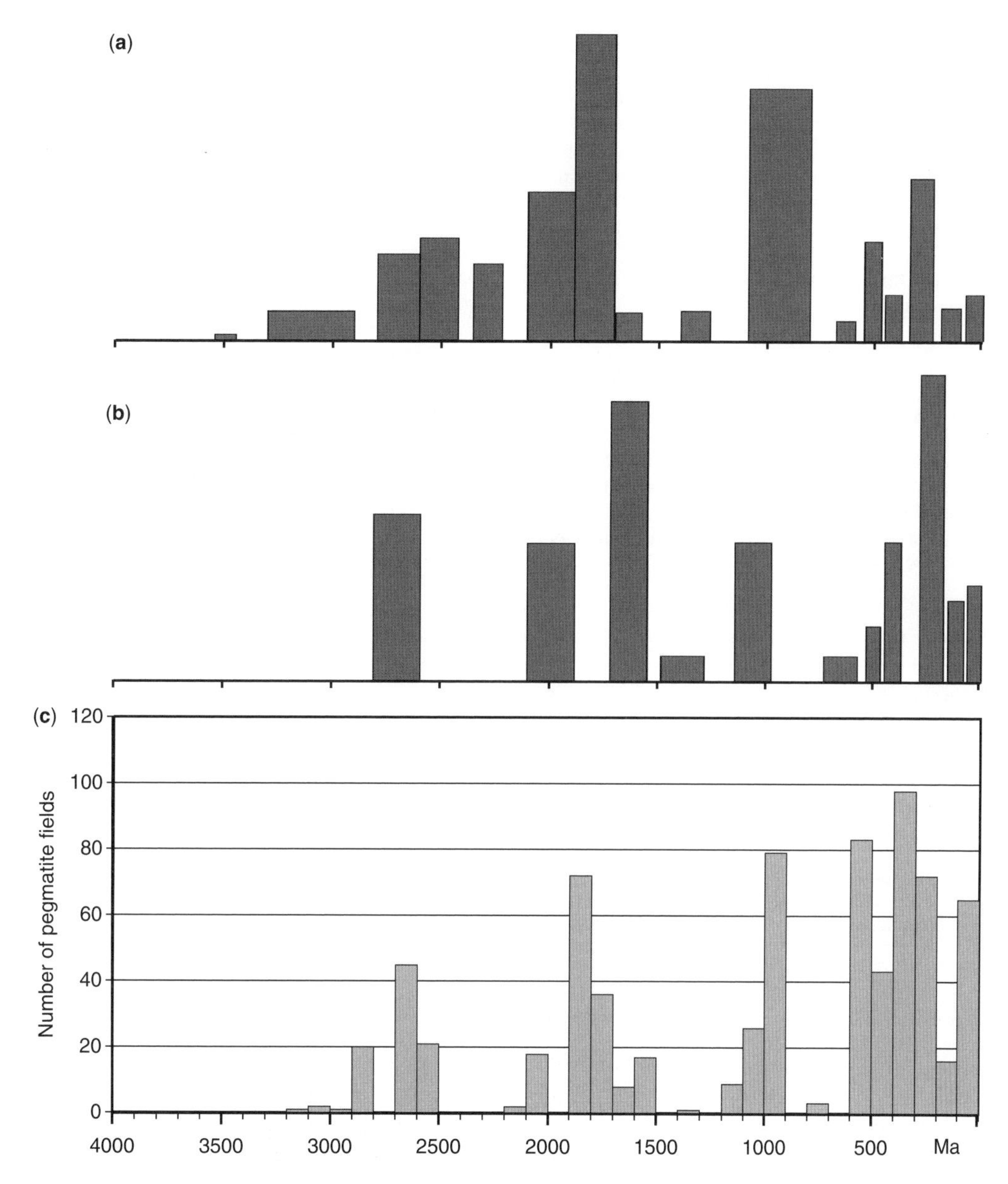

Fig. 2. Comparisons of the geochronological reconstruction for granitic pegmatite deposits throughout the Earth's history: (**a**) a diagram modified from Ovchinnikov *et al.* (1975), (**b**) a diagram modified from Ginzburg *et al.* (1979), (**c**) this study. For comments, see the text.

The discrepancies described and some other minor differences, which an attentive reader can see in Figure 2, are primarily related to the accuracy of the data used. The works cited for comparison are based on research results from the period of geochronological studies when the precision of both tools and basic physical constants was far from the current high level. Besides, in obtaining magmatic ages, data from all methods (K–Ar, Rb–Sr, U–Pb) were used indiscriminately, without paying special attention to closure temperatures of isotopic systems or to likely disturbance and restarting of isotopic clocks. All these could lead to the widening of pulses on the time scale and to false peaks, which are most evident in Ovchinnikov *et al.* (1975). Besides, it is clear that some incorrect data was used. Thus, the age of the Bernic Lake field in Ginzburg *et al.* (1979) is taken to be 2.0 Ga, while currently it is known to be 2.64 Ga (see supplementary material). In the same work, discrepancies for some other provinces are not so large, although still essential. Note that these discrepancies are non-systematic; that is, ages from Ginzburg *et al.* (1979) may be either younger or older than current datings: Wodgina, 2.7/2.84 Ga; Greenbushes, 2.7/2.53 Ga; Mama–Chuya belt, (0.7–0.4)/(0.38–0.33) Ga; and so on.

The onset in the Mesoarchaean: why at that time?

The earliest occurrences of mineralized pegmatites appeared in the Barberton greenstone belt and the adjacent Ancient Gneiss Complex of Swaziland *c.* 3.1 Ga ago (Harris *et al.* 1995; Trumbull 1995). All of them have typical features of the rare-metal pegmatite class, although their mining perspectives have never been highly valued. However, some of the deposits were sources of cassiterite placers in the Tin belt of Swaziland; these placers have been mined occasionally since the end of the 19th century (Maphalala & Trumbull 1998).

In order to understand why pegmatites were formed at this location, precisely at that period close to *c.* 3.1 Ga, one should compare the geology of the region with that of pegmatite provinces that developed prior to and during the same period and that contain no rare-metal-enriched pegmatites. Table 1 presents such comparison for the Isukasia 'barren' block and the Swaziland 'productive' block. These units are similar as both of them contain grey gneiss complexes (TTG: tonalite–trondhjemite–granodiorite complex) and supracrustal rocks (Isua and Barberton greenstone belts, respectively), as well as late- and post-orogenic

Table 1. *Comparison of ancient sialic blocks with and without mineralized pegmatites*

Compared features	Isukasia block with the Isua GB (1–5)	Swaziland block with the Barberton GB (6–13)
Tectonic development	Discontinuous active inner tectonics *c.* 3.85–3.60 Ga; A few episodes of external stresses on a stabilized block during 3.6–2.55 Ga	Discontinuous active inner tectonics *c.* 3.66–3.08 Ga Anorogenic intraplate magmatism during 2.87–2.69 Ga
Main structural complexes involved and produced in orogenesis	TTG; Volcano-sedimentary supracrustals *c.* 3.8–3.7 Ga; Late to post tectonic granitoids and pegmatites *c.* 3.6 and *c.* 2.95 Ga	TTG; Volcano-sedimentary supracrustals *c.* 3.55–3.20 Ga; Late to post tectonic granitoids and pegmatites *c.* 3.10–3.07
Greenstone belts	*c.* 35 km long and up to 2 km wide; Supracrustals up to 0.5 km thick: Volcanogenic/chemogenic/terrigenous ≈10/10/1	*c.* 140 km long and up to 50 km wide; Supracrustals up to 12 km thick: Volcanogenic/chemogenic/terrigenous ≈10/1/7
Meta-terrigenous rocks	Few dozens m thick *c.* 3.7 Ga; Volcanic rocks as an evident provenance only; High-ferruginous low-mature Metamorphosed up to amphibolitic facies	Up to 6 km thick *c.* 3.26–3.20 Ga Essential up to main role of TTG in provenance One half is represented by mature low-ferruginous and low-calcium sediments with a big share of argillaceous varieties Metamorphosed up to amphibolitic facies
Presence of highly evolved granitoids and mineralized pegmatites	No	Yes: *c.* 3.1–3.07 Ga

Abbreviations: GB, greenstone belt, TTG, tonalite–trondhjemite–granodiorite complex ('grey' gneisses).
Chemogenic rocks mostly include chert, BIF and carbonate ones.
1–5: Nutman *et al.* (1984, 2000, 2002); Hanmer *et al.* (2002); Friend & Nutman (2005).
6–13: Maphalala *et al.* (1989); Trumbull (1993); de Ronde & de Wit (1994); Harris *et al.* (1995); Trumbull (1995); Hofmann (2005); Hessler & Lowe (2006); Schoene *et al.* (2008).

granites. However, some differences are clearly apparent. The most important difference is that the Swaziland block contains large reservoirs of terrigenous rocks with a notable amount of high-maturity metasediments. Compared to their source rocks, these metasediments are enriched in K-feldspar and light-coloured mica and depleted in plagioclase and dark-coloured minerals (Hessler & Lowe 2006). The large proportion of these rocks points to deep chemical decay of the provenance rocks caused by weathering of voluminous continental masses. In the Tin belt of Swaziland, the fertile granite of the Sinceni field is the best studied geochemically, so far (Trumbull 1993; Trumbull 1995; Trumbull & Chaussidon 1998). It displays all features of highly evolved granites melted from a source resembling these metasediments.

In southwestern Greenland (not only in the Isukasia block but in the whole Itsaq Gneiss Complex), there is no reservoir of terrigenous rocks comparable with that in the Barberton belt. This area is totally devoid of mineralized pegmatites, although it shows voluminous late-phase pegmatites related to the large (50 km × 18 km) post-orogenic *c.* 2.54 Ga Qôrqut Granite Complex (Brown *et al.* 1981). It was formed by anatexis of the Itsaq gneisses and evolution of resultant melts (Moorbath *et al.* 1981). In the entire Archaean craton of Greenland, it is only *c.* 2.96 Ga late-tectonic granites in the Ivisârtoq greenstone belt of the Kapisilik block (Friend & Nutman 2005) that are accompanied by two small groups of pegmatite dykes with sparse beryl crystals that are of mineralogical interest (Seacher *et al.* 2008). The Ivisârtoq belt incorporates the largest Mesoarchaean supracrustal complex in the craton, but its metasedimentary constituents are not voluminous, which makes it similar to the Isua belt and different from the Barberton belt. The rare occurrence of beryl in pegmatites of the Ivisârtoq belt and total lack of any rare-element minerals in pegmatites of the Isukasia block might result from compositional differences between metasedimentary rocks due to their different origins. However, no comparison has so far been made.

Besides the Itsaq Gneiss Complex, several smaller blocks older than 3.6 Ga are known worldwide. All of them are composed of broadly similar rock complexes (Nutman *et al.* 2001). No mineralized pegmatites are mentioned in the geological literature on these blocks, which provides a good reason for claiming their absence.

Therefore, the generation in the Earth's crust of pegmatites with distinct features of the rare-metal class is restricted to those time intervals and areas in which the first large-scale terrigenous sediment accumulations occurred, that along with other supracrustal and infracrustal rocks, could have been affected by anatectic processes. Even though economically attractive granitic pegmatite deposits in orogenic belts may be located in quite different non-metaterrigenous rocks (amphibolites, anorthosites, marbles, etc.), closer inspection of each particular pegmatite-bearing province reveals considerable masses of metapelitic to metapsammitic rocks. This does not mean that metaterrigenous rocks are the only contributors to the production of fertile melts, but their input of fluid and ore-forming components into anatectic melts must be critical for the completion of the ore-forming process in a pegmatite chamber.

Cyclicity in the metallogeny of granitic pegmatites

The matter of cyclicity in the intensity of generation of mineralized pegmatites in the Earth's crust has to the author's knowledge not been discussed before. However, this kind of cyclicity has been established in the course of this study. The author has identified at least two cyclic trends. The cyclicity is more evident when the data collected is distributed with a step of 50 Ma (Fig. 3). The peaks in pegmatite generation intensity at 2.65–2.60, 1.90–1.85, 1.00–0.95, 0.55–0.50, and 0.30–0.25 Ga are the highest. If the 0.55–0.50 Ga peak is excluded, the rest of the peaks form a quasi-regular cyclic trend with a periodicity of 0.8 ± 0.1 Ga (Series 1). On the other hand, the 0.50 Ga peak together with the lower second-order peaks form another series with nearly the same periodicity: 0.55–0.50, 1.20–1.15, 2.10–2.05, and 2.85–2.80 Ga (Series 2). It is of special interest that peaks of Series 2 correspond to pegmatite fields of Gondwanan continental blocks only. The maxima of Series 1 are more varied, but the input of Laurasian continental masses is the most important. Hence, there is a certain lack of synchronism between these two large groups of continental blocks with regards to the position of pegmatite production peaks on the geological timescale.

If one compares this conclusion with the existing concepts of continental crust growth and supercontinental cycles (Condie 1998, 2002; Kerrich *et al.* 2005), the most active formation of pegmatite deposits occurred during the stages of the most intense growth of the supercontinents. Besides, the peaks at 2.65–2.60 Ga (Kenorland supercontinent) and 1.90–1.85 Ga (Columbia supercontinent) coincide with final phases of the most powerful pulses of growth of juvenile continental crust in the Earth's history. Studies in younger epochs show no pulses of crust growth of the same extent, as the process was wavelike, with smoothed shape of the curve (Condie 2001). This means that the

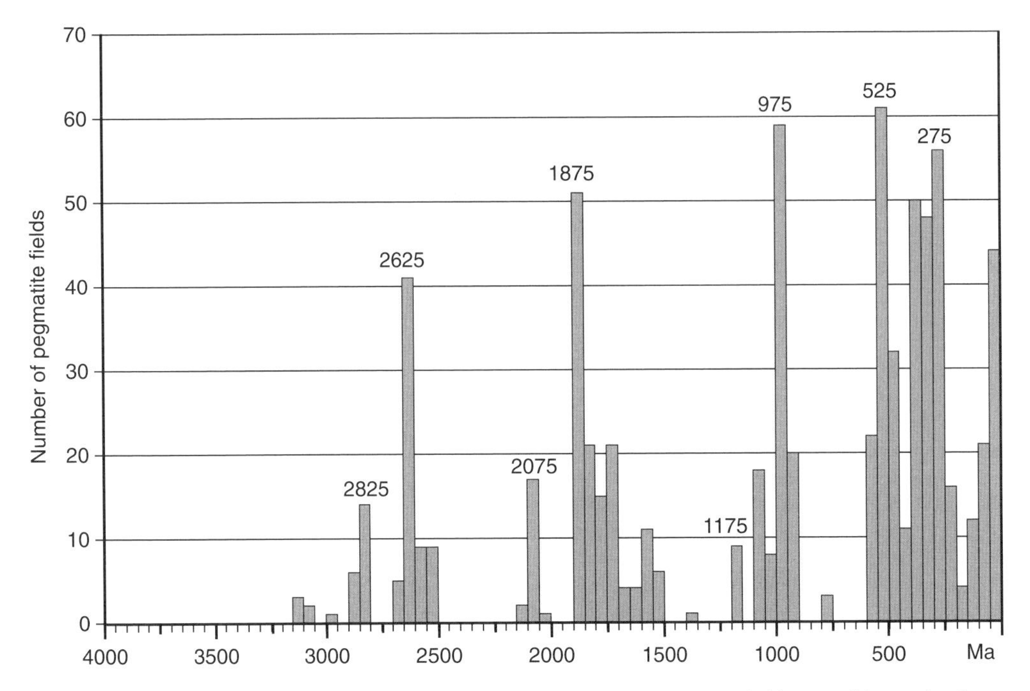

Fig. 3. Two series of cyclicity in metallogeny of granitic pegmatites. Principal peaks (bold numerals) are related to those periods in which pegmatite deposits appeared in all groups of crustal blocks, with the prevalence of deposits in Laurasia-group blocks. Subordinate peaks (italicized numerals) are only related to pegmatite deposits in Gondwana-group blocks. The numerals refer to the middle of corresponding age internals.

formation of the younger supercontinents was not accompanied by intense growth of juvenile continental crust. However, the processes generating mineralized pegmatites were characterized by even stronger pulses in the Neoproterozoic and Phanerozoic. These pulses coincided with the formation of the Rodinia, Gondwana, and Pangaea supercontinents. As a corollary, ancient continental crust and its erosion products must have been in even greater predominance in fertile sources of granitic melts in post-Early Precambrian orogens, as compared with Archaean and Palaeoproterozoic ones.

It should be specially noted that 'empty' time gaps between pegmatite generation pulses became shorter and shorter over the course of time. Ultimately, since 0.6 Ga, such gaps are not observed at all at a data-generalization step of 50 Ma (Fig. 3). With a step of 25 Ma, two such gaps appear in the Phanerozoic interval, while the diagram for the earlier period becomes much more 'sparse' (Fig. 1). This frequency pattern suggests that the total continental crust area had reached a critical value by 0.6 Ga. Then, the interaction of continental blocks in collision belts created orogens at almost any period divisible by 25 Ma, with the resultant formation of pegmatite-hosted mineral deposits.

Mineral deposits affiliated with the main pegmatite classes throughout geological time

The modern classifications of granitic pegmatites (Zagorsky *et al.* 2003; Černý & Ercit 2005) take into account geological settings favorable for pegmatite generating processes, as well as mineralogical and geochemical signatures. They are multi-branched and contain a number of hierarchic ranks such as classes ('formations' in Russian classifications), subclasses, types, and subtypes. Geochronological and geological data on pegmatite deposits amassed to date has allowed the author to trace evolutionary trends for the classes only. In general, the authors analysed the pegmatite classes identical to those from Černý & Ercit (2005). However, some of the classes have changed names: in line with the wording used customarily in the Russian pegmatitic classifications (Ginzburg *et al.* 1979; Zagorsky *et al.* 1999, 2003), the modifier 'rare-element' has been changed for 'rare-metal'.

Besides, the authors have been addressing rare-metal–miarolitic pegmatites, which are important for the analysis carried out in this study. They are not specified in the classification by Černý & Ercit (2005) within either the rare-element class or the miarolitic one. Zagorsky *et al.* (2003) pick out miarolitic varieties in almost all the pegmatite classes and consider these varieties to be in a special, additional classification, as parts of corresponding classes. In this classification the rare-metal–miarolitic pegmatites are placed in the rare-metal class (formation) of the basic classification. The author has followed this definition of the term.

Rare-metal pegmatites first appeared in the Mesoarchaean inside and south of the Barberton greenstone belt *c.* 3.1 Ga ago (see above) and continued to form in later eras (see supplementary material). Only at the very end of the Mesoarchaean in granite-greenstone belts of the Pilbara craton, were the first pegmatites formed, with accumulations of rare metals, reaching the values that were attractive to start hard ore mining for Ta, Sn, and minor Be. The most notable deposit of this kind is Wodgina–Cassiterite Mountain. According to the statistics on the USGS website, the mine has produced up to 25% of the world's primary tantalum over the last five years. Albite–spodumene and albite-type pegmatites prevail among economically attractive bodies in the Pilbara craton (Sweetapple & Collins 2002). Complex-type pegmatites are also known, but they play no essential role in rare-metal reserves of the region.

All types of rare-metal pegmatites have been established for the Neoarchaean as well. Unlike the previous era, this one is earmarked by complex-type deposits, with Tanco (Li, Ta, Cs, Be), Bikita (Li, Cs, Be, Ta), and Greenbushes (Li, Ta, Sn) being the brightest examples. These deposits show extremely high degrees of differentiation of their inner structure (Martin 1964; Partington *et al.* 1995; Černý 2005) and display the world's highest ore grades of Li, Ta, and Cs in the whole exploration history of pegmatite deposits. The Palaeozoic types of rare-metal pegmatites do not differ from the Neoarchaean ones, but no deposits with equally high-grade mineralization have been found within them.

Some rare-metal pegmatite bodies are of mining interest not only for rare elements, but also for gems and high-priced mineral collections specimens from residual miarolitic cavities. In the prevailing number of such pegmatite bodies the latter represent a greater economic interest than the former. These pegmatites appeared for the first time in the terminal Precambrian in the fields of the Eastern Brasilian pegmatite province (Morteani *et al.* 2000; Pedrosa-Soares *et al.* 2011). Thus, at that time, a new intermediate *rare-metal–miarolitic* type of pegmatites appeared. Vugs in the Archaean and Proterozoic rare-metal pegmatites are not abundant and are only of scientific interest. These vugs are not residual cavities: they are small, are not attached to core zones, and were produced by the leaching activity of relatively low-temperature fluids. For instance, Stilling *et al.* (2006) mention rare vugs with only low-temperature mineral lining in the Tanco pegmatite, which are concentrated in the pegmatite's upper and central intermediate zones. The author has not succeeded in finding any published descriptions of high-temperature mineral associations in the vugs of such ancient pegmatites from orogenic settings.

Since the Neoproterozoic–Palaeozoic boundary, rare-metal–miarolitic pegmatites have been routinely formed, and with time they came to dominate in the Phanerozoic belts over the rare-metal pegmatites (supplementary material). Besides, global analysis of the rare-metal pegmatite class shows concurrent gradual degradation of the pegmatites' inner zoning. This is most clearly manifested in the general decrease of the average amount of minerals in pegmatites and in the increasing prevalence of very primitively zoned albite–spodumene type bodies over the better zoned types in all pegmatite provinces (Solodov 1985). The appearance of an exotic variety as aphanitic pegmatite dykes in the Hindu Kush belt (Rossovskiy *et al.* 1976) may be viewed as a climax of the trend. Solodov's conclusion (1985) about the total extinction of the complex-type pegmatites in orogens by the Cenozoic is evidence of general degradation of pegmatite-forming melts in terms of their geochemical evolvement. So, in the course of geological time, there is a distinct general change for the worse, in the chances of the correct conditions to evolve, especially to crystallize pegmatite-forming melts.

Note that the oldest *rare-metal granites* in orogenic belts (Abu Dabbab and their counterparts in the Eastern Desert of Egypt, Taourirt in the Hoggar Mts, etc.) were formed at the waning stages of some orogens in the Early Palaeozoic (Abdalla *et al.* 1998; Kesraoui & Nedjari 2002). These granites are ore-bearing for Ta, Sn $\pm$ Li, Be. Besides, they are quasi-synchronous to rare-metal pegmatites in other parts of the same provinces. Rare-metal granites and pegmatites are very similar in terms of petrology, mineralogy, and geochemistry (Beskin & Marin 2003). The number of rare-metal granites increased manifold in the Hercynides and Mesozoids. The large Alakha (Li, Ta) deposit generated in the Altai orogen at the Triassic–Jurassic boundary is represented by a spodumene granite unknown in any earlier epoch (Kudrin *et al.* 1994). In deposits predating this boundary, spodumene is known only from pegmatites. This granite differs from some of the albite-spodumene pegmatites only by a plug-shaped morphology, smaller-sized minerals (0.n–10 mm, mainly 2–3 mm), and a stronger primitive zoning revealed only from sampling results (Kudrin *et al.* 1994).

Muscovite pegmatite deposits are the main suppliers of sheet muscovite and practically the only source of high-quality (low-defect) large sheets of muscovite. They are usually barren of rare-metal minerals, except for some known cases of scarce accessories. The colour of the sheet muscovite is light brown, reddish brown, and sometimes light green. *Muscovite–rare-metal pegmatite* deposits, which are also mined for the same purpose with a non-systematic co-production of some rare-metal minerals (usually beryl), mostly contain books of greenish or whitish muscovite of a lower quality. The first muscovite pegmatite deposits were formed in the Belomorian and East Sayan belts in the Palaeoproterozoic *c.* 1.87 Ga ago (see supplementary material). They are located in metamorphic formations with a great share of aluminous (two-mica ± garnet ± kyanite/sillimanite) middle-amphibolite-facies paragneisses and schists of kyanite–sillimanite (Barrovian) series in elongate fold belts (Ginzburg & Rodionov 1960; Sal'ye & Glebovitsky 1976). Occurrences of this kind of metamorphic rocks are known in the Archaean structures but are not numerous, and they are relatively small in area (Percival 1979 and references therein). They appear to be related to partial convective overturn adjacent to Archaean rising granitic domes rather than to collision belts (Collins & Van Kranendonk 1999). The size of the Barrovian-type metamorphosed blocks created due to this tectonic scenario and the duration of favorable conditions were probably not sufficient for the muscovite pegmatite deposits to be generated there.

The average quality of the sheet muscovite in the Belomorian belt deposits is the highest in the world (Tkachev *et al.* 1998). The Bihar Mica belt, famous for its deposits, essentially exceeds the Belomorian one and any other of the Palaeoproterozoic provinces in terms of resource abundance, although it is slightly inferior to its older counterparts listed above in terms of the average quality of muscovite sheets: birth defects in mica crystals (colour zoning, 'A'-structure, staining, microintergrowths with other minerals, etc.) are more abundant here. Muscovite pegmatite deposits keep this bias tendency: in general the share of high-grade crystals (in terms of sheet mica quality) in the best Palaeozoic deposits of this class is notably lower than in the Proterozoic ones (Tkachev & Gershenkop 1997; Tkachev *et al.* 1998). Probably the youngest (Permian) deposits of the muscovite pegmatite class are located in the Urals belt (supplementary material). Historic archival exploration data from the deposits show that only few bodies in them contain vanishingly small amounts of high-grade sheet muscovite. The rest of pegmatites contain only low-grade muscovite. No Mesozoic or Cenozoic muscovite pegmatite deposits have been found worldwide so far. Merely small-scale accumulations of light-coloured sheet mica that have been found in their host muscovite–rare-metal pegmatites, requiring a slightly lower lithostatic pressure to be generated as compared to muscovite pegmatites proper. Although Barrovian-type metamorphic complexes in orogens continued to appear even in the Late Cenozoic, and mineralized granitic pegmatites are widespread in them. However, these orogens are not fertile for muscovite pegmatites proper. Hence, there are many reasons for supposing cessation of this pegmatite class deposits in the crust after the Palaeozoic.

For example, numerous pegmatites are known in the Miocene-age Muzkol Metamorphic Complex of the Barrovian type in the Pamirs (Dufour *et al.* 1970), but none of them contain sheet mica zones! The pegmatites with miarolitic mineralization are the most noted in the region (Zagorsky *et al.* 1999). Besides, some bodies contain uneconomic rare-metal mineralization. Small deposits of sheet mica (Zagorsky *et al.* 1999) are hosted by a rock complex of the same type and of similar age in the Neelum River valley, High Himalaya (Fontan *et al.* 2000). These deposits belong to the muscovite–rare-metal class, and, in addition, a number of them contain miarolitic vugs with gems (Zagorsky *et al.* 1999).

The most ancient (*c.* 1.73 Ga) *miarolitic pegmatites* proper, that is, those that are attractive for mining exclusively because they contain crystals lining cavities (mainly residual ones), are related to anorogenic rapakivi granites of the Ukrainian shield (Lazarenko *et al.* 1973). Some other occurrences of ancient miarolitic pegmatites are known in rapakivi granites of the Fennoscandian shield (1.67–1.47 Ga) with most notable ones in the Wyborg batholith (Haapala 1995). It is of specific interest that in post-orogenic settings this class of pegmatites appeared only at the end of the Mesoproterozoic. According to the data collected (supplementary material and references therein), the most ancient pegmatite fields of this category are hosted by the *c.* 1.07 Ga old Katemcy and Streeter granites in the western part of the Llano Uplift in Texas. The post-orogenic nature of the granites is reliably established (Mosher *et al.* 2008). These deposits were mined to extract black quartz (morion) and jewel topaz crystals (Broughton 1973). Of the rare-metal minerals, only cassiterite was recorded. Note that rare-element pegmatites in the NE of the uplift (Ehlmann *et al.* 1964; Landes 1932) are related to the Lone Grove granite with major-element chemistry very similar to the Katemcy pluton but its age is *c.* 20 Ma older (Rougvie *et al.* 1999). Here, pegmatites are located in the granites as well as in the country gneiss of the upper amphibolite facies. Topaz is unknown, and vugs are very scarce, although one of them has a notable size (Landes 1932). Hence, the conditions of melt differentiation and crystallization over a period of *c.* 20 Ma changed in such a manner that petrologically similar granites gave birth to metallogenically different-class pegmatites.

Later on, miarolitic pegmatites have been getting progressively widespread in orogenic belts in course of time. It is possible to claim the same about the intermediate rare-metal–miarolitic class (see above in this section). Since the Late Palaeozoic (*c.* 300 Ma) these classes have been prevailing in mineralized pegmatite fields. At the same time, these pegmatites have been slowly diminishing in size, with their inner zoning becoming progressively less developed, in keeping with the trend mentioned above for rare-metal pegmatites.

Driving forces of the metallogenic evolution of granitic pegmatites in orogenic belts

Evidence of global changes in the conditions of pegmatite crystallization is provided by the analysis of the following features described in a section above (either considered separately or analysed jointly for increased benefit): (i) gradual deterioration (degradation) of pegmatites of the rare-metal class from the Neoarchaean to the Cenozoic; (ii) the restriction of the highest-grade sheet mica deposits to the Proterozoic and the total absence of muscovite-class pegmatites in post-Palaeozoic orogens; (iii) the first appearance of miarolitic pegmatites in the Late Mesoproterozoic and rare-metal–miarolitic ones in the Late Neoproterozoic; and (iv) progressively increasing proportion of both of these classes in mineralized pegmatite fields from the Cambrian to the Neogene. The first wide-scale appearance of rare-metal granites in the beginning of the Palaeozoic and their increasing abundance in the Phanerozoic orogens on the periphery of pegmatite belts and sometimes instead of them are thought to be part of the same sequence of interconnected events.

The degradation of rare-metal and sheet mica deposits may be tentatively explained by the well-established gradual cooling of the Earth and the decrease in the mean value of the lithospheric heat flow (Taylor & McLennan 1985). On the one hand, the cooling impairs the conditions for crystallization and differentiation of pegmatite-forming melts both during preliminary stages and in the final pegmatite chambers, because the decrease in the global heat flow reduces the possibilities for high-T fields (which make for low heat conductivity in pegmatite-hosting rocks at crustal levels (7–12 km) in orogens, favourable for the rare-metal class formation) to persist over a sufficiently long time. On the other hand, sufficiently lasting existence of extensive areas with Barrovian-type middle amphibolite facies conditions in the middle crust of extending orogens (16–22 km) became possible when at *c.* 1.9 Ga in the Palaeoproterozoic the mean heat flow values dropped below the Archaean ones. This is the condition for the formation of the muscovite-class pegmatite deposits, because the growth of large (up to 2–3 m^2) low-defect mica crystals requires long-lasting high partial pressure of H_2O, which cannot be reached in fertile undersaturated melts without a high enough lithostatic pressure in the country rocks (Tkachev *et al.* 1998).

However, it is impossible to explain why miarolitic pegmatites appear precisely in the Grenvillian orogens by invoking the lithospheric cooling alone. The formation of residual miarolitic cavities in pegmatites results from H_2O supersaturation of fertile melts (London 2005). If the vugs clustering in the central (core) pegmatite zone are abundant or scarce but sizeable (n-10 n m^3), this most likely implies great H_2O supersaturation during subliquidus crystallization, that is, intra-chamber boiling of partially crystallized melt. The higher the supersaturation, the larger the bulk share of miarolitic cavities in a pegmatite. In theory, so 'sudden' appearance of miarolitic pegmatites at the end of the Mesoproterozoic and their progressively wider spread in the younger epochs can only result from two factors.

Firstly, it may be supposed that Early Precambrian pegmatite-forming melts were lower in H_2O. Over time, the melts became increasingly enriched in H_2O and reached saturation, and as a result, boiling in the course of intra-chamber crystallization became increasingly common. Secondly, in view of the well-known fact that H_2O solubility in granitic melts decreases with lithostatic pressure (Luth *et al.* 1964; Luth 1976; Huang & Wyllie 1981; Holtz *et al.* 1995), the above evolutionary trend is possible to assume that in any given era pegmatite-forming melts had more or less the same H_2O concentrations, but since the end of the Mesoproterozoic these melts crystallized at increasingly shallower depths. If this is so, the pegmatite formation maximum shifted closer to the surface, that is, to a zone with P–T conditions where the boiling limit was lower, while thermal field gradients in the country rocks of crystallizing pegmatites were higher when at greater depths.

The first assumption (variations in H_2O concentration in melts as a function of age) is not very plausible. The second one is more realistic.

The last two decades have seen an increase in thermochronological studies in orogens (e.g. Hodges 2003) and in mathematical modelling of orogens based on realistic physical parameters (e.g. Mareschal 1994). The results of these studies make it possible to calculate exhumation (uplift) rates of extending orogens on a reasonable basis. Unfortunately, for well-known reasons, these studies have focused on Phanerozoic fold belts which describe the number of rate estimates, summarized for different epochs in Table 2. Nevertheless, all the main epochs are (to a varying extent) characterized by uplift rates of orogen roots in the course of the post-culmination extension.

These data provide weighty support to the second assumption: the likely decrease in crystallization depths of pegmatite-forming melts through geological time. Table 2 clearly shows notable differences in exhumation rates of orogens: all Early Precambrian rates are below 0.35 mm $\times$ a^{-1} with a mean value of *c.* 0.2 mm $\times$ a^{-1} (0.2 km $\times$ Ma^{-1} or 2 km per 10 Ma), whereas, the rates for younger belts are higher by a factor of 3 or more. Some of the exhumation rates in Phanerozoic belts were calculated for relatively short periods of 2–5 Ma. Even when the data is smoothed over longer periods of 10–20 Ma, which is taken to be

Table 2. *Exhumation rates in orogens during post-culmination extension*

Tectonic structure	Calculated rates*, mm a^{-1}	References
Cenozoic		
Himalayan belt, Namche-Barwa syntaxis	3–10	Burg *et al.* (1997)
Himalayan belt, Nanga Parbat syntaxis	*c.* 5	Shroder & Bishop (2000)
Himalayan belt, South Tibet junction	1–5	Ruppel & Hodges (1994)
Himalayan belt, Zanskar	1.0–1.1	Searle *et al.* (1992)
Black Mountains area	2.3–3.2	Holm *et al.* (1992)
Ruby Mountains area	1.33–5.8	Hacker *et al.* (1990)
Omineca belt, Idaho batholith	0.4–1.6	House *et al.* (2002)
Pyrenean belt	*c.* 2	Sinclair *et al.* (2005)
Mesozoic		
Cordilleran belt, Sierra Nevada batholith	0.35–1.33	DeGraaff-Surpless *et al.* (2002)
ibid	0.5–1.0	Vermeesch *et al.* (2006)
Qinling-Dabie-Sulu belt, Dabie zone	1–8	Ayers *et al.* (2002)
Late Neoproterozoic–Palaeozoic		
Altai belt	1.75–1.82	Briggs *et al.* (2007)
Appalachian belt, Acadian orogeny	1–2	Hames *et al.* (1989)
Ibid	*c.* 1.4	Armstrong & Tracy (2000)
Variscan belt, Iberian crystalline massif	0.6–1.3	Martínez *et al.* (1988)
Variscan belt, French Massif Central	0.3–1.5	Scaillet *et al.* (1996)
Variscan belt, Bohemian crystalline massif (north-western part)	1.1–2.5	Zulauf *et al.* (2002)
Variscan belt, Bohemian crystalline massif (eastern part)	2.8–4.3	Kotková *et al.* (2007)
Lützow–Holm Complex, East Antarctica	1.15	Fraser *et al.* (2000)
Late Mesoproterozoic – Early Neoproterozoic		
Grenvillian belt, western part	0.33	Martignole & Reynolds (1997)
Grenvillian belt, eastern part	0.41–1.22	Cox *et al.* (2002)
Sveconorwegian belt, Bamble zone	1.5–1.0	Cosca *et al.* (1998)
Sveconorwegian belt, Idefjorden zone	*c.* 1	Söderlund *et al.* (2008)
Palaeoproterozoic		
Athabasca round-basin area, Hearn-Rae junction (1.80–1.85 Ga interval)	<0.2	Flowers *et al.* (2006)
Svecofennian belt	0.1	Lindh (2005)
Belomorian belt	0.06	Alexeev *et al.* (2003)
Limpopo belt	*c.* 0.3	Zeh *et al.* (2004)
Neoarchaean		
Yellowknife belt	0.15–0.35	Bethune *et al.* (1999)

*All rates are given with a precision shown in the referred works.

a model time-gap between the termination of collision and the start-up of granite magmatism (Thompson 1999 and references therein), they give uplift rates that are still higher than those of the Early Precambrian.

Thus, starting in the Grenvillian epoch, anatectic granitic melts had more opportunities to penetrate from their sites of origin at depths greater than *c.* 15 km (Brown 2001) into the upper levels by means of passive transport along with country rocks. This process was further facilitated by the wider distribution of brittle deformation at these conditions, providing additional magma conduits to the uppermost crust horizons (Thompson 1999). As discussed above, low-pressure settings are very favorable for large-scale crystallization of miarolitic pegmatites or even rare-metal granites, rather than pegmatites. Since the Mesozoic, the uplift rates became so high as to leave insufficient time for muscovite class pegmatites to originate at a favourable depth in kyanite-sillimanite metamorphic complexes. Instead, the other types of pegmatites were generated, including those with miarolitic cavities. The youngest known pegmatites of the abyssal feldspar–rare-element class are the Ordovician ones (supplementary material); the author believes that the considerations above are also applicable in this case.

It is not easy to unambiguously define the factors responsible for the changes in the post-culmination behaviour of orogens. This issue may conceivably be unravelled by analysing the lithosphere thermo-density models presented by Poudjom Djomani *et al.* (2001). These models were developed using the subcontinental lithospheric 4D mapping technique based on mantle xenolith data (O'Reilly & Griffin 1996). According to these models, subcontinental lithospheric mantle (SCLM) of Archaean age (>2.5 Ga) has considerable buoyancy relative to its underlying asthenosphere. The Proterozoic (2.5–1.0) SCLM is slightly thinner and has somewhat lower density parameters than the Archaean one. Nevertheless, it is buoyant relative to the asthenosphere within any reasonably possible thermal fields. As for the Grenvillian–Phanerozoic SCLM (<1 Ga), it is the thinnest, the densest, and has the greatest gradient in terms of the vertical distribution pattern of density. In general, this SCLM is always buoyant only where lithospheric geotherms are elevated, as in the Cenozoic active volcanic provinces. However, as the geothermal gradient relaxes toward a stable conductive profile during orogenic post-culmination extension, SCLM sections thinner than *c.* 100 km become denser than the asthenosphere or, in other words, negatively buoyant, and as a result the whole lithosphere becomes gravitationally unstable because of a heavy lithospheric root. This could promote delamination of the SCLM in all or some of its parts, upwelling of asthenospheric material, and fast uplifting of the crust. Hence, it does not seem to be mere chance that the generation of the first miarolitic pegmatites started exactly at the Mesoproterozoic–Neoproterozoic boundary, when the continental lithosphere in orogens became unstable because of such a pattern of density distribution. In this connection, it is apropos to point out that a recently developed tectonic reconstruction for the Llano Uplift area in the Grenvillian epoch perfectly fits such a kind of scenario (Mosher *et al.* 2008).

The changes in SCLM are conditioned by different levels of depletion depending on variations in the volumes and temperatures of mafic–ultramafic melts in different epochs, which is, in turn, a consequence of the Earth's cooling throughout geological time (Poudjom Djomani *et al.* 2001). Therefore, this evolutionary trend in pegmatite metallogeny (the appearance of miarolitic pegmatites, extinction of abyssal and muscovite pegmatites) is also related to the same factor that was proposed as the most important cause of the changes (general simplification of zoning, widespread vugs) in the deposits of the rare-metal pegmatite class. However in this case it acts through a more complex chain of processes. No doubt, this chain, amongst other things, also played a role in the evolution of the rare-metal pegmatite class.

Conclusions

The collected data on geology and geochronology of different mineralized pegmatite classes have made it possible to correlate the evolutionary trends of pegmatite metallogeny in orogens with the global evolution of the lithosphere. The metallogeny shows two principal trends: (a) a pulsatory pattern with quasi-regular periodicity; and (b) unidirectional development of pegmatite classes from their inception to their extinction.

The pulses or cyclicity in pegmatite generation are in correlation with the extent of crust magmatism as well as supercontinental cycles. The gradual growth of pegmatite generation is in perfect alignment with the existing concepts of continental crust growth from the Archaean to the Cenozoic.

Different factors are responsible for the diversity of pegmatite classes, including compositions of supracrustal country rocks, metamorphic facies series in orogenic belts, and uplift rates during orogenic extension. As these factors changed, the classes of mineralized pegmatites also changed their aspects. These changes controlled life cycles of the classes, from inception to peak through to decline and eventual total extinction. The metallogeny of rare-metal class pegmatites is characterized by the longest life cycle. Generation of these pegmatites began in the Mesoarchaean and persisted through all the later eras, to wane gradually after the Early Precambrian with the eventual strong degradation of their zoning structure, as observed in Cenozoic deposits. Since the terminal Neoproterozoic, rare-metal–miarolitic pegmatites have been generated more and more frequently inside or instead of rare-metal pegmatite deposits. Mineral deposits of the muscovite pegmatite class appeared for the first time in the second half of the Palaeoproterozoic and came to the end of their life cycle at the Palaeozoic–Mesozoic boundary. Miarolitic pegmatite deposits first appeared in anorogenic settings in the Late Palaeoproterozoic, whereas, in post-orogenic plutons they occurred first in Grenvillian-aged structures and dominated throughout the pegmatite metallogeny of many Phanerozoic belts.

All these changes were ultimately induced by the general cooling of the Earth.

This study was supported by grants from the Russian Academy of Sciences and the Russian Ministry of Education and Science (State Contract # 02.515.12.5010). I am deeply grateful to I. Kravchenko-Berezhnoy and N. Kuranova, who helped me to put my ideas into English. I thank M. Cronwright, T. Oberthür, S. Misra, and D. Ray for their assistance in obtaining certain new data on African and Indian pegmatite deposits. Three anonymous reviewers are thanked for critical reading of the manuscript and many helpful comments and suggestions.

Assistance in the acquisition of any new geochronology data from mineralized pegmatites and fertile granites worldwide would be very much appreciated.

References

ABDALLA, H. M., HELBA, H. A. & MOHAMED, F. H. 1998. Chemistry of columbite-tantalite minerals in rare metal granitoids, Eastern Desert, Egypt. *Mineralogical Magazine*, **62**, 821–836.

ALEXEEV, N. L., BALAGANSKY, V. V. ET AL. 2003. Rates of Early Proterozoic orogenic processes: a study of U–Pb and Sm–Nd zircon and garnet systems and metamorphic processes in rocks of the Pon'gom-Navolok Island, Central Belomorian Region. *In*: KOZAKOV, I. K. & KOTOV, A. B. (eds) *Isotope Geochronology for Resolving Problems of Geodynamics and Ore Genesis*. Centre for Information Culture Publishers, St. Petersburg, 60–63 [in Russian].

ARMSTRONG, T. R. & TRACY, R. J. 2000. One-dimensional thermal modelling of Acadian metamorphism in southern Vermont, USA. *Journal of Metamorphic Geology*, **18**, 625–638.

AYERS, J. C., DUNKLE, S., GAO, S. & MILLER, C. E. 2002. Constraints on timing of peak and retrograde metamorphism in the Dabie Shan ultrahigh-pressure metamorphic belt, east-central China, using U–Th–Pb dating of zircon and monazite. *Chemical Geology*, **186**, 315–331.

BALASHOV, Y. A. & GLAZNEV, V. N. 2006. Endogenic cycles and the problem of crustal growth. *Geochemistry International*, **44**, 131–140.

BESKIN, S. M. & MARIN, YU. B. 2003. About evolution of the rare-metal granite mineral- and ore-forming process during the geological history. *Zapiski Vserossiyskogo Mineralogicheskogo Obschestva (Proceedings of the Russian Mineralogical Society)*, **132**, 1–14 [in Russian].

BETHUNE, K. M., VILLENEUVE, M. E. & BLEEKER, W. 1999. Laser $^{40}Ar/^{39}Ar$ thermochronology of Archaean rocks in Yellowknife domain, southwestern Slave province: Insights into the cooling history of an Archaean granite–greenstone terrane. *Canadian Journal of Earth Sciences*, **36**, 1189–1206.

BRIGGS, S. M., YIN, A., MANNING, C. E., CHEN, Z. L., WANG, X. F. & GROVE, M. 2007. Late Paleozoic tectonic history of the Ertix Fault in the Chinese Altai and its implications for the development of the Central Asian Orogenic System. *Geological Society of America Bulletin*, **119**, 944–960.

BROUGHTON, P. L. 1973. Precious topaz deposits of the Llano Uplift area, central Texas. *Rocks & Minerals*, **48**, 147–156.

BROWN, M. 2001. Orogeny, migmatites and leucogranites: a review. *Proceedings of Indian Academy of Sciences, Earth Planetary Sciences*, **110**, 313–336.

BROWN, M., FRIEND, C. R. L., MCGREGOR, V. R. & PERKINS, W. T. 1981. The late-Archaean Qôrqut granite complex of southern west Greenland. *Journal of Geophysical Research*, **86**, 10617–10632.

BURG, J.-P., DAVY, P. ET AL. 1997. Exhumation during crustal folding in the Namche-Barwa syntaxis. *Terra Nova*, **9**, 53–56.

ČERNÝ, P. 1991*a*. Rare-element granite pegmatites: Part I. Anatomy and internal evolution of pegmatite deposits. *Geoscience Canada*, **18**, 49–67.

ČERNÝ, P. 1991*b*. Rare-element granite pegmatites: Part II. Regional to global environments and petrogenesis. *Geoscience Canada*, **18**, 68–81.

CERNY, P. 1991*c*. Fertile granites of Precambrian rare-element pegmatite fields: is geochemistry controlled by tectonic setting or source lithologies? *Precambrian Research*, **51**, 429–468.

ČERNÝ, P. 2005. The Tanco rare-element pegmatite deposit, Manitoba: regional context, internal anatomy, and global comparisons. *In*: LINNEN, R. L. & SAMSON, I. M. (eds) *Rare-element Geochemistry and Mineral Deposits*. Geological Association of Canada, Short Course Notes, **17**, 127–158.

ČERNÝ, P. & ERCIT, T. S. 2005. Classification of granitic pegmatites. *Canadian Mineralogist*, **43**, 2005–2026.

COLLINS, W. J. & VAN KRANENDONK, M. J. 1999. Model for the development of kyanite during partial convective overturn of Archaean granite-greenstone terranes: the Pilbara Craton, Australia. *Journal of Metamorphic Geology*, **17**, 145–156.

CONDIE, K. C. 1998. Episodic continental growth and supercontinents: a mantle avalanche connection? *Earth and Planetary Science Letters*, **163**, 97–108.

CONDIE, K. C. 2001. Continental growth during formation of Rodinia at 1.35–0.9 Ga. *Gondwana Research*, **4**, 5–16.

CONDIE, K. C. 2002. The supercontinent cycle: are there two patterns of cyclicity? *Journal of African Earth Sciencies*, **35**, 179–183.

COSCA, M. A., MEZGER, K. & ESSENE, E. J. 1998. The Baltica-Laurentia connection: Sveconorwegian (Grenvillian) metamorphism, cooling, and unroofing in the Bamble Sector, Norway. *Journal of Geology*, **106**, 539–552.

COX, R. A., INDARES, A. & DUNNING, G. R. 2002. Temperature– time paths in the high-P Manicouagan Imbricate zone, eastern Grenville Province: evidence for two metamorphic events. *Precambrian Research*, **117**, 225–250.

DEGRAAFF-SURPLESS, K., GRAHAM, S. A., WOODEN, J. L. & MCWILLIAMS, M. O. 2002. Detrital zircon provenance analysis of the Great Valley Group, California: Evolution of an arc-forearc system. *Geological Society of America Bulletin*, **114**, 1564–1580.

DE RONDE, C. E. J. & DE WIT, M. J. 1994. Tectonic history of the Barberton Greenstone Belt, South Africa: 490 million years of Archaean evolution. *Tectonics*, **13**, 983–1005.

DUFOUR, M. S., POPOVA, V. A. & KRIVETS, T. N. 1970. *Alpine Metamorphic Complex of the Eastern Central Pamirs*. LGU Publishing House, Leningrad [in Russian].

EHLMANN, A. J., WALPER, J. L. & WILLIAMS, J. 1964. A new, Barringer Hill-type, rare-earth pegmatite from the Central Mineral Region, Texas. *Economic Geology*, **59**, 1348–1360.

FAURE, G. & MENSING, T. M. 2005. *Isotopes: Principles and Applications*. John Wiley & Sons, New Jersey.

FERSMAN, A. E. 1931. *Pegmatites: Their Scientific and Practical Importance. V.1. Granitic Pegmatites*. USSR Academy of Sciences Publishing House, Leningrad [in Russian].

FERSMAN, A. E. 1940. *Pegmatites. V.1. Granitic Pegmatites (3rd edition: corrected and supplemented)*. USSR Academy of Sciences Publishing House, Moscow-Leningrad [in Russian].

FLOWERS, R. M., MAHAN, K. H., BOWRING, S. A., WILLIAMS, M. L., PRINGLE, M. S. & HODGES, K. V. 2006. Multistage exhumation and juxtaposition of lower continental crust in the western Canadian Shield: Linking high-resolution U–Pb and $^{40}Ar/^{39}Ar$ thermochronometry with pressure-temperature-deformation paths: *Tectonics*, **25**, TC4003, doi: 10.1029/2005TC001912.

FONTAN, D., SCHOUPPE, M. A., HUNZIKER, J., MARTINOTTI, G. & VERKAEREN, J. 2000. Metamorphic evolution, $^{40}Ar-^{39}Ar$ chronology and tectonic model for the Neelum valley, Azad Kashmir, NE Pakistan. *In*: KHAN, M. A., TRELOAR, P. J., SEARLE, M. P. & JAN, M. Q. (eds) *Tectonics of the Nanga Parbat Syntaxis and of the Western Himalaya*. Geological Society, London, Special Publications, **170**, 431–453.

FRASER, G., MCDOUGALL, I., ELLIS, D. J. & WILLIAMS, I. S. 2000. Timing and rate of isothermal decompression in Pan-African granulites from Rundvågshetta, East Antarctica. *Journal of Metamorphic Geology*, **18**, 441–454.

FRIEND, C. R. L. & NUTMAN, A. P. 2005. New pieces to the Archaean jigsaw puzzle in the Nuuk region, southern West Greenland: steps in transforming a simple insight into a complex regional tectonothermal model. *Journal of the Geological Society*, **162**, 147–162.

GINZBURG, A. I. & RODIONOV, G. G. 1960. On the depth of formation of granitic pegmatites. *Geologia Rudnykh Mestorozhdeniy (Geology of Ore Deposits)*, **1**, 45–54 [in Russian].

GINZBURG, A. I., TIMOFEYEV, I. N. & FELDMAN, L. G. 1979. *Principles of Geology of the Granitic Pegmatites*. Nedra, Moscow [in Russian].

HAAPALA, I. 1995. Metallogeny of the rapakivi granites. *Mineralogy and Petrology*, **54**, 141–160.

HACKER, B. R., YIN, A., CHRISTIE, J. M. & SNOKE, A. W. 1990. Differential stress, strain rate, and temperatures of mylonitization in the Ruby Mountains, Nevada: Implications for the rate and duration of uplift. *Journal of Geophysical Research*, **95**, 8569–8580.

HAMES, W. E., TRACY, R. J. & BODNAR, R. J. 1989. Postmetamorphic unroofing history deduced from petrology, fluid inclusions, thermochronometry, and thermal modeling: an example from southwestern New England. *Geology*, **17**, 727–730.

HANMER, S., HAMILTON, M. A. & CROWLEY, J. L. 2002. Geochronological constraints on Paleoarchaean thrust nappe and Neoarchaean accretionary tectonics in southern West Greenland. *Tectonophysics*, **350**, 255–271.

HARRIS, P. D., ROBB, L. J. & TOMKINSON, M. J. 1995. The nature and structural setting of rare-element pegmatites along the northern flank of the Barberton greenstone belt, South Africa. *South Africa Journal of Geology*, **98**, 82–94.

HESSLER, A. M. & LOWE, D. R. 2006. Weathering and sediment generation in the Archaean: An integrated study of the evolution of siliciclastic sedimentary rocks of the 3.2 Ga Moodies Group, Barberton Greenstone Belt, South Africa. *Precambrian Research*, **151**, 185–210.

HODGES, K. V. 2003. Geochronology and thermochronology in orogenic systems. *In*: RUDNICK, R. L. (ed.) *Treatise on Geochemistry*, The Crust, **3**, Elsevier, New York, 263–292.

HOFMANN, A. 2005. The geochemistry of sedimentary rocks from the Fig Tree Group, Barberton greenstone belt: Implications for tectonic, hydrothermal and surface processes during mid-Archaean times. *Precambrian Research*, **143**, 23–49.

HOLM, D. K., SNOW, J. K. & LUX, D. R. 1992. Thermal and barometric constraints on the intrusive and unroofing history of the Black Mountains: Implications for timing, initial dip, and kinematics of detachment faulting in the Death Valley region, California. *Tectonics*, **11**, 507–522.

HOLTZ, F., BEHRENS, H., DINGWELL, D. B. & JOHANNES, W. 1995. Water solubility in haplogranitic melts: Compositional, pressure and temperature dependence. *American Mineralogist*, **80**, 94–108.

HOUSE, M. A., BOWRING, S. A. & HODGES, K. V. 2002. Implications of middle Eocene epizonal plutonism for the unroofing history of the Bitterroot metamorphic core complex, Idaho-Montana. *Geological Society of America Bulletin*, **114,** 448–461.

HUANG, W. L. & WYLLIE, P. J. 1981. Phase relations of S-type granite with H_2O to 35 kbar: muscovite granite from Harney Peak, South Dakota. *Journal of Geophysical Research*, **86**, 10515–10529.

KERRICH, R., GOLDFARB, R. J. & RICHARDS, J. 2005. Metallogenic provinces in an evolving geodynamic framework. *Economic Geology* 100th Anniversary Volume, 1097–1136.

KESRAOUI, M. & NEDJARI, S. 2002. Contrasting evolution of low-P rare metal granites from two different terranes in the Hoggar area, Algeria. *Journal African Earth Sciences*, **34**, 247–257.

KOTKOVÁ, J., GERDES, A., PARRISH, R. R. & NOVÁK, M. 2007. Clasts of Variscan high-grade rocks within Upper Viséan conglomerates – constraints on exhumation history from petrology and U–Pb chronology. *Journal of Metamorphic Geology*, **25**, 781–801.

KRATZ, K. O. (ed.) 1984. *Principles of the Metallogeny of the Precambrian Metamorphic Belts*. Nauka, Leningrad [in Russian].

KUDRIN, V. S., STAVROV, O. D. & SHURIGA, T. N. 1994. New spodumene type of tantalum-bearing rare metal granites. *Petrologia*, **2**, 88–95 [in Russian].

LANDES, K. K. 1932. The Baringer Hill pegmatite. *American Mineralogist*, **17**, 381–390.

LANDES, K. K. 1935. Age and distribution of pegmatites. *American Mineralogist*, **20**, 81–105, 153–175.

LAZARENKO, E. K., PAVLISHIN, V. I., LATYSH, V. T. & SOROKIN, Y. G., 1973. *Mineralogy and Genesis of the Chamber Pegmatites of Volynia*. Vischa Shkola, Lvov [in Russian].

LINDH, A. 2005. Origin of chemically distinct granites in a composite intrusion in east-central Sweden: geochemical and geothermal constraints. *Lithos*, **80**, 249–266.

LONDON, D. 2005. Granitic pegmatites: an assessment of current concepts and directions for the future. *Lithos*, **80**, 281–303.

LONDON, D. 2008. *Pegmatites*. The Canadian Mineralogist, Québec, Special Publication, **10**.

LUTH, W. C. 1976. Granitic rocks. *In*: BAILEY, D. K. & MACDONALD, R. (eds) *The Evolution of Crystalline Rocks*. Academic Press, London, 335–417.

LUTH, W. C., JAHNS, R. H. & TUTTLE, O. F. 1964. The granite system at pressures of 4 to 10 kilobars. *Journal of Geophysical Research*, **69**, 759–773.

MAKRYGINA, V. A., MAKAGON, V. M., ZAGORSKY, V. E. & SMAKIN, B. M. 1990. *Granitic Pegmatites. V.1: Mica-bearing Pegmatites*. 'Nauka', Novosibirsk [in Russian].

MAPHALALA, R. M. & TRUMBULL, R. B. 1998. A geochemical and Rb/Sr isotopic study of Archaean pegmatite dykes in the Tin Belt of Swaziland. *South African Journal of Geolology*, **101**, 53–65.

MAPHALALA, R. M., KRÖNER, A. & KRAMERS, J. D. 1989. Rb–Sr ages for Archaean granitoids and tin-bearing pegmatites in Swaziland, southern Africa. *Journal of African Earth Sctences*, **9**, 749–757.

MARESCHAL, J.-C. 1994. Thermal regime and post-orogenic extension in collision belts. *Tectonophysics*, **238**, 471–484.

MARTIGNOLE, J. & REYNOLDS, P. 1997. $^{40}Ar/^{39}Ar$ thermochronology along a western Quebec transect of the Grenville Province, Canada. *Journal of Metamorphic Geology*, **15**, 283–296.

MARTIN, Y. J. 1964. The Bikita tinfield. *Southern Rhodesia Geological Survey Bulletin*, **58**, 114–143.

MARTÍNEZ, F. J., JULIVERT, M., SEBASTIAN, A., ARBOLEYA, M. L. & GIL IBARGUCHI, J. I. 1988. Structural and thermal evolution of high-grade areas in the northwestern parts of the Iberian Massif. *American Journal of Science*, **288**, 969–996.

MOORBATH, S., TAYLOR, P. N. & GOODWIN, R. 1981. Origin of granitic magma by crustal remobilisation: Rb–Sr and Pb/Pb geochronology and isotope geochemistry of the late Archaean Qôrqut Granite Complex of southern West Greenland. *Geochimica et Cosmochimica Acta*, **45**, 1051–1060.

MORTEANI, G., PREINFALK, A. & HORN, A. H. 2000. Classification and mineralization potential of the pegmatites of the Eastern Brazilian Pegmatite Province. *Mineralium Deposita*, **35**, 638–655.

MOSHER, S., LEVINE, J. S. F. & CARLSON, W. D. 2008. Mesoproterozoic plate tectonics: a collisional model for the Grenville-aged orogenic belt in the Llano uplift, central Texas. *Geology*, **36**, 55–58.

NUTMAN, A., ALLAART, J., BRIDGWATER, D., DIMROTH, E. & ROSING, M. 1984. Stratigraphic and geochemical evidence for the depositional environment of the Early Archaean Isua supracrustal belt, southern West Greenland. *Precambrian Research*, **25**, 365–396.

NUTMAN, A. P., BENNETT, V. C., FRIEND, C. R. L. & MCGREGOR, V. R. 2000. The early Archaean Itsaq Gneiss Complex of southern West Greenland: the importance of field observations in interpreting age and isotopic constraints for early terrestrial evolution. *Geochimica et Cosmochimica Acta*, **64**, 3035–3060.

NUTMAN, A. P., FRIEND, C. R. L. & BENNETT, V. C. 2001. Review of the oldest (4400–3600 Ma) geological record: glimpses of the beginning. *Episodes*, **24**, 93–101.

NUTMAN, A. P., FRIEND, C. R. L. & BENNETT, V. C. 2002. Evidence for 3650–3600 Ma assembly of the northern end of the Itsaq Gneiss Complex, Greenland: implication for early Archaean tectonics. *Tectonics*, **21**, 10.1029/2000TC001203.

O'REILLY, S. Y. & GRIFFIN, W. L. 1996. 4-D lithospheric mapping: a review of the methodology with examples. *Tectonophysics*, **262**, 3–18.

OVCHINNIKOV, L. N., VORONOVSKIY, S. N. & OVCHINNIKOVA, L. B. 1975. Radiogeochronology of granitic pegmatites. *Doklady of the USSR Academy of Sciencies*, **223**, 1202–1205 [in Russian].

OVCHINNIKOV, L. N., VORONOVSKIY, S. N. & OVCHINNIKOVA, L. B. 1976. Radiogeochronology of granitic pegmatites. *In*: *Studies on Geological Petrology*. Nauka, Moscow, 319–326 [in Russian].

PARTINGTON, G. A., MCNAUGHTON, N. J. & WILLIAMS, I. S. 1995. A review of the geology, mineralization, and geochronology of the Greenbushes pegmatite, Western Australia. *Economic Geology*, **90**, 616–635.

PEDROSA-SOARES, A. C., DE CAMPOS, C. P. ET AL. 2011. Late Neoproterozoic–Cambrian granitic magmatism in the Araçuaí orogen (Brazil), the Eastern Brazilian Pegmatite Province and related mineral resources. *In*: SIAL, A. N., BETTENCOURT, J. S., DE CAMPOS, C. P. & FERREIRA, V. P. (eds) *Granite-Related Ore Deposits*. Geological Society, London, Special Publications, **350**, 25–51.

PERCIVAL, J. A. 1979. Kyanite-bearing rocks from the Hackett River area, N.W.T.: Implications for Archaean geothermal gradients. *Contributions to Mineralogy and Petrology*, **69**, 177–184.

Poudjom Djomani, Y. H., O'Reilly, S. Y., Griffin, W. L. & Morgan, P. 2001. The density structure of subcontinental lithosphere through time. *Earth and Planetary Science Letters*, **184**, 605–621.

Rossovskiy, L. N., Chmyrev, V. M. & Salakh, A. S. 1976. Genetic relationship of aphanitic spodumene dikes to lithium-pegmatite veins. *Doklady of the USSR Academy of Sciences, Earth Science Section*, **226**, 170–172.

Rougvie, J. R., Carlson, W. D., Copeland, P. & Connelly, J. N. 1999. Late thermal evolution of Proterozoic rocks in the northeastern Llano Uplift, central Texas. *Precambrian Research*, **94**, 49–72.

Ruppel, C. & Hodges, K. V. 1994. Pressure-temperature-time paths from two-dimensional thermal models: prograde, retrograde, and inverted metamorphism. *Tectonics*, **13**, 17–44.

Sal'ye, M. E. & Glebovitsky, V. A. 1976. *Metallogenic Specialisation of Pegmatites on the East of the Baltic Shield*. Nauka, Leningrad [in Russian].

Scaillet, S., Cuney, M., Le Carlier de Veslud, C., Cheilletz, A. & Royer, J. J. 1996. Cooling patterns and mineralization history of the Saint Sylvestre and Western Marche leucogranite plutons, French Massif Central. II. Thermal modelling and implications for the mechanisms of U-mineralization. *Geochimica et Cosmochimica Acta*, **60**, 4673–4688.

Schneiderhöhn, H. 1961. *Die Erzlagerstatten der Erde. Bd. 2. Die Pegmatite*. Gustav Fisher Verlag, Stuttgart.

Schoene, B., De Wit, M. J. & Bowring, S. A. 2008. Mesoarchaean assembly and stabilization of the eastern Kaapvaal craton: A structural-thermochronological perspective. *Tectonics*, **27**, TC5010, doi: 10.1029/2008TC002267, 1–27.

Seacher, K., Steenfelt, A. & Garde, A. A. 2008. Pegmatites and their potential for mineral exploration in Greenland. *Geology and Ore*, **10**, 2–12.

Searle, M. P., Waters, D. J., Rex, D. C. & Wilson, R. N. 1992. Pressure, temperature and time constraints on Himalayan metamorphism from eastern Kashmir and western Zanskar. *Journal of the Geological Society*, **149**, 753–773.

Shmakin, B. M., Zagorsky, V. E. & Makagon, V. M. 2007. *Granitic Pegmatites, v.4: Rare-earth Pegmatites. Pegmatites of Unusual Composition*. 'Nauka', Novosibirsk [in Russian].

Shroder, J. F., Jr. & Bishop, M. P. 2000. Unroofing of the Nanga Parbat Himalaya. *In*: Khan, M. A., Treloar, P. J., Searle, M. P. & Jan, M. Q. (eds) *Tectonics of the Nanga Parbat Syntaxis and the Western Himalaya*. Geological Society, London, Special Publications, **170**, 163–179.

Sinclair, H. D., Gibson, M., Naylor, M. & Morris, R. G. 2005. Asymmetric growth of the Pyrenees revealed through measurement and modeling of orogenic fluxes. *American Journal of Science*, **305**, 369–406.

Söderlund, U., Hellström, F. A. & Kamo, S. L. 2008. Geochronology of high-pressure mafic granulite dykes in SW Sweden: tracking the P–T–t path of metamorphism using Hf isotopes in zircon and baddeleyite. *Journal of Metamorphic Geology*, **26**, 539–560.

Solodov, N. A. 1985. *Metallogeny of Rare-metal Formations*. Nedra, Moscow [in Russian].

Stilling, A., Černý, P. & Vanstone, P. J. 2006. The Tanco pegmatite at Bernic Lake, Manitoba. XVI. Zonal and bulk compositions and their petrogenetic significance. *Canadian Mineralogist*, **44**, 599–623.

Sweetapple, M. T. & Collins, P. L. F. 2002. Genetic framework for the classification and distribution of Archaean rare metal pegmatites in the North Pilbara craton, Western Australia. *Economic Geology*, **97**, 873–895.

Taylor, S. R. & McLennan, S. M. 1985. *The Continental Crust: Its Composition and Evolution*. Blackwell, Oxford.

Thompson, A. B. 1999. Some time–space relationships for crustal melting and granitic intrusion at various depths. *In*: Castro, A., Fernandez, C. & Vigneresse, J. L. (eds) *Understanding Granites: Integrating New and Classical Techniques*. Geological Society, London, Special Publications, **168**, 7–25.

Tkachev, A. V. & Gershenkop, A. Sh. 1997. *Mineral Raw Materials*. Mica. Handbook. ZAO 'Geoinformmark', Moscow [in Russian].

Tkachev, A. V., Sapozhnikova, L. N., Zhukova, I. A. & Zhukov, N. A. 1998. Location and generation conditions of the sheet muscovite deposits with large reserves and high quality of raw materials. *Otechestvennaya Geologia (Domestic Geology)*, **4**, 35–39 [in Russian].

Trumbull, R. B. 1993. A petrological and Rb/Sr isotopic study of an early Archaean fertile granite-pegmatite system: the Sinceni Pluton in Swaziland. *Precambrian Research*, **61**, 89–116.

Trumbull, R. B. 1995. Tin mineralization in the Archaean Sinceni rare element pegmatite field, Kaapvaal Craton, Swaziland. *Economic Geology*, **90**, 648–657.

Trumbull, R. B. & Chaussidon, M. 1998. Chemical and boron isotopic composition of magmatic and hydrothermal tourmalines from the Sinseni granite-pegmatite system in Swaziland. *Chemical Geology*, **153**, 125–137.

Vermeesch, P., Miller, D. D., Graham, S. A., De Grave, J. & McWilliams, M. O. 2006. Multimethod detrital thermochronology of the Great Valley Group near New Idria, California. *Geological Society of America Bulletin*, **118**, 210–218.

Zagorsky, V. E., Makagon, V. M., Shmakin, B. M., Makrygina, V. A. & Kuznetzova, L. G. 1997. *Granitic Pegmatites, v.2: Rare-metal Pegmatites*. 'Nauka', Novosibirsk [in Russian].

Zagorsky, V. E., Peretyazhko, I. S. & Shmakin, B. M. 1999. *Granitic Pegmatites, v.3: Miarolitic Pegmatites*. 'Nauka', Novosibirsk [in Russian].

Zagorsky, V. E., Makagon, V. M. & Shmakin, B. M. 2003. Systematics of granitic pegmatites. *Russian Geology and Geophysics*, **44**, 422–435.

Zeh, A., Klemd, R., Buhlmann, S. & Barton, J. M. 2004. Pro- and retrograde P–T evolution of granulites of the Beit Bridge Complex (Limpopo Belt, South Africa): constraints from quantitative phase diagrams and geotectonic implications. *Journal of Metamorphic Geology*, **22**, 79–95.

Zulauf, G., Dorr, W., Fiala, J., Kotkova, J., Maluski, H. & Valverde-Vaquero, P. 2002. Evidence for high-temperature diffusional creep preserved by rapid cooling of lower crust (North Bohemian shear zone, Czech Republic). *Terra Nova*, **14**, 343–354.

Late Neoproterozoic–Cambrian granitic magmatism in the Araçuaí orogen (Brazil), the Eastern Brazilian Pegmatite Province and related mineral resources

A. C. PEDROSA-SOARES[1]*, CRISTINA P. DE CAMPOS[2], CARLOS NOCE[1], LUIZ CARLOS SILVA[3], TIAGO NOVO[1], JORGE RONCATO[1], SÍLVIA MEDEIROS[4], CRISTIANE CASTAÑEDA[1], GLÁUCIA QUEIROGA[1], ELTON DANTAS[5], IVO DUSSIN[1] & FERNANDO ALKMIM[6]

[1]*Universidade Federal de Minas Gerais, IGC–CPMTC, Campus Pampulha, 31270-901 Belo Horizonte, MG, Brazil*

[2]*Department of Earth and Environmental Sciences – LMU Theresienstrasse 41/III – 80333, Munich, Germany*

[3]*Serviço Geológico do Brasil–CPRM, Belo Horizonte, MG, Brazil*

[4]*Universidade Estadual do Rio de Janeiro, UERJ–Faculdade de Geologia, Rio de Janeiro, RJ, Brazil*

[5]*Universidade de Brasília, UnB–IG–Laboratório de Geocronologia, Asa Norte, Brasília, DF, Brazil*

[6]*Universidade Federal de Ouro Preto, DEGEO, Campus do Cruzeiro, Ouro Preto, Brazil*

**Corresponding author (e-mail: pedrosa@pq.cnpq.br)*

Abstract: The Araçuaí orogen extends from the eastern edge of the São Francisco craton to the Atlantic margin, in southeastern Brazil. Orogenic igneous rocks, formed from *c.* 630 to *c.* 480 Ma, cover one third of this huge area, building up the Eastern Brazilian Pegmatite Province and the most important dimension stone province of Brazil. G1 supersuite (630–585 Ma) mainly consists of tonalite to granodiorite, with mafic to dioritic facies and enclaves, representing a continental calc-alkaline magmatic arc. G2 supersuite mostly includes S-type granites formed during the syn-collisional stage (585–560 Ma), from relatively shallow two-mica granites and related gem-rich pegmatites to deep garnet-biotite granites that are the site of yellow dimension stone deposits. The typical G3 rocks (545–525 Ma) are non-foliated garnet-cordierite leucogranites, making up autochthonous patches and veins. At the post-collisional stage (530–480 Ma), G4 and G5 supersuites were generated. The S-type G4 supersuite mostly consists of garnet-bearing two-mica leucogranites that are the source of many pegmatites mined for tourmalines and many other gems, lithium (spodumene) ore and industrial feldspar. G5 supersuite, consisting of high-K–Fe calc-alkaline to alkaline granitic and/or charnockitic to dioritic/noritic intrusions, is the source of aquamarine-topaz-rich pegmatites but mainly of a large dimension stone production.

The Late Neoproterozoic–Cambrian Araçuaí orogen encompasses the entire region between the São Francisco craton and the Atlantic continental margin, north of latitude 21°S, in eastern Brazil (Fig. 1a). Synthesis on the definition, stratigraphy, magmatism, tectonics and evolution of the Araçuaí orogen are found in Pedrosa-Soares & Wiedemann-Leonardos (2000), Pedrosa-Soares *et al.* (2001*a*, 2007, 2008), De Campos *et al.* (2004), Silva *et al.* (2005) and Alkmim *et al.* (2006).

The most remarkable feature of this crustal segment is the huge amount of different plutonic igneous rocks of Late Neoproterozoic up to Cambro–Ordovician ages, depicting a long lasting (*c.* 630–480 Ma) succession of granite production events. Granitic rocks cover one third of the orogenic region, and built up the outstanding Eastern Brazilian Pegmatite Province (Fig. 1b) and the most important dimension stone province of Brazil. This is due to the exposure of shallow to deep crustal

From: SIAL, A. N., BETTENCOURT, J. S., DE CAMPOS, C. P. & FERREIRA, V. P. (eds) *Granite-Related Ore Deposits*. Geological Society, London, Special Publications, **350**, 25–51.
DOI: 10.1144/SP350.3 0305-8719/11/$15.00

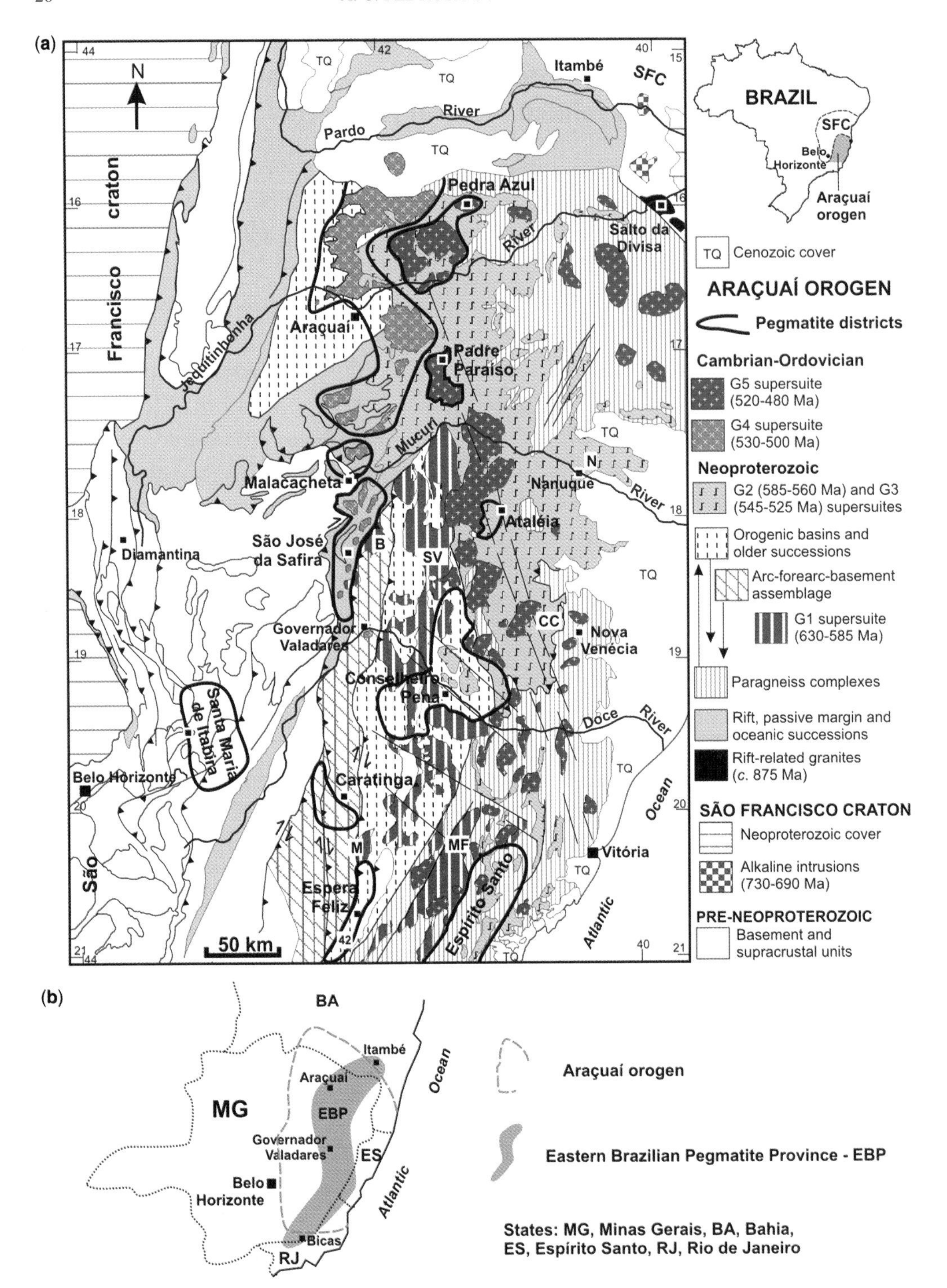

Fig. 1. Simplified geological map of the Araçuaí orogen and adjacent cratonic region, highlighting the Neoproterozoic and Cambrian granite supersuites (geology modified from Pedrosa-Soares *et al.* 2008; pegmatite districts modified from Netto *et al.* 2001 and Pedrosa-Soares *et al.* 2001*b*). SFC, São Francisco craton. Location of U–Pb zircon or whole-rock Sm–Nd analysed samples: B, Brasilândia; CC, Carlos Chagas; M, Manhuaçu; MF, Muniz Freire; N, Nanuque; SV, São Vitor.

levels along an area over 350 000 km^2. Therefore, those granites record the whole evolutionary history of the Araçuaí orogen, from the subduction-controlled pre-collisional stage up to the post-collisional gravitational collapse.

The authors present the state-of-art of the granite genesis events of the Araçuaí orogen and related mineral deposits with emphasis on the Eastern Brazilian Pegmatite Province. This information is complemented by new geochemical, geothermo-barometric and geochronological data.

Based on field relations, structural features, geochemical and geochronological data, granites from this orogen were formerly grouped into six suites (G1, G2, G3S, G3I, G4 and G5) by Pedrosa-Soares & Wiedemann-Leonardos (2000), Pedrosa-Soares *et al.* (2001*a*) and Silva *et al.* (2005). Additional data supported a regrouping into five suites (G1 to G5 from De Campos *et al.* 2004 and Pedrosa-Soares *et al.* 2008). In the Brazilian geological literature, the term suite has usually been applied to single batholiths and smaller bodies, as well as to local petrological associations, named after a confusing plethora of geographical names. This is the reason why, in this work, we use the designations G1 to G5 supersuites, instead of suites and geographical names, to avoid misunderstandings. The grouping of diverse rock units into a supersuite is strictly based on petrological and geochemical similarities, and is constrained by zircon U–Pb ages. Therefore, supersuites include suites, batholiths, stocks and other bodies, which local names will be referred to in the following sections. These supersuites can be easily recognized along extensive areas, recording different evolutionary stages of the Araçuaí orogen.

The Brasiliano orogenic event in the Araçuaí orogen has been subdivided into four geotectonic stages (Pedrosa-Soares *et al.* 2008), namely pre-collisional (*c.* 630–585 Ma), syn-collisional (*c.* 585–560 Ma), late collisional (*c.* 560–530 Ma) and post-collisional (*c.* 530–480 Ma). However, in order to better constrain the complexity of the magmatism, protracted transitions, from one geotectonic stage to another, have been taken into consideration, since changes in processes and timing control magma genesis. Accordingly, the G1 supersuite is pre-collisional because it represents the building of a calc-alkaline magmatic arc, formed in response to subduction of oceanic lithosphere, from *c.* 630 to *c.* 585 Ma. The G2 supersuite is syn-collisional as it was mainly generated by partial melting of metasedimentary piles associated with the major crustal thickening caused by contractional thrusting and folding, from *c.* 585 to *c.* 560 Ma. G3 supersuite is late collisional to post-collisional (*c.* 545–525 Ma). Late collisional refers to the transitional stage from the waning of convergent forces to the extensional relaxation of the orogen, generally accompanied by delamination and convective removal of lithospheric mantle. The post-collisional stage is related to the climax of the gravitational collapse of the orogen, which is coeval to asthenosphere ascent. G4 (*c.* 530–500 Ma) and G5 (*c.* 520–480 Ma) supersuites are post-collisional, and include plutons that cut and disturb the regional tectonic trend, as well as concordant bodies intruded along structures of distinct ages (Pedrosa-Soares & Wiedemann-Leonardos 2000; De Campos *et al.* 2004; Pedrosa-Soares *et al.* 2001*a*, 2008). Lateral escape of rock masses along major strike–slip shear zones also took place from the late collisional to post-collisional stages, providing preferred sites for magma emplacement, mainly in the southern region of the Araçuaí orogen (De Campos *et al.* 2004; Alkmim *et al.* 2006).

G1 supersuite

The following synthesis on G1 supersuite is based on data from Söllner *et al.* (1991); Nalini-Junior *et al.* (2000*a*, 2005, 2008); Bilal *et al.* (2000); Noce *et al.* (2000, 2006); Pedrosa-Soares & Wiedemann-Leonardos (2000); Pedrosa-Soares *et al.* (2001*a*, 2008); Pinto *et al.* (2001) ; Whittington *et al.* (2001); Silva *et al.* (2002, 2005, 2007); De Campos *et al.* (2004); Martins *et al.* (2004); Vauchez *et al.* (2007); Vieira (2007); Gomez (2008); Novo (2009); Petitgirard *et al.* (2009), and references therein. This supersuite includes suites, batholiths and stocks locally named Brasilândia, Derribadinha, Divino, Estrela-Muniz Freire, Galiléia, Guarataia, Manhuaçu, Mascarenhas-Baixo Guandu, Muriaé, São Vitor, Teófilo Otoni, Valentim and others.

The G1 supersuite mainly consists of tonalite to granodiorite stocks and batholiths, with mafic to diorite facies and autoliths, regionally deformed during the Brasiliano orogeny but with locally well-preserved magmatic features (Figs 1a & 2). This supersuite also includes metamorphosed orthopyroxene-bearing rocks, ranging in composition from monzogabbro to quartz monzonite (Novo 2009). Supracrustal correlatives of the G1 supersuite are metamorphosed pyroclastic and volcaniclastic rocks of dacite to rhyolite composition and volcanic arc signature, dated around 585 Ma (Vieira 2007).

Data from almost two hundred samples, from several plutonic and volcanic G1 bodies, outline a predominant medium- to high-K calc-alkaline (Fig. 3), metaluminous (Fig. 4), pre-collisional signature (Fig. 5), representing a magmatic arc formed on an active continental margin setting, from *c.* 630 Ma to *c.* 585 Ma.

This magmatic arc shows a hybrid isotopic signature, in agreement with Sm–Nd isotope data

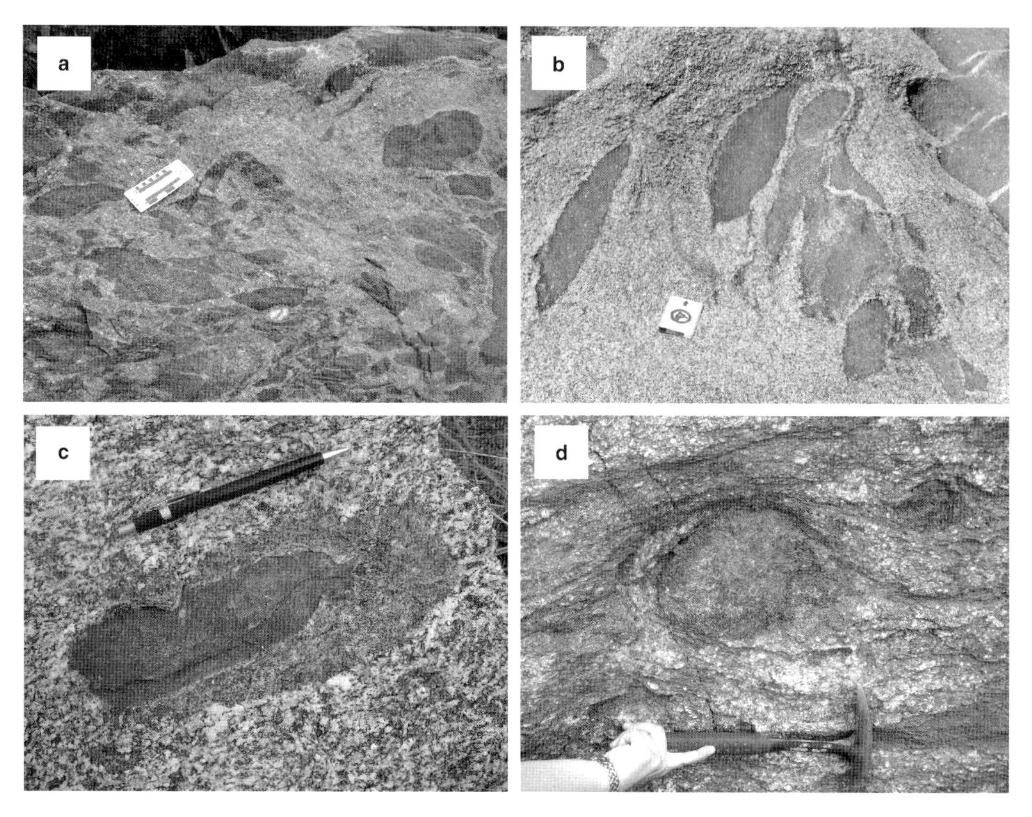

Fig. 2. Features of the G1 supersuite: (**a**) the Manhuaçu tonalite crowded with mafic autoliths, both well-preserved from the regional deformation. (**b**) stretched mafic enclaves along the regional solid-state foliation of the Galiléia suite. (**c**) mafic autholith partially assimilated by the Galiléia tonalite. (**d**) highly deformed Derribadinha tonalitic orthogneiss with stretched and rotated amphibolite enclaves.

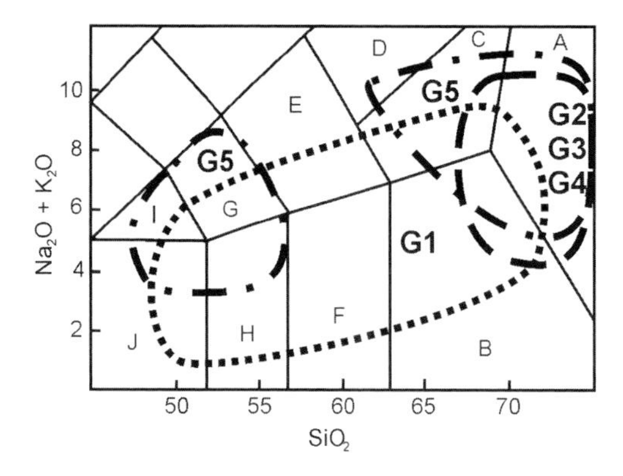

Fig. 3. Major compositional fields of the G1 (dotted line), G2, G3 and G4 (dashed line), and G5 (dashed–dotted line) supersuites of the Araçuaí orogen, in the $Na_2O + K_2O$ *vs.* SiO_2 diagram (data references in text). A, granite; B, granodiorite and tonalite; C, quartz monzonite; D, syenite; E, monzonite; F, diorite; G, monzodiorite; H, gabbro-diorite; I, monzogabbro; J, gabbro. Data from orthopyroxene-bearing rocks of G1, G2 and G5 supersuites are also represented in the diagram.

(epsilon Nd = −5 to −13; T_{DM} model ages = 1.2 to 2.1 Ga), as well as U–Pb ages of inherited zircon grains that suggest significant contribution of a Palaeoproterozoic basement, with involvement of mantle components related to subduction of Neoproterozoic oceanic lithosphere.

Actually, Neoproterozoic ophiolites occur to the west of the G1 magmatic arc, suggesting subduction to the east (in relation to Fig. 1a), but the northern sector of the Araçuaí orogen remained ensialic (Pedrosa-Soares *et al.* 1998, 2001*a*, 2008; Queiroga *et al.* 2007). Such a confined basin (i.e. an inland-sea basin like a partially oceanized gulf) implies in subduction of a relatively small amount of oceanic lithosphere, supporting the quoted epsilon Nd negative values and the new Sm–Nd isotopic data presented below.

New isotopic Sm–Nd data

The Neoproterozoic mantle contribution is obvious in G1 bodies that show large amounts of gabbroic/noritic facies and/or autoliths, such as Brasilândia,

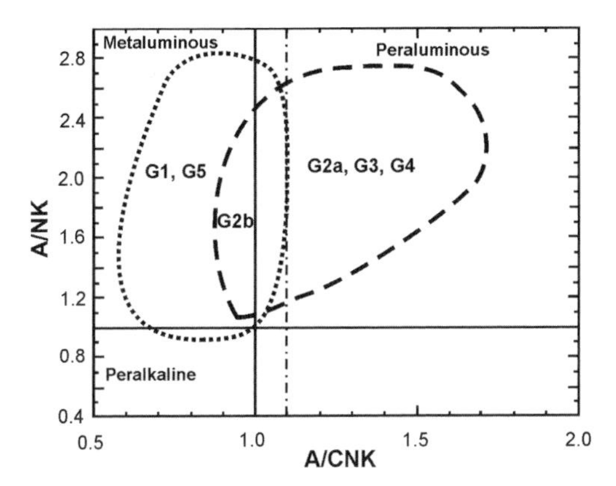

Fig. 4. General relations of G1 and G5 (dotted line), and G2, G3 and G4 (dashed line) supersuites of the Araçuaí orogen in the alumina saturation diagram (data references in text): A/CNK = $Al_2O_3/(Na_2O + K_2O)$; A/NK = $Al_2O_3/(Na_2O + K_2O + CaO)$. G2a, peraluminous rocks; G2b, metaluminous rocks. The dashed–dotted line (A/CNK = 1.1) limits the I-type (A/CNK < 1.1) and S-type (A/CNK > 1.1) fields from Chappel & White (2001).

Divino, Galiléia, Manhuaçu and Valentin (Nalini-Junior *et al.* 2000*b*; Martins *et al.* 2004; Noce *et al.* 2006; Novo 2009; H. Söllner & C. De Campos, pers. comm. 2008; and data presented below).

In this work we present new isotopic Sm–Nd data for the Brasilândia stock and São Vitor batholith (Table 1). The Brasilândia stock (*c.* 595 Ma; Noce *et al.* 2000), located to the east of São José da Safira (B in Fig. 1a), is a deformed tonalite intrusion rich in metadiorite to metagabbro facies and autoliths. The São Vitor batholith (*c.* 585 Ma; Whittington *et al.* 2001) has been reported as relatively

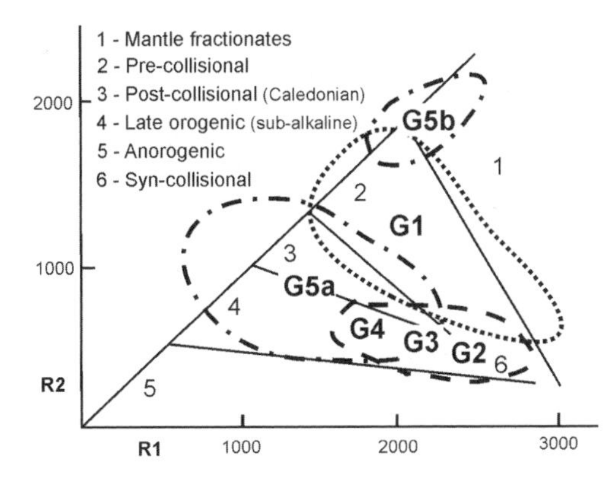

Fig. 5. General relations of G1 (dotted line), G2, G3 and G4 (dashed line), and G5 (dotted–dashed line) supersuites of the Araçuaí orogen in the R1–R2 diagram of De La Roche *et al.* (1980). G5a, high- and very high-K rocks; G5b, tholeiitic rocks.

poor of mafic enclaves (Pinto *et al.* 2001). However, an outcrop located around 50 km from both the NE of Governador Valadares and east of São José da Safira (SV in Fig. 1a) is crowded with well-preserved diorite autoliths. Both Brasilândia and São Vitor outcrops studied show clear evidence of magma mixing processes.

This evidence together with the new Sm–Nd isotopic data (epsilon Nd = −6.9 to −9.1; T_{DM} model ages = 1.62 to 2.09 Ga; Table 1) support the idea that the gabbro-norite to diorite facies and autoliths record the source of Neoproterozoic mantle magma involved in the crustal melting of an essentially Palaeoproterozoic basement, during the generation of the G1 supersuite (see also Nalini-Junior *et al.* 2000*b*, 2005; Martins *et al.* 2004).

New U–Pb geochronological data

U–Pb zircon ages for the G1 supersuite constrain magma crystallization between *c.* 630 and *c.* 585 Ma. Nevertheless, U–Pb ages from zircon overgrowths and zircon crystals from neosomes of migmatized G1 bodies, as well as U–Pb ages from monazite and some Pb–Pb evaporation ages in zircons, ranging from *c.* 575 to *c.* 560 Ma fall in the time interval of the syn-collisional stage that imprinted the main regional foliation under high temperature metamorphic conditions, causing syn-cinematic metamorphic recrystallization and partial melting on G1 rocks (e.g. Söllner *et al.* 1991; Nalini-Junior *et al.* 2000*b*; Noce *et al.* 2000; Whittington *et al.* 2001; Silva *et al.* 2002, 2005; De Campos *et al.* 2004; Vieira 2007; Novo 2009; Petitgirard *et al.* 2009).

New U–Pb zircon ages for the Manhuaçu stock and Estrela-Muniz Freire batholith are presented in this work. Both new geochronological datasets presented here further support time constraints previously obtained and published for the G1 magmatic event.

The Manhuaçu stock is a newly recognized G1 body (Noce *et al.* 2006) located between Caratinga and Espera Feliz, in the southern region of the Araçuaí orogen (M in Fig. 1a). It consists of biotite-hornblende tonalite crowded with mafic autoliths (Fig. 2a). The analysed sample was taken from the tonalite groundmass. Twenty-seven U–Pb laser ablation–inductively coupled plasma–mass spectrometry (LA–ICP–MS) spot analyses were carried out on 15 short-prismatic zircon crystals (Table 2). Eleven crystals display inherited cores and five of them yielded Mesoproterozoic and Palaeoproterozoic ages (Fig. 6). Analysis for the remaining inherited cores plot around 600 Ma close to the tonalite magmatic age, pointing towards a severe lead-loss. Spot analysis of magmatic zircon grains or overgrowths plot on a

Table 1. *Sm–Nd isotopic data for samples from the G1 supersuite (Brasilândia stock and São Vitor batholith, locations in Fig. 1a)*

Rock, G1 body or suite	Sm (ppm)	Nd (ppm)	$^{147}Sm/^{144}Nd$	$^{143}Nd/^{144}Nd$ (±2SE)	εNd(590 Ma)	T_{DM} (Ga)
Monzodiorite facies, Brasilândia	7.997	41.14	0.1175	0.511978 ± 6	−6.92	1.67
Gabbro autolith, Brasilândia	10.103	46.078	0.1325	0.511924 ± 5	−9.11	2.09
Gabbro-diorite facies, Brasilândia	9.884	47.61	0.1255	0.512012 ± 5	−6.86	1.77
Quartz-rich tonalite, São Vitor	5.460	28.225	0.1169	0.511899 ± 6	−8.42	1.79
Diorite autolith, São Vitor	4.173	18.449	0.1367	0.512023 ± 7	−7.49	2.00
Tonalite, São Vitor	5.589	30.135	0.1121	0.511955 ± 6	−6.96	1.62

discordia line with a poorly constrained upper intercept age of 600 ± 32 Ma. A concordia age was calculated from the eight near-concordant spots at 597 ± 3 Ma (Fig. 6), which is assumed as the best estimate for the magmatic crystallization age of the Manhuaçu tonalite.

The Estrela-Muniz Freire batholith is one of the easternmost records of the G1 supersuite in the Araçuaí orogen (MF in Fig. 1). This batholith mainly comprises medium- to coarse-grained granodioritic to tonalitic gneisses, locally migmatized, with prominent augen structure and stretched mafic enclaves (De Campos *et al.* 2004 and references therein). Twenty five clean zircon crystals, free of inherited cores, extracted from a foliated tonalite, yielded a U–Pb LA–ICP–MS concordia age of 588 ± 4 Ma (Table 3, Fig. 7). This value is assumed to be the magmatic crystallization age of this portion of the Estrela–Muniz Freire batholith (MF in Fig. 1). This age corroborates previous zircon U–Pb dating for this batholith (Söllner *et al.* 1991).

G2 supersuite

This supersuite mostly includes S-type granites formed during the syn-collisional stage (*c.* 585–560 Ma) of the Araçuaí orogen, such as the units called Ataléia, Carlos Chagas, Montanha, Nanuque, Pescador, Urucum and Wolf (e.g. Nalini-Junior *et al.* 2000*a*, *b*; Pedrosa-Soares & Wiedemann-Leonardos 2000; Pinto *et al.* 2001; Silva *et al.* 2002, 2005; Noce *et al.* 2000, 2006; Castañeda *et al.* 2006; Pedrosa-Soares *et al.* 2006*a*, *b*, 2008; Baltazar *et al.* 2008; Roncato 2009; and references therein). However, G2 supersuite also comprises I-type melts generated from the syn-collisional migmatization of the Palaeoproterozoic basement (e.g. De Campos *et al.* 2004; Noce *et al.* 2006, 2007; Novo 2009). In fact, data from the authors quoted, outline a restricted compositional range (Fig. 3), mostly peraluminous (Fig. 4) with a syn-collisional geochemical signature (Fig. 5).

From the Doce River to the north, the G2 supersuite makes up a huge batholith mostly composed of peraluminous S-type granites (Fig. 1a). Few of them are typical muscovite-bearing leucogranites, crystallized in relatively shallow crustal levels (e.g. the Urucum suite located to the north of Conselheiro Pena; Nalini-Junior *et al.* 2000*a*, *b*). However, most G2 granites occurring to the north of the Doce River vary in composition from prevailing garnet–biotite granite to minor garnet-rich granodiorite to tonalite (Pinto *et al.* 2001; Castañeda *et al.* 2006; Pedrosa–Soares *et al.* 2006*a*, *b*; Baltazar *et al.* 2008; Roncato 2009). These garnet-rich granodiorite and tonalite represent autochthonous to semi-autochthonous anatectic melts mainly from plagioclase-rich peraluminous gneisses, like those cropping out in the Nova Venécia region (Roncato 2009). These paragneisses reached peak metamorphic conditions around 820°C at 6.5 kbar (Munhá *et al.* 2005) and were the source of very large amounts of granitic melts produced during the syn-collisional stage.

To the south of the Doce river, where deeper crustal levels and the tholeiitic to calc-alkaline Palaeoproterozoic basement are widely exposed, the G2 supersuite also includes anatectic metaluminous (I-type) melts represented by hornblende-bearing orthogneisses and orthopyroxene-bearing charnockitic rocks (e.g. De Campos *et al.* 2004; Horn 2006; Noce *et al.* 2006, 2007; Novo 2009). Along this region, the most common G2 peraluminous rocks are garnet–biotite granite to granodiorite that form relatively small bodies.

Similar to the G1 supersuite, G2 rocks generally display the regional solid-state foliation. However, locally they may present very well-preserved magmatic fabrics. The huge Carlos Chagas batholith, located west of Nova Venécia, from north of the Doce river to the surroundings of Nanuque

Table 2. *Summary of LA–ICP–MS U–Pb zircon data for sample CN–341, Manhuaçu stock (Fig. 1a). Spot analysis performed at the Laboratory of Geochronology, Federal University of Rio Grande do Sul, Brazil. 1, Sample and standard are corrected after Pb and Hg blanks; 2, $^{207}Pb/^{206}Pb$ and $^{206}Pb/^{238}U$ are corrected after common Pb presence. Common Pb assuming $^{206}Pb/^{238}U$ and $^{207}Pb/^{235}U$ concordant age; 3, $^{235}U = 1/137.88 * U_{total}$; 4, Standard GJ–1; 5, $Th/U = ^{232}Th/^{238}U * 0.992743$; 6, All errors in the table are calculated 1 sigma (% for isotope ratios, absolute for ages)*

Spot	Ratios							Age (Ma)											
	$^{207}Pb/^{235}U$	±	$^{206}Pb/^{238}U$	±	Rho 1	$^{207}Pb/^{206}Pb$	±	$^{206}Pb/^{238}U$	±	$^{207}Pb/^{235}U$	±	$^{207}Pb/^{206}Pb$	±	$^{232}Th/^{238}U$	% Disc	f206	Th (ppm)	U (ppm)	Pb (ppm)
1a	0.82988	2.22	0.09909	1.54	0.70	0.06074	1.60	609	9	614	14	630	10	0.46	3	0.0002	45.9	540.4	63.2
1b	0.82034	2.48	0.09944	1.44	0.58	0.05983	2.02	611	9	608	15	598	12	0.08	−2	0.0005	69.2	170.3	19.1
2	0.80026	2.16	0.09748	1.30	0.60	0.05954	1.72	600	8	597	13	587	10	0.05	−2	0.0001	40.5	989.3	96.3
3a	0.80743	2.40	0.09813	1.30	0.54	0.05968	2.01	603	8	601	14	592	12	0.32	−2	0.0003	241.3	752.6	97.0
3b	4.56897	1.89	0.30801	1.39	0.74	0.10758	1.28	1731	24	1744	33	1759	23	0.48	2	0.0003	91.1	192.4	73.5
4a	0.82285	2.87	0.09967	2.11	0.74	0.05987	1.94	612	13	610	17	599	12	0.25	−2	0.0002	213.4	844.4	88.4
4b	0.82059	3.65	0.09984	2.00	0.55	0.05961	3.05	613	12	608	22	589	18	0.41	−4	0.0007	42.2	106.0	10.7
5a	0.78666	3.23	0.0945	2.57	0.80	0.06037	1.96	582	15	589	19	617	12	0.06	6	0.0006	14.1	262.1	25.0
5b	2.63147	2.02	0.22474	1.42	0.70	0.08492	1.43	1307	19	1309	26	1314	19	0.46	1	0.0003	83.8	189.5	45.1
6	0.76451	2.38	0.09266	1.62	0.68	0.05984	1.75	571	9	577	14	598	10	0.42	4	0.0000	517.5	1243.9	127.0
7a	0.79119	2.41	0.0943	1.32	0.55	0.06083	2.02	581	8	592	14	633	13	0.01	8	0.0012	12.4	950.8	98.2
7b	0.81707	3.39	0.0995	1.59	0.47	0.05957	2.99	611	10	606	21	588	18	0.01	−4	0.0006	27.2	2247.8	74.8
9a	0.82890	4.48	0.10218	1.84	0.41	0.05883	4.08	627	12	613	27	561	23	0.32	−12	0.0015	42.9	136.2	15.1
9b	2.12422	3.50	0.19352	1.37	0.39	0.07961	3.22	1140	16	1157	40	1187	38	1.13	4	0.0012	107.8	113.3	30.3
10a	0.82807	3.50	0.10079	1.69	0.48	0.05959	3.06	619	10	613	21	589	18	0.38	−5	0.0002	54.3	146.3	11.8
10b	0.78391	2.45	0.09508	1.58	0.64	0.05980	1.87	585	9	588	14	596	11	0.12	2	0.0002	31.6	234.5	24.7
11a	0.75805	3.89	0.09432	2.87	0.74	0.05829	2.63	581	17	573	22	541	14	0.36	−7	0.0010	71.7	263.2	6.6
11b	0.92756	2.50	0.1082	1.87	0.75	0.06217	1.66	662	12	666	17	680	11	0.15	3	0.0005	105.1	705.0	24.9
13a	0.79499	3.45	0.0960	2.69	0.78	0.06004	2.16	591	16	594	20	605	13	0.02	2	0.0001	42.7	1816.3	56.7
13b	4.65872	1.60	0.2949	1.01	0.63	0.11456	1.24	1666	17	1760	28	1873	23	1.52	11	0.0005	486.5	338.4	52.0
21a	0.80233	2.34	0.09678	1.43	0.61	0.06013	1.86	595	8	598	14	608	11	0.35	2	0.0014	74.3	212.7	25.1
21b	0.85379	3.17	0.10313	1.92	0.61	0.06005	2.52	633	12	627	20	605	15	0.33	−5	0.0014	30.9	93.6	11.1
22a	0.81574	3.42	0.09886	1.88	0.55	0.05985	2.85	608	11	606	21	598	17	0.59	−2	0.0008	48.1	64.5	8.9
22b	0.77027	3.75	0.09354	1.93	0.51	0.05972	3.22	576	11	580	22	594	19	0.58	3	0.0005	43.2	74.3	9.7
23	0.76677	2.91	0.0933	1.73	0.59	0.05958	2.34	575	10	578	17	588	14	0.12	2	0.0004	65.0	475.3	12.4
28a	0.76699	4.46	0.09246	2.54	0.57	0.06016	3.66	570	14	578	26	609	22	0.33	6	0.0008	139.5	568.5	12.7
28b	4.60593	2.39	0.2781	1.35	0.56	0.12012	1.98	1582	21	1750	42	1958	39	0.39	19	0.0018	117.6	557.0	30.9

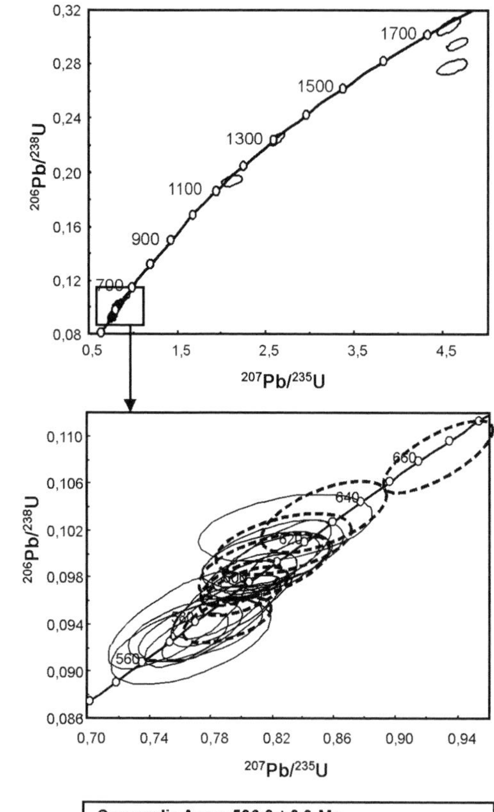

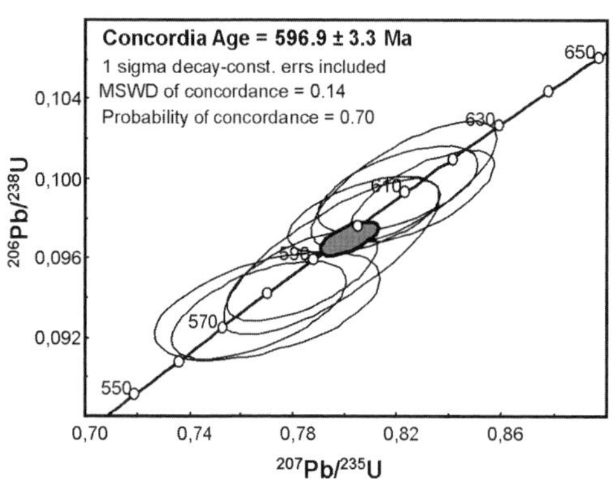

Fig. 6. U–Pb concordia diagrams for the Manhuaçu tonalite groundmass (sample CN-341, concordia age = 596.9 ± 3.3 Ma; MSWD = 0.14; probability of concordance = 0.70). Spot analysis on zircon crystals by LA-ICP-MS, Geochronology Lab., Federal University of Rio Grande do Sul, Brazil.

(Fig. 1a), provides a superb example of the heterogeneous deformation imprinted on granitic bodies during the syn-collisional stage (Figs 8 & 9). This batholith mainly includes granites of the Carlos Chagas, Montanha and Nanuque suites that only differ from each other due to slight differences in mafic mineral content. Therefore, the Carlos Chagas batholith is quite homogeneous in composition, consisting of very coarse- to medium-grained garnet–biotite granite crowded with K-feldspar phenocrysts (Pinto *et al.* 2001; Castañeda *et al.* 2006; Pedrosa-Soares *et al.* 2006*a*, *b*; Baltazar *et al.* 2008; Roncato 2009). Nevertheless, the Carlos Chagas batholith is structurally heterogeneous and shows both small and large areas displaying clear magmatic features, which were preserved from the syn-collisional deformation that widely imprinted the regional solid-state foliation to the G2 granites (Fig. 8).

In the Carlos Chagas batholith, the regional foliation is marked by the ductile deformation of quartz, feldspars and garnet, and by recrystallization of stretched garnet and oriented sillimanite (Fig. 9). This aluminosilicate formed from the breakdown of biotite that tends to disappear in high strain zones. Our quantitative geothermometric data suggest that the garnet–sillimanite-bearing assemblage is syncinematic to the solid-state foliation and recrystallized around 660 °C (Table 4). This temperature is compatible with the stability of the magmatic mineral assemblage of granitic composition so that no syncinematic retrograde reaction has been observed. Furthermore, the solid-state deformation imprinted to the Carlos Chagas batholith immediately follows the magmatic crystallization so that both deformed and non-deformed facies have consistently yielded zircon U–Pb ages around 575 Ma (Silva *et al.* 2002, 2005; Vauchez *et al.* 2007; Roncato 2009).

Also according to U–Pb ages, the oldest G2 granites have been dated around 585–582 Ma and the youngest ones appear to be *c.* 560 Ma old, but the climax of G2 granite generation seems to have taken place around 575 Ma (Söllner *et al.* 1991; Nalini-Junior *et al.* 2000*a*; Noce *et al.* 2000; Silva *et al.* 2002, 2005; Vauchez *et al.* 2007; Roncato 2009).

The autochtonous plagioclase–garnet-rich (Ataléia) facies of G2 granites, as well as other G2 granites, commonly shows inherited zircon grains with magmatic U–Pb ages between 630 Ma and 590 Ma, and metamorphic overgrowths of *c.* 575 Ma (Silva *et al.* 2002, 2005; Roncato 2009). Despite their clear association with the regional plagioclase-rich paragneisses, that can be their melting sources, these plagioclase–garnet-rich granites (*s.l.*) have been interpreted as being G1 bodies by some authors (De Campos *et al.* 2004; Vieira 2007).

Extensive bodies of foliated garnet-two-mica granite, located along the Jequitinhonha River valley to the east of Araçuaí (Fig. 1), yielded zircon U–Pb ages around 540 Ma (Whittington *et al.* 2001; Silva *et al.* 2007). This garnet–two-mica granite shows a persistent solid-state

Table 3. *Summary of LA–ICP–MS U–Pb zircon data for the Muniz Freire batholith (sample OPU–1412; MF in Fig. 1a). Spot analysis performed at the Laboratory of Geochronology, University of Brasília, Brazil, according to methodology described in Della Giustina* et al. *(2009). All errors in the table are calculated 1 sigma (% for isotope ratios, absolute for ages)*

Spot	$^{206}Pb/^{204}Pb$ ratio	$^{207}Pb/^{206}Pb$ ratio	±	$^{207}Pb/^{235}U$ ratio	±	$^{206}Pb/^{238}U$ ratio	±	$^{207}Pb/^{206}Pb$ age (Ma)	±	$^{207}Pb/^{235}U$ age (Ma)	±	$^{206}Pb/^{238}U$ age (Ma)	±	Rho	Conc (%)
Z1	265	0.05917	2.61302	0.74725	0.03557	0.09292	0.00442	541.8	17.2	566.6	20.5	572.8	26.0	0.57	99.9
Z3	887	0.05937	1.88023	0.78329	0.02405	0.09435	0.00290	611.0	8.3	587.4	13.6	581.2	17.0	0.83	100.1
Z4	477	0.05963	2.10423	0.78583	0.03025	0.09585	0.00369	583.8	13.2	588.8	17.1	590.1	21.7	0.28	100.0
Z5	520	0.05958	0.37564	0.78597	0.05227	0.09558	0.00636	590.3	20.1	588.9	29.3	588.5	37.3	0.58	100.0
Z6	546	0.05971	0.21052	0.78191	0.02967	0.09625	0.00365	564.0	12.8	586.6	16.8	592.4	21.4	0.71	99.9
Z7	1015	0.05996	6.39302	0.80346	0.09301	0.09785	0.01133	587.3	27.0	598.8	51.1	601.8	66.2	0.40	100.0
Z8	4310	0.06022	2.47862	0.81147	0.03199	0.09937	0.00392	575.3	4.0	603.3	17.8	610.7	22.9	0.94	99.9
Z9	3526	0.05986	4.80301	0.79460	0.06362	0.09722	0.00778	577.0	11.7	593.8	35.4	598.1	45.6	0.88	99.9
Z10	1373	0.05967	1.51777	0.78086	0.02174	0.09603	0.00267	566.0	6.9	586.0	12.3	591.1	15.7	0.64	99.9
Z11	2354	0.06024	5.92345	0.82247	0.08681	0.09957	0.01051	600.1	36.8	609.4	47.3	611.9	61.3	0.50	100.0
Z12	4823	0.05904	7.36983	0.75986	0.10145	0.09228	0.01232	593.3	22.4	573.9	56.9	569.0	72.3	0.93	100.1
Z13	3571	0.06007	3.00000	0.85984	0.04684	0.09895	0.00539	708.7	17.6	630.0	25.3	608.3	31.5	0.14	100.3
Z14	425	0.06024	3.07476	1.00495	0.04929	0.10106	0.00496	989.1	11.6	706.3	24.7	620.6	29.0	0.84	101.4
Z15	8167	0.05970	7.91294	0.78607	0.10698	0.09626	0.01310	575.3	16.7	588.9	59.1	592.4	76.6	0.78	99.9
Z28	6532	0.05969	4.06494	0.79979	0.05672	0.09630	0.00683	611.9	16.1	596.7	31.5	592.7	40.0	0.76	100.1
Z17	13217	0.05902	1.76462	0.76164	0.02123	0.09218	0.00257	600.7	6.4	575.0	12.2	568.4	15.1	0.84	100.1
Z18	25136	0.05993	3.41522	0.79613	0.04342	0.09761	0.00532	572.5	3.0	594.6	24.3	600.4	31.2	0.94	99.9
Z19	3139	0.05913	3.02168	0.76632	0.04442	0.09283	0.00538	598.7	18.7	577.6	25.2	572.3	31.7	0.76	100.1
Z20	2024	0.06023	1.05640	0.85547	0.15492	0.09981	0.01808	679.5	25.6	627.6	81.4	613.3	105.1	0.33	100.2
Z21	1533	0.05978	1.63828	0.79573	0.02130	0.09678	0.00259	590.1	12.1	594.4	12.0	595.5	15.2	0.55	100.0
Z22	26204	0.05982	1.20364	0.79044	0.01584	0.09696	0.00194	571.6	5.7	591.4	8.9	596.6	11.4	0.68	99.9
Z23	469	0.05964	1.52525	0.95030	0.02170	0.09728	0.00222	952.9	62.3	678.3	11.2	598.5	13.0	0.04	101.3
Z24	6295	0.05973	8.63920	0.80049	0.11796	0.09654	0.01423	608.4	8.1	597.1	64.4	594.1	83.1	0.51	100.0
Z25	1354	0.05970	2.19985	0.78305	0.02799	0.09625	0.00344	567.1	6.8	587.2	15.8	592.4	20.2	0.82	99.9

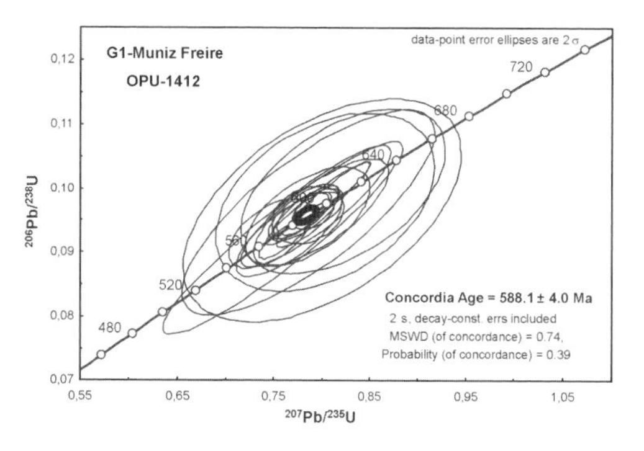

Fig. 7. U–Pb concordia diagram for the Muniz Freire tonalite (concordia age = 588.1 ± 4.0 Ma; MSWD = 0.74; probability of concordance = 0.39). Analysis performed on zircon crystals by LA–ICP–MS, Geochronology Lab., University of Brasília, Brazil.

foliation concordant with the regional structural trend, and hosts patches and veins (neossomes) of cordierite-garnet leucogranite, similar to G3 leucogranite, overprinting its foliation. Consequently, those bodies of garnet-two-mica granite were considered to belong to the G2 supersuite by Pedrosa-Soares & Wiedemann-Leonardos (2000) and Pedrosa-Soares *et al.* (2001*a*, 2008). The presence of primary muscovite indicates that these granites crystallized at relatively shallow levels, especially when compared to G2 garnet–biotite granites formed at deeper crustal levels. In this case, the emplacement of G2 granites would have lasted until the late collisional stage. Alternatively, Whittington *et al.* (2001) suggested that the solid-state foliation shown by the garnet-two-mica granite could be related to the gravitational collapse of the Araçuaí orogen (Marshak *et al.* 2006). Besides, in restricted geochronological terms, the foliated garnet-two-mica granite was considered to belong to the G3 supersuite (Pinto 2008; Silva *et al.* 2007). If this is the case, neossomes of cordierite–garnet leucogranite overprinting the foliation of the garnet-two-mica granite remains an open question. A possible explanation could be related to the presence of huge G5 intrusions close to garnet-two-mica granite bodies. In this case the cordierite–garnet neossomes could have resulted from dehydrating migmatization in contact aureoles (see section on G3 supersuite). Actually, those foliated garnet-two-mica granite bodies provide interesting problems to solve.

G3 supersuite

One important feature, commonly observed in outcrop, is the re-melting process of G2 S-type granites, leading to the formation of G3 leucogranites from the late collisional to the post-collisional stages. The G3 supersuite is much less voluminous than the other granitic supersuites of the Araçuaí orogen and generally occurs intimately associated with the G2 supersuite. G3 bodies have quite a few formal names, such as Água Branca, Água Boa, Almenara, Barro Branco, Itaobim and Poranga (Pedrosa-Soares & Wiedemann-Leonardos 2000; Pedrosa-Soares *et al.* 2001*a*, 2006*a*, *b*, 2008; De Campos *et al.* 2004; Silva *et al.* 2005, 2007; Castañeda *et al.* 2006; Baltazar *et al.* 2008; Drumond & Malouf 2008; Gomes 2008; Heineck *et al.* 2008; Junqueira *et al.* 2008; Paes *et al.* 2008; Pinto 2008).

The most typical rocks in the G3 supersuite are leucogranites with variable garnet and/or cordierite and/or sillimanite contents. They are free of the regional solid-state foliation, and mainly occur as autochthonous patches and veins to semi-autochthonous small plutons, hosted by the parent G2 granites (Fig. 10). These G3 rocks are sub-alkaline (rich in K-feldspar; Fig. 3), peraluminous (Fig. 4), late orogenic (Fig. 5) S-type granites (De Campos *et al.* 2004; Castañeda *et al.* 2006; Pedrosa-Soares *et al.* 2006*a*, *b*; Queiroga *et al.* 2009; Roncato *et al.* 2009).

The typical G3 leucogranites are particularly abundant to the north of Doce River, where they are associated with several G2 granites (Fig. 10). However, south of the Doce River, where deeper crustal levels are generally exposed, the G3 supersuite also includes charnockite to enderbite, especially along the coastal region of Espírito Santo (De Campos *et al.* 2004).

As mentioned before, cordierite–garnet leucogranites are also formed in contact aureoles of G5 intrusions, and both the intrusive and host rocks show very similar U–Pb monazite ages (Whittington *et al.* 2001). Indeed, these leucogranites do not belong to the G3 supersuite. They are products of contact metamorphism and partial melting of mica-bearing host rocks, catalysed by the high temperatures released from the crystallizing G5 intrusion (Whittington *et al.* 2001; Queiroga *et al.* 2009; Roncato *et al.* 2009).

Zircon and monazite U–Pb ages suggest that most G3 cordierite–garnet leucogranites crystallized between *c.* 540 Ma and *c.* 525 Ma (Whittington *et al.* 2001; Noce *et al.* 2004; Pedrosa-Soares *et al.* 2006*a*, *b*; Silva *et al.* 2005, 2007; and new data presented below).

New U–Pb geochronological data

An outcrop along a road cut located close to Nanuque (N in Fig. 1a) was sampled in order to compare the ages of G3 and G2 granites in the same outcrop. The G2 sample was collected from a foliated porphyroclastic garnet–biotite granite of

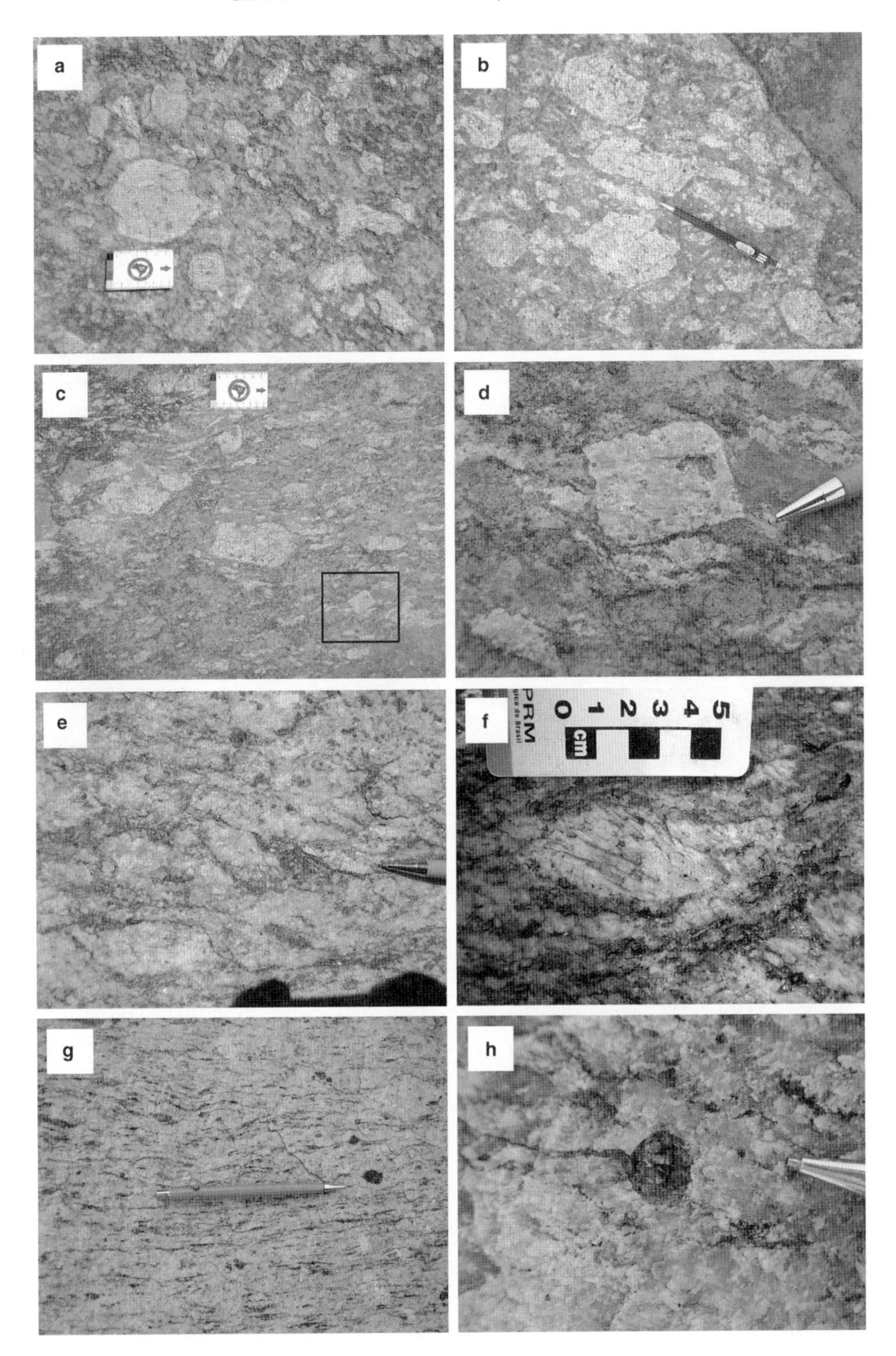

Fig. 8. Structural features of the G2 supersuite, illustrated by the Carlos Chagas suite in Nova Venécia region (Fig. 1): (**a**) magmatic isotropic structure with disordered K-feldspar phenocrysts. (**b**) igneous flow structure marked by oriented K-feldspar phenocrysts. (**c**) incipient solid-state deformation along the igneous flow (**d**) detail of photo C showing a slightly deformed K-feldspar phenocryst. (**e**) solid-state foliation, parallel to the igneous flow, with development of augen-structure. (**f**) detail showing a sigmoidal K-feldspar porphyroclast. (**g**) strongly stretched fabric with well-developed solid-state foliation. (**h**) detail showing deformed rotated garnet with recrystallization tail.

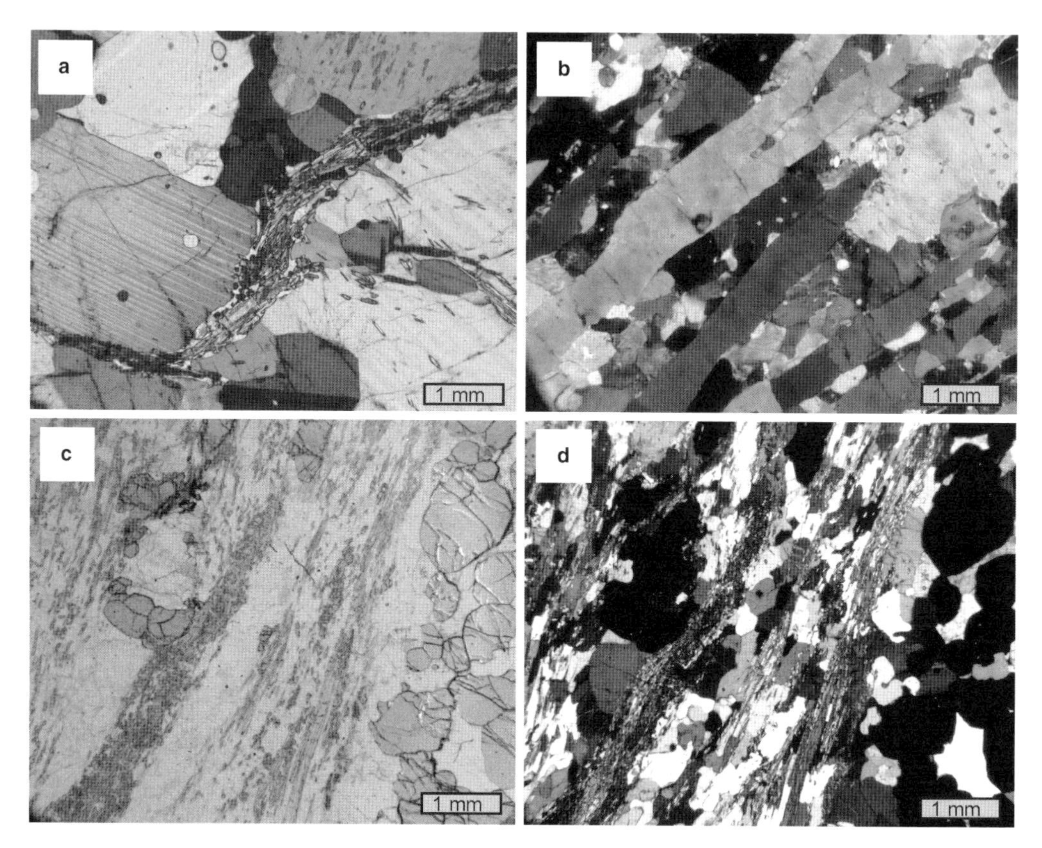

Fig. 9. Microscopic features of the Carlos Chagas deformed granite. (**a**) incipient development of solid-state foliation marked by oriented recrystallization of biotite–sillimanite trails in the quartz-feldspar matrix (polarized light). (**b**) outstanding solid-state foliation marked by recrystallization of quartz ribbons and stretched feldspars (polarized light). (**c**) penetrative solid-state foliation marked by recrystallization of sillimanite (minor biotite) trails between garnet-rich bands. (**d**) the same of C in polarized light.

the Nanuque suite, which yielded a zircon U–Pb SHRIMP age of 573 ± 5 Ma (previously published by Silva *et al.* 2002).

The G3 sample was taken from an isotropic vein of garnet–cordierite leucogranite, free of the regional foliation, showing gradational contacts with the host G2 granite (Fig. 11). U–Pb thermal ionization mass spectrometry (TIMS) analytical procedures were performed in the Geochronological Research Center of São Paulo University, according to conventional routines. The zircon U–Pb results furnished a very large mean square weighted deviation (MSWD) (31), suggesting that it may represent a single data population but with large detectable geological scatter (Fig. 11, Table 5). Despite of analytical imprecision, the zircon age of 532 ± 11 Ma is validated by a number of well-constrained zircon and monazite U–Pb (LA–ICP–MS and TIMS) ages, in the range between *c.* 540 and 525 Ma, obtained from similar G3 leucogranites located to the north (Whittington *et al.* 2001; Silva *et al.* 2007) and south (Noce *et al.* 2004; Pedrosa-Soares *et al.* 2006*b*) of Nanuque area.

G4 supersuite

This supersuite occurs along the central-northern sector of the Araçuaí orogen, where intermediate to shallow crustal levels of amphibolite to greenschist metamorphic facies are exposed (Fig. 1a), and includes suites, batholiths and plutons called by local names such as Campestre, Caraí, Córrego do Fogo, Itaporé, Laje Velha, Mangabeiras, Piauí, Quati, Santa Rosa and Teixeirinha (Pedrosa 1997*a*, *b*; Pedrosa & Oliveira 1997; Basílio *et al.* 2000; Pinto *et al.* 2001; Heineck *et al.* 2008; Paes *et al.* 2008).

The G4 supersuite consists of balloon-like zoned plutons composed of biotite granite cores and roots,

Table 4. *Geothermobarometric data for samples of the G2 and G3 supersuites, calculated with THERMOCALC software (e.g. Powell & Holland 2006). G2 samples are foliated garnet–biotite–sillimanite leucogranites with mylonitic textures (Carlos Chagas suite) and G3 samples are mica-free non-foliated leucogranites, collected in the Nova Venécia region (Fig. 1a)*

Unit-sample	Mineral assemblage	Associated structure	Mineral chemistry (mol) used in Thermocalc	T (°C)	P (kb)
G2–48	qtz + kfs + pl + grt + bt + sil	Solid-state foliation	py 0.00057, gr 0.0000060, alm 0.64, spss 0.000015, phl 0.0138, ann 0.076, east 0.016, an 0.22, ab 0.82	683	–
G2–50	qtz+ kfs + pl + grt + bt + sil	Solid-state foliation	py 0.0051, gr 0.000056, alm 0.44, spss 0.000016, phl 0.059, ann 0.031, east 0.046, an 0.43, ab 0.71	666	–
G2–54	qtz+ kfs + pl + grt + bt + sil	Solid-state foliation	py 0.0093, gr 0.000110, alm 0.41, spss 0.000017, phl 0.042, ann 0.035, east 0.039, an 0.48, ab 0.68	642	–
G3–55B	qtz + pl + kfs + grt + sil	Isotropic, igneous	py 0.020, gr 0.000076, alm 0.33, spss 0.0000064, phl 0.046, ann, 0.021, east 0.057, an 0.75, ab 0.92	814	5.0
G3–55C	qtz + pl + kfs + cd + sil	Isotropic, igneous	py 0.0028, gr 0.0000149, alm 0.45, phl 0.028, ann 0.055, east 0.026, crd 0.38, fcrd 0.18, an 0.29, ab 0.79	819	4.9

grading into two-mica and muscovite–garnet leucogranite towards the borders, capped by pegmatoid cupolas. G4 plutons commonly show xenoliths and roof-pendants of the country rocks. The emplacement mechanism forced the regional foliation to accommodate around the intrusions, forming post-cinematic curvilinear structures clearly outlined in remote sensing images. Generally, G4 plutons show igneous flow structures and towards the intrusion border this orientation can be parallel to the regional solid-state foliation. Mineral assemblages formed by contact metamorphism and the mineralization of petalite, instead of spodumene, in some pegmatites indicate depths of emplacement between 5 to 15 km. G4 granites are generally peraluminous (Figs 3 & 4), but the biotite-rich facies can be slightly metaluminous (Pedrosa-Soares *et al.* 1987). They represent the sub-alkaline, post-collisional (late orogenic; Fig. 5) S-type event of granite genesis related to the gravitational collapse of the Araçuaí orogen (Pedrosa-Soares & Wiedemann-Leonardos 2000; Pedrosa-Soares *et al.* 2001*a*, 2008; Marshak *et al.* 2006).

Zircon U–Pb data constrain the magmatic crystallization age of the G4 supersuite from *c.* 530 Ma to *c.* 500 Ma (Whittington *et al.* 2001; Paes *et al.* 2008; Silva *et al.* 2005, 2007). The Ibituruna intrusion, a thick quartz syenite sill located in the southern vicinities of Governador Valadares (Fig. 1a) and dated at 534 ± 5 Ma (zircon U–Pb SHRIMP; Petitgirard *et al.* 2009), probably represents the southernmost G4 body of considerable size exposed and preserved from erosion in the Araçuaí orogen.

G5 supersuite

This supersuite represents the most outstanding post-collisional magmatic event related to the gravitational collapse of the Araçuaí orogen (Pedrosa-Soares & Wiedemann-Leonardos 2000). Most G5 bodies occur to the east and north of the pre-collisional magmatic arc (G1 supersuite), following a NE trend to the south of 20° S parallel, and NW trends to the north of this latitude (Fig. 1a).

The G5 supersuite includes suites, batholiths, complex zoned plutons, sills and dykes called by many local names, such as Aimorés, Caladão, Cotaxé, Guaratinga, Lagoa Preta, Lajinha, Medina, Padre Paraíso, Pedra Azul, Pedra do Elefante, Rubim, Santa Angélica, Salomão, Santo Antônio do Jacinto and Várzea Alegre (e.g. Silva *et al.* 1987; Wiedemann 1993; Pinto *et al.* 2001; Wiedemann *et al.* 1995, 2002; Celino *et al.* 2000; Pedrosa-Soares & Wiedemann 2000; De Campos *et al.* 2004; Castañeda *et al.* 2006; Pedrosa-Soares *et al.* 2006*a*, *b*; Queiroga *et al.* 2009).

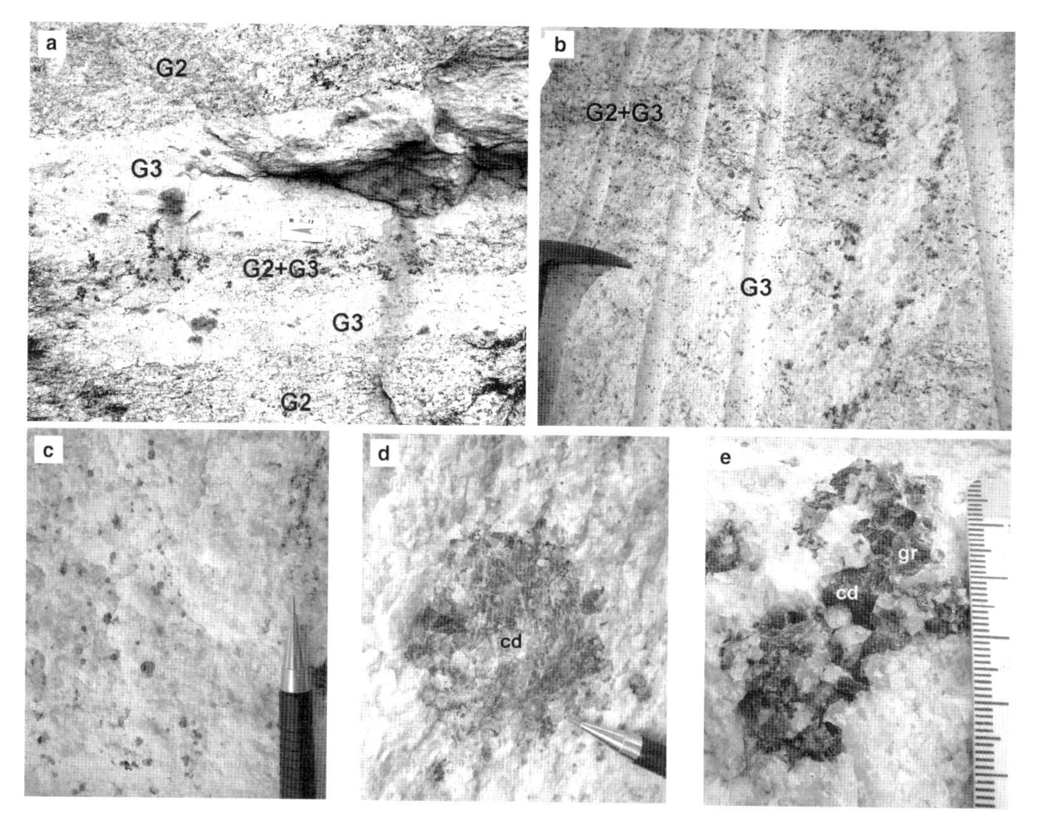

Fig. 10. Photos from G3 leucogranites associated with G2 granites in the Carlos Chagas batholith: (**a**) veins of G3 leucogranite hosted by G2 Ataléia granite separated by a mixture band (G2 + G3) of rather incipient G3 melting. (**b**) typical outcrop of G3 leucogranite with G2 biotite schlieren and restite (G2 + G3). (**c**) G3 leucogranite with fine-grained garnet. (**d**) G3 leucogranite with poikilitic cordierite (cd). (**e**) G3 leucogranite with garnet (gr) and cordierite (cd).

The different crustal levels exposed along the Araçuaí orogen reveal distinct parts, sizes and characteristics of the G5 bodies. In general, crustal depth increases from north to south and from west to east, so that apparently small G5 bodies tend to be exposed in southern and eastern regions of the orogen (Fig. 1a). In fact, such apparently small G5 bodies are deeply eroded plutons, whereas the huge G5 batholiths located in the northern region outline amalgamated intrusions exposed in relatively upper crustal levels.

To the south of the Doce River and in the Nova Venécia region, the G5 bodies range in composition from gabbro-norite to granite, and many of them also include enderbite to charnockite facies, indicating crystallization under high CO_2 fluid pressure. In this region, the host rocks are mainly high grade paragneisses, and rocks of the G1 and G2 supersuites. Deep erosion levels, together with vertical exposures of over 500 m, disclose the internal structure of G5 plutons. The most outstanding features revealed are the roots of diapirs and their inverse zoning, displaying interfingering of mafic to intermediate rocks in the core and syeno-monzonitic to granitic borders, together with widespread evidence of magma mixing (e.g. Bayer *et al.* 1987; Schmidt-Thomé & Weber-Diefenbach 1987; Wiedemann 1993; Mendes *et al.* 1999, 2005; Medeiros *et al.* 2000, 2001, 2003; Wiedemann *et al.* 2002; De Campos *et al.* 2004; Pedrosa-Soares *et al.* 2006*a*, *b*). Metaluminous to peraluminous, high-K calc-alkaline, I-type granitoids (e.g. Horn & Weber-Diefenbach 1987; Wiedemann 1993; Mendes *et al.* 1999) progressively evolve into more markedly alkaline to peralkaline rocks (Ludka *et al.* 1998). These post-collisional melts originated from contrasting sources, involving important mafic contributions from an enriched mantle, partial re-melting from a mainly metaluminous continental crust and dehydration melting from slightly peraluminous rocks (e.g. Wiedemann *et al.* 1995; Ludka *et al.* 1998; Mendes *et al.* 1999; De Campos *et al.* 2004).

To the north of 19° S parallel, G5 exposures tend to be larger and often reach batholithic sizes

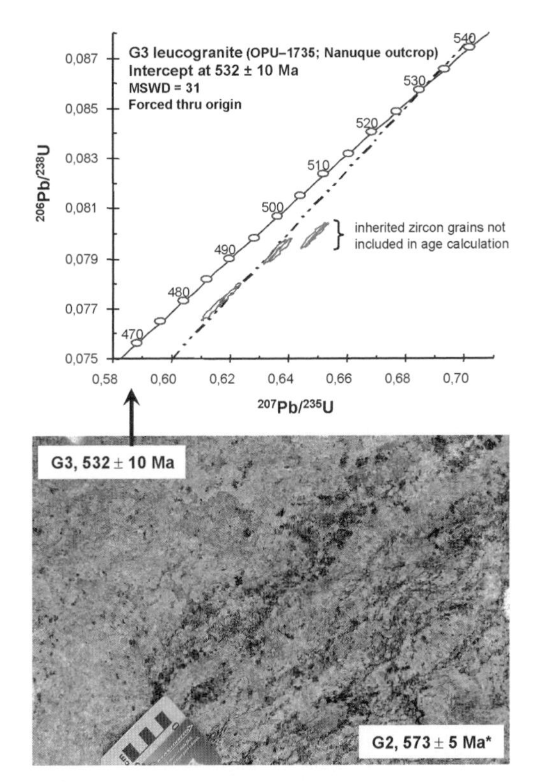

Fig. 11. U–Pb concordia diagram for a sample (OPU-1735) of non-foliated G3 leucogranite from the Nanuque region (upper intercept age = 532 ± 10 Ma; MSWD = 31). Analysis performed on zircon crystals by using TIMS equipment, Geochronology Research Center, University of São Paulo, Brazil. (*) Photo also illustrates the G2 foliated granite dated by Silva *et al.* (2002).

(Fig. 1a). Their host rocks are mainly G2 granites and high grade paragneisses. Coalescent intrusive plutons of mega-porphyritic granites grading into granodiorites, and their charnockite equivalents, form large polydiapiric structures, with the metamorphic foliation of the host rocks wrapped around them. Magma mingling and mixing evidence is widespread. The orientation of crystals by igneous flow is, generally, well developed. Due to the relatively high erosion level mafic cores are absent. Migmatization of the mica-bearing host rocks, forming cordierite-garnet leucogranitic neossomes, can be widespread in aureoles of contact metamorphism close to G5 bodies. G5 granites located to the north of the 19° S parallel show a high-K and high-Fe calc-alkaline to alkaline, I- to A2-type, post-collisional signature (Fernandes 1991; Faria 1997; Achtschin 1999; Celino *et al.* 2000; Whittington *et al.* 2001; Pinto *et al.* 2001; Sampaio *et al.* 2004; Pinto 2008).

A great number of geochemical data from many G5 plutons, located in different sectors of the Araçuaí orogen, reveal an essentially bimodal composition (Figs 3 & 5). This supersuite mainly includes metaluminous to slightly peraluminous (Fig. 4), calc-alkaline to alkaline, I- to A--types, post-collisional granites (Fig. 5), originated in the lowermost continental crust with an important mantle contribution (Bayer *et al.* 1987; Fernandes 1991; Wiedemann 1993; Faria 1997; Achtschin 1999; Mendes *et al.* 1999, 2005; Celino *et al.* 2000; Medeiros *et al.* 2000, 2001, 2003; De Campos *et al.* 2004; Martins *et al.* 2004; Castañeda *et al.* 2006; Pedrosa-Soares *et al.* 2006*a*, *b*; Pinto 2008; Silva *et al.* 2007; Queiroga *et al.* 2009; Roncato *et al.* 2009).

Zircon and monazite U–Pb ages and zircon Pb–Pb evaporation ages constrain the evolution of the G5 supersuite from *c.* 520 to *c.* 480 Ma (Söllner *et al.* 1991, 2000; Noce *et al.* 2000; Medeiros *et al.* 2000; Whittington *et al.* 2001; De Campos *et al.* 2004; Mendes *et al.* 2005; Castañeda *et al.* 2006; Pedrosa-Soares *et al.* 2006*a*, *b*). Rb–Sr and Sm–Nd isotopic data suggest an enriched-mantle reservoir for G5 basic to intermediate rocks (Medeiros *et al.* 2000, 2001, 2003; De Campos *et al.* 2004; Martins *et al.* 2004; Mendes *et al.* 2005).

Mineral resources

Pegmatite and dimension stone deposits are the main mineral resources directly related to the Neoproterozoic–Cambrian granitic magmatism in the region encompassed by the Araçuaí orogen (Correia-Neves *et al.* 1986; Chiodi-Filho 1998; Pinto & Pedrosa-Soares 2001; Dardenne & Schobbenhaus 2003). The most important deposits are associated with the G2 to G5 magmatic events. Besides ordinary building materials, no important mineral resource is hitherto known in the G1 supersuite. However, the recent characterization of dacitic volcanic rocks opens new targets for base metal prospecting on the pre-collisional magmatic arc (Vieira 2007).

The Eastern Brazilian Pegmatite Province

Pegmatite gemstones became officially known in Brazil's history since the last few decades of the 17th century, when green tourmalines were found in eastern Minas Gerais by Fernão Dias Paes Leme, one of the most famous leaders of Brazilian colonizers. In fact, pegmatite gems were found in Brazil before Fernão Dias expedition. Although his challenge was to discover emerald deposits (he was called 'The Emerald Hunter'), he found tourmaline, a mineral already known by former colonizers

Table 5. *Summary of zircon U–Pb TIMS data for the G3 leucogranite (OPU–1735) of Nanuque region (N in Fig. 1a). *1, not corrected for blank or non-radiogenic Pb; *2, Radiogenic Pb corrected for blank and initial Pb; U corrected for blank; Total U and Pb concentrations corrected for analytical blank; Ages given in Ma using Ludwig Isoplot/Ex program; decay constants recommended by Steiger & Jäger (1977)*

Sample no.	Weight (mg)	U (ppm)	Pb (ppm)	$^{206}Pb/^{204}Pb^{*1}$	$^{207}Pb/^{235}U^{*2}$	Error (%)	$^{206}Pb/^{238}U^{*2}$	Error (%)	Error correlation	$^{238}Pb/^{206}Pb$	Error (%)	$^{207}Pb/^{206}Pb$	Error (%)	$^{206}Pb/^{238}U$ Age (Ma)	$^{207}Pb/^{235}U$ Age (Ma)	$^{207}Pb/^{206}Pb$ Age (Ma)
1712	0.097	591.6	44.5	6115.4	0.637368	0.494	0.0793493	0.483	0.97876	12.6025056	0.48	0.0582567	0.101	492	501	539
1713	0.109	684.7	50.0	6687.0	0.619746	0.495	0.0775557	0.486	0.982181	12.8939588	0.49	0.0579561	0.093	482	490	528
1714	0.086	569.7	43.3	1064.9	0.614244	0.533	0.0770086	0.515	0.967843	12.9855627	0.52	0.0578496	0.134	478	486	524
1711	0.091	718.3	53.9	6002.6	0.649652	0.504	0.0799318	0.498	0.989128	12.5106653	0.5	0.0589468	0.074	496	508	565
1715	0.090	578.3	44.2	2851.0	0.648465	0.525	0.0799018	0.516	0.983197	12.5153626	0.52	0.0588612	0.096	496	508	562

and prospectors. Pioneer naturalists and geologists, of first decades of the 19th century, such as Eschwege, Spix and Martius, and Saint-Hilaire, referred to pegmatite gem deposits located in regions of the Jequitinhonha and Doce river valleys.

However, only in the 20th century, particularly during and after the Second World War, pegmatites became important mineral deposits in Brazil, owing to efforts to increase the production of mica, beryl and quartz for the allied countries military industry. This mining development was accompanied by pioneering geological studies and several new mineral specimens were discovered. Accordingly, Brazilian pegmatitic populations were grouped into the Eastern, Northern and Southern Brazilian pegmatite provinces by Paiva (1946).

The Eastern Brazilian Pegmatite Province encompasses a very large region of about 150 000 km^2, from Bahia to Rio de Janeiro, but more than 90% of its whole area is located in eastern Minas Gerais and southern Espírito Santo, in terrains of the Araçuaí orogen where countless pegmatites crystallized from 630–*c.* 480 Ma (Fig. 1). At least one thousand pegmatites have been mined in this province since the beginning of the 20th century, for gems (aquamarine, tourmalines, topaz, quartz varieties and others), Sn, Li and Be ores, industrial minerals (mainly feldspars and muscovite), collection and rare minerals (phosphate and lithium minerals, giant quartz and feldspar crystals, oxides of Ta, Nb, U, and other minerals), dimension stone, and minerals for esoteric purposes. Only Sá (1977), Issa-Filho *et al.* (1980), Pedrosa-Soares *et al.* (1990), Netto *et al.* (2001) and Ferreira *et al.* (2005) record mining activities on more than 800 different pegmatites, from the Rio Doce to the Jequitinhonha river valleys.

However, virtually all pegmatites that became targets for mineral exploration were, and most still are, exploited by means of primitive techniques in small and chaotic ‘mines’ (i.e. digs): the so-called *garimpo* (plural: *garimpos*) and their workers, the *garimpeiros* (singular: *garimpeiro*). Therefore, because of its long history of predatory mining, the Eastern Brazilian Pegmatite Province has become a scenario of economic declining. In the whole province, there is really only one organized mine, the CBL Cachoeira mine (*Companhia Brasileira de Lítio* or Brazilian Lithium Company), that has produced spodumene for lithium ore in the Araçuaí pegmatite district (Romeiro & Pedrosa-Soares 2005). The quarrying for pegmatite dimension stone began early in the 2000s and, in many cases, has been much more economically attractive than the traditional *garimpos*. Many quarries for dimension stone are located on pegmatites and pegmatoid cupolas of intrusive plutons in the Araçuaí and Conselheiro Pena districts.

Since Paiva (1946) the limits and subdivisions of the Eastern Brazilian Pegmatite Province have been redefined and refined, according to more detailed geological maps and analytical data (e.g. Correia-Neves *et al.* 1986; Pedrosa-Soares *et al.* 1990, 2001*b*; Morteani *et al.* 2000; Netto *et al.* 2001; Pinto *et al.* 2001; Pinto & Pedrosa-Soares 2001, and references therein). Two pegmatite districts of the province occur outside the Araçuaí orogen: namely Itambé, located in the São Francisco craton, and Bicas-Mar de Espanha, located in the Ribeira orogen just to the south of the Araçuaí orogen (Fig. 1).

In the Araçuaí orogen, the province can be subdivided into eleven districts, encompassing the most important pegmatite populations, based on their main mineral resources, pegmatite sizes, types and classes, and relations to parent and host rocks (Table 6, Fig. 1a). Most pegmatites of the Araçuaí, Ataléia, Conselheiro Pena, Espera Feliz, Padre Paraíso, Pedra Azul and São José da Safira districts are residual melts from granites. Anatectic pegmatites prevail in the Caratinga, Santa Maria de Itabira and Espírito Santo districts. They are mainly formed from the partial melting of paragneisses. In economic terms, the residual pegmatites are much more important than the anatectic pegmatites and their parent rocks belong to the syn-collisional G2 (Conselheiro Pena district) and post-collisional G4 (São José da Safira and Araçuaí districts) and G5 (Ataléia, Espera Feliz, Padre Paraíso and Pedra Azul districts) supersuites. External pegmatites (i.e. enveloped by country rocks) are the most important mineral deposits in the Araçuaí, Ataléia, Conselheiro Pena and São José da Safira districts, but internal pegmatites (i.e. hosted by the parent granite) largely predominate in the Espera Feliz, Padre Paraíso and Pedra Azul districts. Most anatectic pegmatites formed during the collisional stage of the Araçuaí orogen. They are commonly associated with migmatitic to granulitic paragneisses, and may be deposits of kaolin, K-feldspar, mica, corundum and quartz, mainly in the Caratinga and Espírito Santo districts (e.g. Sá 1977; Issa-Filho *et al.* 1980; Correia-Neves *et al.* 1986; Pedrosa-Soares *et al.* 1987, 1990, 2001*b*; Cassedanne 1991; Moura 1997; Achtschin 1999; Morteani *et al.* 2000; Castañeda *et al.* 2001; Gandini *et al.* 2001; Netto *et al.* 2001; Kahwage & Mendes 2003; Ferreira *et al.* 2005; Romeiro & Pedrosa-Soares 2005; Pinho-Tavares *et al.* 2006; Horn 2006).

Besides pegmatites *s.s.*, the Santa Maria de Itabira (or Nova Era–Itabira–Ferros) and Malacacheta districts include hydrothermal gem deposits which have been also called ‘pegmatites’ because of their coarse-grained quartz-feldspar composition. Such hydrothermal quartz-feldspar veins hosted by ultramafic schists and banded iron formations are

Table 6. *Districts of the Eastern Pegmatite Province in the Araçuaí orogen (modified from Correia-Neves* et al. *1986, Morteani* et al. *2000, Netto* et al. *2001, and Pedrosa-Soares* et al. *2001b). 1, pegmatite size in relation to thickness: very small, < 0.5 m; small, 0.5–5 m; medium, 5–15 m; large, 15–50 m; and very large, > 50 m thick (cf. Issa-Filho* et al. *1980). 2, pegmatite type and class based on definitions synthesized by Cerný (1991)*

District name	Main mineral resources and rare minerals	Pegmatite size[1], type, class[2]	Parent and host rock
Pedra Azul	aquamarine, topaz, quartz	very small to small, residual, rare element	G5 granite
Padre Paraíso	aquamarine, topaz, quartz, goshenite, chrysoberyl	very small to small, residual, rare element	G5 granite and charnockite
Araçuaí	spodumene, ornamental granite, gem varieties of tourmaline, beryl and quartz, industrial feldspar, schorl, ambligonite, albite, petalite, cleavelandite, apatite, rare phosphates, cassiterite, columbite-tantalite, bismuthinite	very large to very small, residual, rare element	G4 granite; mica schist, metawacke, quartzite, meta-ultramafic rock
Ataléia	aquamarine, chrysoberyl	very small to small, residual, rare element	G5 granite
São José da Safira	industrial feldspar, gem tourmalines, beryl ore, muscovite, aquamarine, garnet, albite, cleavelandite, apatite, heliodor, Mn-tantalite, bertrandite, microlite, zircon	very large to medium, residual, rare element to muscovite	G4 granite; mica schist, metawacke, quartzite, meta-ultramafic rock
Conselheiro Pena	industrial feldspar, gem varieties of tourmaline, beryl and quartz, beryl ore, albite, cleavelandite, triphylite, brasilianite, barbosalite and other rare phosphates, spodumene, kunzite, lepidolite,	very large to medium, residual, rare element	G2 granite; mica schist, metawacke, quartzite, meta-ultramafic rock
Malacacheta	muscovite, beryl, chrysoberyl (alexandrite formed in hydrothermal systems)	small to medium, residual, rare element	G4 granite; mica schist, meta-ultramafic rock
Santa Maria de Itabira	emerald, alexandrite, aquamarine, amazonite	hydrothermal systems and anatectic pegmatites	ultramafic schist, iron formation, migmatite
Caratinga	kaolin, corundum, beryl	small to large, anatectic, ceramic	migmatitic paragneiss
Espera Feliz	aquamarine, topaz, quartz	very small to small, residual, rare element	G5 granite
Espírito Santo	kaolin, quartz; aquamarine, topaz	very small to medium; anatectic (ceramic) and residual (rare element)	migmatitic paragneiss and G5 granite

significant deposits of emerald, alexandrite and/or aquamarine. At least one mineralization episode is late Neoproterozoic to Cambrian in age, but there is no solid evidence of intrusive granites directly associated to these deposits, even though chemical contributions from granite sources should be expected (Ribeiro-Althoff *et al.* 1997; Basílio *et al.* 2000; Gandini *et al.* 2001; Mendes & Barbosa 2001; Pinto & Pedrosa-Soares 2001; Preinfalk *et al.* 2002).

Geochemical and mineralogical specializations in the Eastern Brazilian Pegmatite Province

Geochemical and mineralogical specializations of residual pegmatites depend on the magma genesis and chemistry of the parent granites, as well as on the crystallization conditions of the granite–pegmatite systems (e.g. Cerný 1991). Accordingly, populations of residual pegmatites of the Eastern Brazilian Pegmatite Province show general geochemical specializations in relation to the regional granite supersuites, but their mineralizations can vary according to their emplacement depth (i.e. PT conditions of crystallization). The pegmatites rich in Na, B, Be, Li, P, Ta and/or Cs are associated with S-type two-mica granites of the G2 and G4 supersuites, emplaced in relatively shallow crustal levels. On the other hand, the pegmatites rich in Fe, Be and F, but poor in B and Na, are residual melts from the I- and A2-types deep-seated plutons of the G5 supersuite. Some of the most important pegmatite districts of the province, briefly described below, record examples of distinct specializations, according to different conditions of genesis and crystallization (Fig. 12).

Pegmatites of the Conselheiro Pena district represent residual melts from syn–collisional, S-type, two-mica granites of the Urucum suite (G2 supersuite), crystallized around 582 Ma (Nalini *et al.* 2000*a*, *b*). The main host rocks are amphibolite facies garnet-mica schists with intercalations of calc-silicate rock and metawacke. Granites and pegmatites crystallized under PT conditions of 750–600 °C and 4–5 kbar (Nalini *et al.* 2000*a*, *b*, 2008), which are consistent with garnet–staurolite–sillimanite-bearing assemblages, free of andalusite and kyanite, generated by contact metamorphism. The large to very large pegmatites show complex zoning and carry a very diversified mineralogy (Table 6), including albite, schorl, beryl, elbaite, morganite, spodumene, lepidolite, kunzite and many phosphate minerals like apatite, brazilianite, triphylite, barbosalite, moraesite, ruifrancoite, frondelite, tavorite, eosphorite, hureaulite, reddingite, variscite, vivianite, frondelite and

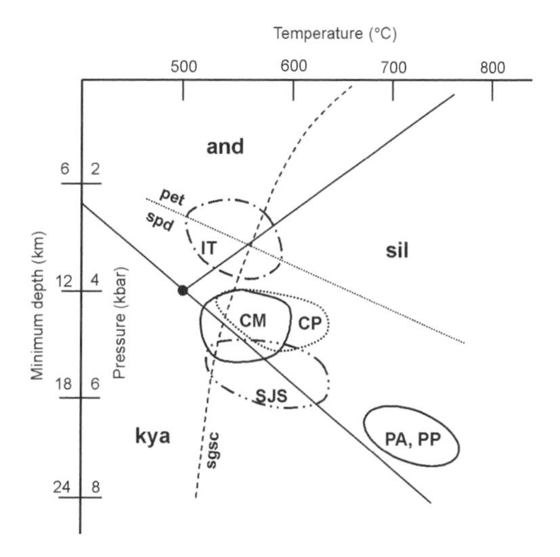

Fig. 12. PT crystallization conditions for some pegmatite populations of the Eastern Brazilian Pegmatite Province: CP, Conselheiro Pena district (dotted curve); CM and IT, Coronel Murta and Itinga (dashed–dotted curve) pegmatite fields of the Araçuaí district; SJS, São José da Safira (double–dotted–dashed curve) district; PA and PP, Pedra Azul and Padre Paraíso districts and, andalusite; kya, kyanite; sil, sillimanite; pet, petalite; spd, spodumene; sgsc, water-saturated granite *solidus* curve.

others (e.g. Lindberg & Pecora 1958; Issa-Filho *et al.* 1980; Cassedanne & Cassedanne 1981; Cassedanne & Baptista 1999; Netto *et al.* 2001; Scholz 2006; Scholz *et al.* 2008; Menezes 2009). Petalite is absent from these pegmatites (Fig. 12). Accordingly, the Conselheiro Pena district shows a Na–B–P–Be–Li geochemical specialization, being classified in the lithium–caesium–tantalum (LCT) family of the rare element pegmatite class (cf. Cerný 1991).

The São José da Safira district includes residual pegmatites related to the post-collisional, S-type granites of the Santa Rosa suite (G4 supersuite), and also beryl-muscovite-rich pegmatites apparently without relation to a parent granite (but, much probably related to non-exposed G4 intrusive granites). The main host rocks are garnet-mica schists with variable contents of staurolite, kyanite and sillimanite, garnet–biotite paragneisses and quartzites. Most pegmatites are very large to large, complex zoned bodies with abnormal contents of muscovite, albite, schorl, elbaite and beryl varieties, and quite rare minerals like bertrandite, Mn-tantalite and microlite (e.g. Federico *et al.* 1998; Castañeda *et al.* 2000; Dutrow & Henry 2000; Gandini *et al.* 2001; Netto *et al.* 2001). Most pegmatites of the São José da Safira district belong to the rare element class and show a Na–B–Be–Li geochemical specialization. Compared to other pegmatitic districts associated with S-type granites in this region, São José da Safira crystallized in deeper crustal levels (Fig. 12).

The most important pegmatites of the Araçuaí district are medium to very large, external, residual bodies from post-collisional two-mica granites of the S-type G4 supersuite (Fig. 1a, Table 6). However, very small and small residual pegmatites hosted by the parent S-type biotite granites can be rich in aquamarine of high gem quality. Contrasting pegmatite populations of the Araçuaí district are the lithium-rich Itinga field, located to the east of Araçuaí, and the boron-rich Coronel Murta field, located to the north of that city (Fig. 12).

Very anomalous contents of lithium minerals, such as spodumene, petalite, lepidolite and/or ambligonite, characterize many pegmatites of the Itinga field (e.g. Sá 1977; Correia-Neves *et al.* 1986; Romeiro & Pedrosa-Soares 2005). Two main pegmatite groups can be distinguished in the Itinga field: a swarm of non-zoned (homogeneous) pegmatitic bodies very rich in spodumene, but free of tourmaline and petalite, mined by CBL (Brazilian Lithium Company; Romeiro & Pedrosa-Soares 2005); and complex zoned pegmatites rich in Na, B, Li, Sn, Ta and Cs, mineralized in spodumene, petalite, lepidolite, ambligonite–montebrasite, albite, cleavelandite, elbaite (e.g. the watermelon elbaite with colourless core and light pink to green rims), cassiterite, tantalite and polucite (e.g. Sá 1977; Correia-Neves *et al.* 1986; Castañeda *et al.* 2000, 2001). In the Itinga pegmatite field, the host rocks are biotite schists with variable contents of andalusite, cordierite and sillimanite, formed during the regional metamorphism and recrystallized by contact metamorphism. Low pressure metamorphic silicates (andalusite and cordierite) together with the presence of petalite in some pegmatites and quantitative geothermobarometric data indicate a relatively shallow crustal level (6 to 12 km) for the Itinga field (Correia-Neves *et al.* 1986; Costa 1989; Pedrosa-Soares *et al.* 1996). The Itinga pegmatite field belongs to the rare element class, showing a clear affinity with the LCT pegmatite family. Their pegmatite groups, however, show two distinct geochemical specializations: Li–Na–P for the spodumene-rich pegmatite swarm of CBL mine, and Na–B–Li–Sn–Ta–Cs for other bodies.

The Coronel Murta field shows a myriad of external and internal pegmatites related to G4 granites (e.g. Pedrosa-Soares *et al.* 1987, 1990, 2001*a*; Pedrosa-Soares 1997*a*, b; Pedrosa-Soares & Oliveira 1997; Castañeda *et al.* 2000, 2001; Pinho-Tavares *et al.* 2006). The external pegmatites are hosted by mica schist, metawacke and quartzite. Garnet, staurolite, kyanite and sillimanite occur in the assemblages of metapelitic country rocks, characterizing an intermediate pressure regime for the regional metamorphism (Pedrosa-Soares *et al.* 1996). Staurolite and sillimanite also recrystallized in aureoles of contact metamorphism, but andalusite and cordierite are absent. These evidences together with quantitative geothermobarometric data suggest that pegmatites of the Coronel Murta field emplaced in crustal levels (12–16 km) deeper than those of the Itinga field (Fig. 12). Accordingly, most external pegmatites of the Coronel Murta field are complex zoned bodies of the rare element class and LCT family, with a diversified mineralogy, including elbaite, schorl, albite, beryl, cleavelandite, morganite, topaz, spodumene, lepidolite, ambligonite–montebrasite, topaz, bismuthinite, herderite and other phosphates, characterizing a Na–B–Be–Li–P–Bi–F geochemical specialization.

Most pegmatites of the Padre Paraíso and Pedra Azul districts represent internal residual melts crystallized inside their parent G5 granites and charnockites. These are relatively simple post-collisional, I- to A2-type, high-K and high-Fe pegmatite–granite systems, emplaced in deep crustal levels (Table 6, Fig. 12). These pegmatites are rich in biotite (instead of muscovite), aquamarine and topaz, but are poor or free of tourmaline, and poor in albite. Their Fe-rich geochemical signature favours the formation of deep blue aquamarine crystals (Achtschin 1999; Gandini *et al.* 2001; Ferreira *et al.* 2005; Kahwage & Mendes 2003). They are pegmatites of the rare element class with Fe–Be–F geochemical specialization.

The Espírito Santo–Minas dimension stone province

We call Espírito Santo – Minas dimension stone province the vast region of the eastern Araçuaí orogen and its continuation just to the south of the 21° S parallel (in terrains of the Ribeira orogen), encompassing the Espírito Santo State, and the regions of eastern Minas Gerais and southernmost Bahia (Fig. 1). This province supplies more than 65% of the whole exported dimension stones produced in Brazil, being most of them exploited from quarries located in the Espírito Santo State (Baltazar *et al.* 2008; Abirochas 2009).

The Espírito Santo – Minas dimension stone province comprises many hundreds of quarries opened mainly on massifs of the G2, G3, G4 and G5 supersuites, but also on host rocks like the high grade paragneisses, cordierite granulites, marbles, amphibolites and biotite schists (e.g. Chiodi-Filho 1998; Machado-Filho 1998; Costa *et al.* 2001; Castañeda *et al.* 2006; Costa & Pedrosa-

Soares 2006; Pedrosa-Soares *et al.* 2006*b*; Baltazar *et al.* 2008; Abirochas 2009). The ornamental granites are commercially classified into general groups, namely yellow, white, green, black, exotic, pink and grey. Despite the supergenic factors that control some rock colours, there are correlations linking the commercial dimension stone groups and the granite supersuites of the Araçuaí orogen, as described below.

The G2 supersuite is the site of many quarries for yellow and light grey ornamental granites, which are mainly opened on massifs of garnet–biotite leucogranites of the Carlos Chagas suite. Yellow ornamental granites have been by far economically the most important dimension stones of the G2 supersuite. Generally, their yellow tints result from two combined factors: the low content of mafic minerals together with the slight chemical weathering owing to infiltration of pluvial water along the foliation surfaces (Fig. 13). The downwards weathering fronts can be clearly observed in many quarries gradually separating the relatively thin (metres to a few decametres) yellow-coloured granite cover from the underneath unaltered light grey to white

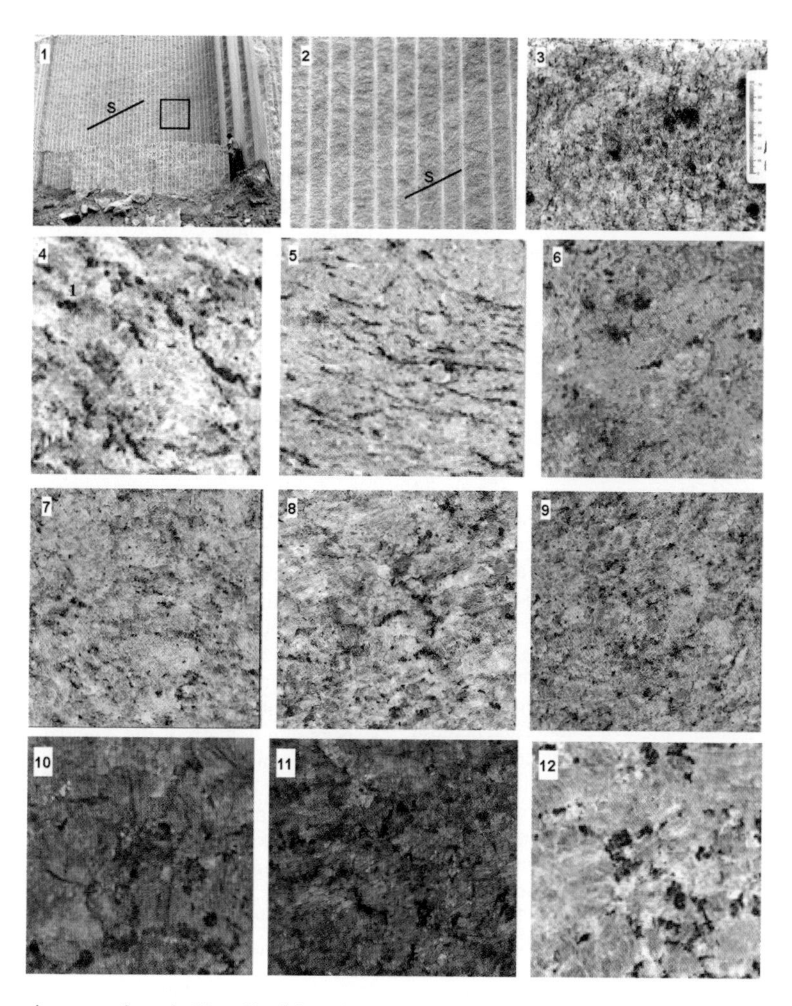

Fig. 13. Dimension stones from the Nova Venécia region (Fig. 1a): (**1**) quarry open pit shows weathering front, roughly parallel to the regional foliation (S), separating the yellow ornamental granite, at the top, from the unaltered light grey leucogranite (G2 Carlos Chagas suite); (**2**) detail of photo 1, showing sharp contact between the yellow and light grey parts of the same leucogranite; (**3**) non-polished yellow ornamental granite (Carlos Chagas suite); (**4**, **5** and **6**) polished plates of yellow G2 granites according to increasing chemical weathering (4, Santa Cecília; 5, Topázio; and 6, Gold 500 ornamental granites); (**7**, **8** and **9**) colour varieties of the Giallo Veneziano granite (G2 Carlos Chagas suite); (**10**, **11** and **12**) plates of ornamental granites extracted from the same G5 pluton, according to increasing weathering (10, Green Jade; 11, Green Gold; 12, Beige Vermont ornamental granites).

leucogranite (Castañeda *et al.* 2006; Pedrosa-Soares *et al.* 2006*b*). A possible exception is the internationally known Giallo Veneziano Granite, a poorly- to non-foliated very coarse-grained granite of the G2 Carlos Chagas suite (*c.* 576 Ma, zircon U–Pb SHRIMP age; Roncato 2009), exploited by the Granasa Company in a mine located to the west of Nova Venécia. After 30 years of mining activities, cutting on average 3000 m^3/month of yellow granite squared blocks sized *c.* 3.0 × 2.2 × 1.7 m, the mine passed to show both yellow and light salmon-coloured granites without unambiguous weathering fronts separating them, more than 200 m deep from the top (Fig. 13). This huge amount of homogeneous yellow granite, together with the probable coalescence of it with the unaltered salmon-coloured facies in relatively high depth suggest that slight metasomatic alteration by rising fluids also took place to form the thick yellow top zone of the Giallo Veneziano pluton, well before the recent tropical weathering (Pedrosa-Soares *et al.* 2006*b*; Granasa 2009).

G3 leucogranites are quarried mainly for white ornamental granites, especially those with disseminated small garnet crystals, and poor to free of cordierite and mica. However, G3 bodies large enough to be mined are rather rare (Castañeda *et al.* 2006; Pedrosa-Soares *et al.* 2006*b*; Baltazar *et al.* 2008).

Since the end of the 1990s quarries for coarse-grained white to yellow dimension stone have been opened on G4 pegmatites and pegmatoid cupolas, and became a very important mineral resource for the Araçuaí region (Fig. 1a). The fine-grained muscovite leucogranite, very abundant in the region to the north of Araçuaí, is also an important target for dimension stone exploration (Pedrosa-Soares 1997*a*, *b*).

Economically speaking, the major importance of the G5 supersuite is its large dimension stone production for a great variety of colours and textural patterns, including light to dark green, peacock-green, yellowish green (green-gold), greenish black, black, light brown (beige), yellow, pink and grey dimension stones. The green to black varieties are exploited from charnockite to gabbro-norite facies of G5 plutons, and most of these quarries are located in the northern Espírito Santo State (Fig. 1a). Nova Venécia region). Yellow to pink coarse-grained granites with rapakivi and antirapakivi textures are abundant in the G5 batholith of the Pedra Azul-Medina region (northern Araçuaí orogen, Fig. 1a). Slight chemical weathering also plays an important role to cause colour variations in G5 plutons, without significant loss of physical properties of the dimension stone. This is the case of the yellowish green (green-gold) and light brown (beige) dimension stones from G5 bodies (Fig. 13). These ornamental rocks can be found in the same quarry, where fronts of chemical weathering caused colour changes from the unaltered green granite (charnockite) to the goldish green variety, which retain most green feldspar crystals, and to light brown varieties in which the feldspar-rich groundmass completely lost its green colour (Pedrosa-Soares *et al.* 2006*b*).

Conclusions

In the scenario of Western Gondwana (Brito-Neves *et al.* 1999; Rogers & Santosh 2004), the Araçuaí orogen represents an example of orogen developed from the closure of a confined basin only partially floored by oceanic crust (Pedrosa-Soares *et al.* 2001*a*, 2008; Alkmim *et al.* 2006). Despite the subduction of only a restricted amount of oceanic lithosphere, a significant volume of pre-collisional calc-alkaline rocks built up a continental magmatic arc, during the pre-collisional stage. Subsequent crustal shortening promoted PT conditions for the generation of a huge amount of syn-collisional granites. Following crustal growth in response to accretion of magmas and tectonic shortening, the extensional relaxation and gravitational collapse of the orogen took place, owing to the declining and ceasing of the tangential convergent forces, accompanied by asthenosphere ascent and crustal re-melting that generated a myriad of igneous plutonic rocks, from the late collisional to post-collisional stages. The erosion levels exposed along the Araçuaí orogen are especially adequate to exhibit rocks from all the magmatic events, from relatively shallow to the deep crust, providing an outstanding example of a long lasting event (*c.* 630–480 Ma) of orogenic granite generation, well-organized in time and space.

However, the uppermost levels of the Neoproterozoic–Cambrian crust were eroded along the extensive granite domain of the Araçuaí orogen so that expected mineral deposits, like volcanic-related hydrothermal systems do not exist so far (or are hitherto unknown). Therefore, the mineral resources associated with the granite forming events of the Araçuaí orogen are typical of the intermediate to lower crusts, like the extensive Eastern Brazilian Pegmatite Province and countless varieties of dimension stones.

The authors acknowledge financial support provided by Brazilian government agencies CNPq (Conselho Nacional de Desenvolvimento Científico e Tecnológico), CAPES (Coordenação de Aperfeiçoamento de Pessoal de Nível Supeior), FINEP (Financiadora de Estudos e Projetos) and FAPEMIG (Fundação de Amparo à Pesquisa de Minas Gerais), and the Geological Survey of Brazil (CPRM). Our gratitude to the anonymous reviewers of this manuscript, and to A. Sial, J. Bettencourt and V. Ferreira.

References

ABIROCHAS 2009. *Associação Brasileira da Indústria de Rochas Ornamentais*. http://www.abirochas.com.br

ACHTSCHIN, A. B. 1999. *Caracterização geológica, mineralógica e geoquímica dos pegmatitos do Distrito de Padre Paraíso, Minas Gerais, e suas variedades de berilo*. MSc thesis, Instituto de Geociências, Universidade Federal de Minas Gerais, Belo Horizonte.

ALKMIM, F. F., MARSHAK, S., PEDROSA-SOARES, A. C., PERES, G. G., CRUZ, S. C. & WHITTINGTON, A. 2006. Kinematic evolution of the Araçuaí–West Congo orogen in Brazil and Africa: Nutcracker tectonics during the Neoproterozoic assembly of Gondwana. *Precambrian Research*, **149**, 43–63.

BALTAZAR, O. F., ZUCHETTI, M., OLIVEIRA, S. A. & SILVA, L. C. 2008. *Folhas São Gabriel Da Palha E Linhares*. Programa Geologia do Brasil, CPRM–Serviço Geológico do Brasil, Rio de Janeiro.

BASÍLIO, M., PEDROSA-SOARES, A. C. & EVANGELISTA, H. J. 2000. Depósitos de alexandrita de Malacacheta, Minas Gerais. *Geonomos*, **8**, 47–54.

BAYER, P., SCHMIDT-THOMÉ, R., WEBER-DIEFENBACH, K. & HORN, H. A. 1987. Complex concentric granitoid intrusions in the Coastal Mobile Belt, Espírito Santo, Brazil: the Santa Angélica pluton – an example. *Geologische Rundshau*, **76**, 357–371.

BILAL, E., HORN, H. ET AL. 2000. Neoproterozoic granitoid suites in southeastern Brazil. *Revista Brasileira de Geociências*, **30**, 51–54.

BRITO-NEVES, B. B., CAMPOS-NETO, M. C. & FUCK, R. 1999. From Rodinia to Western Gondwana: An approach to the Brasiliano–Pan African cycle and orogenic collage. *Episodes*, **22**, 155–199.

CASSEDANNE, J. P. 1991. Tipologia das Jazidas Brasileiras de Gemas. *In*: SCHOBBENHAUS, C. & COELHO, C. S. (eds) *Principais Depósitos Minerais Do Brasil*, Brasília, Departamento Nacional da Produção Mineral, **4A**, 17–52.

CASSEDANNE, J. P. & BAPTISTA, A. 1999. Famous mineral localities: the Sapucaia pegmatite, Minas Gerais, Brazil. *Mineralogical Record*, **30**, 347–365.

CASSEDANNE, J. P. & CASSEDANNE, J. O. 1981. Minerals of the Lavra do Ênio pegmatite. *Mineralogical Record*, **4**, 207–213.

CASTAÑEDA, C., OLIVEIRA, E., PEDROSA-SOARES, A. C. & GOMES, N. S. 2000. Infrared study of O–H sites in tourmalines from the elbaite-schorl serie. *American Mineralogist*, **85**, 1503–1507.

CASTAÑEDA, C., MENDES, J. C. & PEDROSA-SOARES, A. C. 2001. Turmalinas. *In*: CASTAÑEDA, C., ADDAD, J. & LICCARDO, A. (eds) *Gemas De Minas Gerais*, Belo Horizonte, Sociedade Brasileira de Geologia, 152–179.

CASTAÑEDA, C., PEDROSA-SOARES, A. C., BELÉM, J., GRADIM, D., DIAS, P. H., MEDEIROS, S. & OLIVEIRA, L. 2006. *Folha Ecoporanga*. Programa Geologia do Brasil, CPRM–Serviço Geológico do Brasil, Rio de Janeiro.

CELINO, J. J., BOTELHO, N. F. & PIMENTEL, M. M. 2000. Genesis of neoproterozoic granitoid magmatism in the eastern araçuaí fold belt, eastern Brazil: field, geochemical and Sr–Nd isotopic evidence. *Revista Brasileira de Geociências*, **30**, 135–139.

CERNÝ, P. 1991. Rare-element granitic pegmatites. Part I: Anatomy and internal evolution of pegmatite deposits. *Geoscience Canada*, **18**, 49–67.

CHAPELL, B. & WHITE, A. 2001. Two contrasting granite types: 25 years later. *Australian Journal of Earth Sciences*, **48**, 489–499.

CHIODI-FILHO, C. 1998. Aspectos técnicos e econômicos do setor de rochas ornamentais. *Rochas e Equipamentos*, **51**, 84–139.

CORREIA-NEVES, J. M., PEDROSA-SOARES, A. C. & MARCIANO, V. R. 1986. A Província Pegmatitica Oriental do Brasil à luz dos conhecimentos atuais. *Revista Brasileira de Geociencias*, **16**, 106–118.

COSTA, A. G. 1989. Evolução petrológica para uma seqüência de rochas metamórficas regionais do tipo baixa pressão na região de Itinga, NE de Minas Gerais. *Revista Brasileira de Geociências*, **19**, 440–448.

COSTA, A. G. & PEDROSA-SOARES, A. C. 2006. *Catálogo De Rochas Ornamentais Da Região Norte Do Espírito Santo*. Programa Geologia do Brasil, CPRM–Serviço Geológico do Brasil, Rio de Janeiro.

COSTA, A. G., CAMPELLO, M. & PIMENTA, V. B. 2001. Rochas ornamentais e de revestimento de Minas Gerais: Principais ocorrências, caracterização e aplicações na indústria da construção civil. *Geonomos*, **8**, 9–13.

DARDENNE, M. A. & SCHOBBENHAUS, C. 2003. Depósitos minerais no tempo geológico e épocas metalogenéticas. *In*: BIZZI, L. A., SCHOBBENHAUS, C., VIDOTTI, R. M. & GONÇALVES, J. H. (eds) *Geologia, Tectônica E Recursos Minerais Do Brasil*. CPRM–Serviço Geológico do Brasil, Brasília, 365–448.

DE CAMPOS, C. M., MENDES, J. C., LUDKA, I. P., MEDEIROS, S. R., MOURA, J. C. & WALLFASS, C. 2004. A review of the Brasiliano magmatism in southern Espírito Santo, Brazil, with emphasis on post-collisional magmatism. *Journal of the Virtual Explorer*, **17**, http://virtualexplorer.com.au/journal/2004/17/campos.

DE LA ROCHE, H., LETERRIER, J., GRANDCLAUDE, P. & MARCHAI, M. 1980. Classification of volcanic and plutonic rocks using R1–R2 diagram and major-element analyses. Its relationships with current nomenclature. *Chemical Geology*, **29**, 183–210.

DELLA GIUSTINA, M. E. S., OLIVEIRA, C., PIMENTEL, M. M. & BUHN, B. 2009. Neoproterozoic magmatism and high-grade metamorphism in the Goiás Massif: New LA–MC–ICMPS U–Pb and Sm–Nd data and implications for collisional history of the Brasília Belt. *Precambrian Research*, **172**, 67–79.

DRUMOND, J. B. & MALOUF, R. 2008. *Folha Almenara*. Programa Geologia do Brasil, CPRM–Serviço Geológico do Brasil, Rio de Janeiro.

DUTROW, B. & HENRY, D. 2000. Complexly zoned fibrous tourmaline, Cruzeiro mine, Minas Gerais, Brazil: a record of evolving magmatic and hydrothermal fluids. *The Canadian Mineralogist*, **38**, 131–143.

FARIA, L. F. 1997. *Controle e tipologia de mineralizações de grafita flake do nordeste de Minas Gerais e sul da Bahia: uma abordagem regional*. MSc thesis, Instituto de Geociências, Universidade Federal de Minas Gerais, Belo Horizonte.

Federico, M., Andreozzi, G., Lucchesi, S., Graziani, G. & Mendes, J. C. 1998. Compositional variation of tourmaline in the granitic pegmatite dykes of the Cruzeiro Mine, Minas Gerais, Brazil. *The Canadian Mineralogist*, **36**, 415–431.

Fernandes, M. L. 1991. *Geologia, petrografia e geoquímica de rochas granitóides da região de Pedra Azul, MG*. MSc thesis, Instituto de Geociências, Universidade Federal do Rio de Janeiro.

Ferreira, M., Fonseca, M. A. & Pires, F. 2005. Pegmatitos mineralizados em água-marinha e topázio do Ponto do Marambaia, Minas Gerais: tipologia e relações com o Granito Caladão. *Revista Brasileira de Geociências*, **35**, 463–473.

Gandini, A. L, Achtschin, A. B., Marciano, V. R., Bello, R. F. & Pedrosa-Soares, A. C. 2001. Berilo. *In*: Castañeda, C., Addad, J. E. & Liccardo, A. (eds) *Gemas De Minas Gerais, Belo Horizonte, Sociedade Brasileira De Geologia Núcleo Minas Gerais*, 100–127.

Gomes, A. C. 2008. *Folha Rio Do Prado*. Programa Geologia do Brasil, CPRM–Serviço Geológico do Brasil, Rio de Janeiro.

GRANASA 2009. *Granitos Nacionais Ltda*. http://www.granasa.com.br/

Heineck, C., Raposo, F., Malouf, R. & Jardim, S. 2008. *Folha Jequitinhonha*. Programa Geologia do Brasil, CPRM–Serviço Geológico do Brasil, Rio de Janeiro.

Horn, A. H. 2006. *Folha Espera Feliz. Programa Geologia Do Brasil*, CPRM–Serviço Geológico do Brasil, Rio de Janeiro.

Horn, H. A. & Weber-Diefenbach, K. 1987. Geochemical and genetic studies of three inverse zoned intrusive bodies of both alkaline and subalkaline composition in the Araçuaí–Ribeira mobile belt (Espírito Santo, Brazil). *Revista Brasileira de Geociências*, **17**, 488–497.

Issa-Filho, A., Moura, O. & Fanton, J. 1980. Reconhecimento de pegmatitos da província oriental brasileira entre Aimorés e Itambacuri, MG. *31st Congresso Brasileiro de Geologia, Balneário de Camboriú*, Anais, **3**, 1552–1563.

Junqueira, P., Gomes, A. C., Raposo, F. & Paes, V. C. 2008. *Folha Joaíma*. Programa Geologia do Brasil, CPRM–Serviço Geológico do Brasil, Rio de Janeiro.

Kahwage, M. & Mendes, J. C. 2003. O berilo gemológico da Província Pegmatítica Oriental do Brasil. *Geochimica Brasiliensis*, **17**, 13–25.

Lindberg, M. L. & Pecora, W. T. 1958. Phosphate minerals from the Sapucaia pegmatite mine, Minas Gerais. *Boletim da Sociedade Brasileira de Geologia*, **7**, 5–14.

Ludka, I. P., Wiedemann, C. & Töpfner, C. 1998. On origin of incompatible elements in the Venda Nova pluton, State of Espírito Santo, southeast Brazil. *Journal South American Earth Sciences*, **11**, 473–486.

Machado-Filho, M. 1998. *Granito Azul do Espírito Santo, um granulito rico em cordierita usado como rocha ornamental*. MSc thesis, Instituto de Geociências, Universidade Federal do Rio de Janeiro.

Marshak, S., Alkmim, F. F., Whittington, A. & Pedrosa-Soares, A. C. 2006. Extensional collapse in the Neoproterozoic Araçuaí orogen, eastern Brazil: A setting for reactivation of asymmetric crenulation cleavage. *Journal of Structural Geology*, **28**, 129–147.

Martins, V. T. S., Teixeira, W., Noce, C. M. & Pedrosa-Soares, A. C. 2004. Sr and Nd characteristics of Brasiliano/Pan African granitoid plutons of the Araçuaí orogen, southeastern Brazil: Tectonic implications. *Gondwana Research*, **7**, 75–89.

Medeiros, S., Wiedemann, C. & Mendes, J. C. 2000. Post-collisional magmatism in the Araçuaí–Ribeira Mobile belt: geochemical and isotopic study of the Várzea Alegre intrusive complex, ES, Brazil. *Revista Brasileira de Geociências*, **30**, 30–34.

Medeiros, S., Wiedemann, C. & Vriend, S. 2001. Evidence of mingling between contrasting magmas in a deep plutonic environment: the example of Várzea Alegre in the Pan African–Brasiliano mobile belt in Brazil. *Anais Academia Brasileira de Ciências*, **73**, 99–119.

Medeiros, S., Mendes, J. C., McReath, I. & Wiedemann, C. 2003. U–Pb and Rb–Sr dating and isotopic signature of the charnockitic rocks from Várzea Alegre intrusive complex, Espírito Santo, Brazil. *In*: *4th South American Symposium on Isotope Geology*, Salvador. Short papers, **2**, 609–612.

Mendes, J. C. & Barbosa, M. S. 2001. Esmeralda. *In*: Castañeda, C., Addad, J. E. & Liccardo, A. (eds) *Gemas De Minas Gerais, Belo Horizonte*. Sociedade Brasileira De Geologia Núcleo Minas Gerais, 128–151.

Mendes, J. C., Wiedemann, C. & McReath, I. 1999. Conditions of formation of charnockitic magmatic rocks from the Várzea Alegre massif, Espírito Santo, southeast Brazil. *Revista Brasileira de Geociências*, **29**, 47–54.

Mendes, J. C., Medeiros, S., McReath, I. & De Campos, C. 2005. Cambro–Ordovician Magmatism in SE Brazil: U–Pb and Rb–Sr ages combined with Sr and Nd isotopic data of charnockitic rocks from the Várzea Alegre Complex. *Gondwana Research*, **8**, 1–9.

Menezes, L. 2009. Famous mineral localities: The Sapo Mine, Ferruginha District, Conselheiro Pena, Minas Gerais, Brazil. *Mineralogical Record*, **40**, 273–292.

Morteani, G., Preinfalk, C. & Horn, A. H. 2000. Classification and mineralization potential of the pegmatites of the Eastern Brazilian Pegmatite Province. *Mineralium Deposita*, **35**, 638–655.

Moura, O. 1997. Depósitos de feldspato e mica de Pomarolli, Urucum e Golconda, Minas Gerais. *In*: Schobbenhaus, C., Queiroz, E. & Coelho, C. (eds) *Principais Depósitos Minerais Do Brasil*. DNPM/CPRM, Brasília, **4B**, 363–371.

Munhá, J. M., Cordani, U., Tassinari, C. & Palácios, T. 2005. Petrologia e termocronologia de gnaisses migmatíticos da Faixa de Dobramentos Araçuaí (Espírito Santo, Brasil). *Revista Brasileira de Geociências*, **35**, 123–134.

Nalini-Junior, H. A., Bilal, E. & Correia Neves, J. M. 2000*a*. Syncollisional peraluminous magmatism in the Rio Doce region: mineralogy, geochemistry and isotopic data of the Urucum suite (eastern Minas Gerais State, Brazil). *Revista Brasileira de Geociências*, **30**, 120–125.

NALINI-JUNIOR, H. A., BILAL, E., PAQUETTE, J. L., PIN, C. & MACHADO, R. 2000*b*. Géochronologie U–Pb et géochimie isotopique Sr–Nd des granitoides neoproterozoiques des suites Galileia et Urucum, vallée du Rio Doce, Sud-Est du Brésil. *Comptes Rendus Academie Science Paris*, **331**, 459–466.

NALINI-JUNIOR, H. A., MACHADO, R. M. & BILAL, E. 2005. Geoquímica e petrogênese da Suíte Galiléia: exemplo de magmatismo tipo–I, metaluminoso, pré-colisional, neoproterozóico da região do Médio Vale do Rio Doce. *Revista Brasileira de Geociências*, **35** (4– supplement), 23–34.

NALINI-JUNIOR, H. A., MACHADO, R. M., ENDO, I. & BILAL, E. 2008. A importância da tectônica transcorrente no alojamento de granitos pré a sincolisionais na região do vale do médio Rio Doce: o exemplo das suítes graníticas Galiléia e Urucum. *Revista Brasileira de Geociências*, **38**, 741–752.

NETTO, C., ARAÚJO, M. C., PINTO, C. P. & DRUMOND, J. B. 2001. *Pegmatitos*. Projeto Leste, CPRM, Programa Levantamentos Geológicos Básicos do Brasil. CODEMIG, Belo Horizonte.

NOCE, C. M., MACAMBIRA, M. B. & PEDROSA-SOARES, A. C. 2000. Chronology of Neoproterozoic–Cambrian granitic magmatism in the Araçuaí Belt, Eastern Brazil, based on single zircon evaporation dating. *Revista Brasileira de Geociências*, **30**, 25–29.

NOCE, C. M, PEDROSA-SOARES, A. C., PIUZANA, D., ARMSTRONG, R., LAUX, J. H., CAMPOS, C. & MEDEIROS, S. R. 2004. Ages of sedimentation of the kinzigitic complex and of a late orogenic thermal episode in the Araçuaí orogen, northern Espírito Santo State, Brazil: Zircon and monazite U–Pb SHRIMP and ID–TIMS data. *Revista Brasileira de Geociências*, **349**, 587–592.

NOCE, C. M., COSTA, A. G., PIUZANA, D., VIEIRA, V. S. & CARVALHO, C. 2006. *Folha Manhuaçu*. Programa Geologia do Brasil, CPRM–Serviço Geológico do Brasil, Rio de Janeiro.

NOCE, C. M., PEDROSA-SOARES, A. C., SILVA, L. C., ARMSTRONG, R. & PIUZANA, D. 2007. Evolution of polyciclic basement complexes in the Araçuaí orogen, based on U–Pb SHRIMP data: Implications for Brazil–Africa links in Paleoproterozoic time. *Precambrian Research*, **159**, 60–78.

NOVO, T. 2009. *Significado geotectônico das rochas charnockíticas da região de Carangola–MG: implicações para a conexão Araçuaí–Ribeira*. MSc thesis, Instituto de Geociências, Universidade Federal de Minas Gerais, Belo Horizonte.

PAES, V. C., HEINECK, C. & MALOUF, R. 2008. *Folha Itaobim*. Programa Geologia do Brasil, CPRM–Serviço Geológico do Brasil, Rio de Janeiro.

PAIVA, G. 1946. Províncias pegmatíticas do Brasil. *Boletim DNPM–DFPM*, **78**, 13–21.

PEDROSA-SOARES, A. C. 1997*a*. Geologia da Folha Araçuaí. *In*: GROSSI-SAD, J. H., LOBATO, L. M., PEDROSA-SOARES, A. C. & SOARES-FILHO, B. S. (eds) *Projeto Espinhaço*. CODEMIG, Belo Horizonte, 715–852.

PEDROSA-SOARES, A. C. 1997*b*. Geologia da Folha Jenipapo. *In*: GROSSI-SAD, J. H., LOBATO, L. M., PEDROSA-SOARES, A. C. & SOARES-FILHO, B. S. (eds) *Projeto Espinhaço*. CODEMIG, Belo Horizonte, 1053–1198.

PEDROSA-SOARES, A. C. & OLIVEIRA, M. J. R. 1997. Geologia da Folha Salinas. *In*: GROSSI-SAD, J. H., LOBATO, L. M., PEDROSA-SOARES, A. C. & SOARES-FILHO, B. S. (eds) *Projeto Espinhaço*. CODEMIG, Belo Horizonte, 419–541.

PEDROSA-SOARES, A. C. & WIEDEMANN-LEONARDOS, C. M. 2000. Evolution of the Araçuaí Belt and its connection to the Ribeira Belt, Eastern Brazil. *In*: CORDANI, U., MILANI, E., THOMAZ-FILHO, A. & CAMPOS, D. A. (eds) *Tectonic Evolution Of South America*. São Paulo, Sociedade Brasileira de Geologia, 265–285.

PEDROSA-SOARES, A. C., MONTEIRO, R., CORREIA-NEVES, J. M., LEONARDOS, O. H. & FUZIKAWA, K. 1987. Metasomatic evolution of granites, northeast Minas Gerais, Brazil. *Revista Brasileira de Geociências*, **17**, 512–518.

PEDROSA-SOARES, A. C., CORREIA-NEVES, J. M. & LEONARDOS, O. H. 1990. Tipologia dos pegmatitos de Coronel Murta–Virgem da Lapa, Médio Jequitinhonha, Minas Gerais. *Revista Escola de Minas*, **43**: 44–54.

PEDROSA-SOARES, A. C., LEONARDOS, O. H., FERREIRA, J. C. & REIS, L. B. 1996. Duplo regime metamórfico na Faixa Araçuaí: Uma re-interpretação à luz de novos dados. *In*: *39th Congresso Brasileiro de Geologia*, Salvador, Anais, **6**, 5–8.

PEDROSA-SOARES, A. C., VIDAL, P., LEONARDOS, O. H. & BRITO-NEVES, B. B. 1998. Neoproterozoic oceanic remnants in eastern Brazil: Further evidence and refutation of an exclusively ensialic evolution for the Araçuaí–West Congo orogen. *Geology*, **26**, 519–522.

PEDROSA-SOARES, A. C., NOCE, C. M., WIEDEMANN, C. M. & PINTO, C. P. 2001*a*. The Araçuaí–West Congo orogen in Brazil: An overview of a confined orogen formed during Gondwanaland assembly. *Precambrian Research*, **110**, 307–323.

PEDROSA-SOARES, A. C., PINTO, C. P. ET AL. 2001*b*. A Província Gemológica Oriental do Brasil. *In*: CASTAÑEDA, C., ADDAD, J. E. & LICCARDO, A. (eds) *Gemas De Minas Gerais*. Belo Horizonte, Sociedade Brasileira de Geologia, 16–33.

PEDROSA-SOARES, A. C., CASTAÑEDA, C. ET AL. 2006*a*. Magmatismo e tectônica do Orógeno Araçuaí no extremo leste de Minas Gerais e norte do Espírito Santo. *Geonomos*, **14**, 97–111.

PEDROSA-SOARES, A. C., QUEIROGA, G. ET AL. 2006*b*. *Folha Mantena*. Programa Geologia do Brasil, CPRM–Serviço Geológico do Brasil, Rio de Janeiro.

PEDROSA-SOARES, A. C., NOCE, C. M. ET AL. 2007. Orógeno Araçuaí: síntese do conhecimento 30 anos após Almeida 1977. *Geonomos*, **15**, 1–16.

PEDROSA-SOARES, A. C., ALKMIM, F. F. ET AL. 2008. Similarities and differences between the Brazilian and African counterparts of the Neoproterozoic Araçuaí–West Congo orogen. *In*: PANKHURST, R. J., TROUW, R. A. J., BRITO NEVES, B. B. & DE WIT, M. J. (eds) *West Gondwana: Pre-Cenozoic Correlations Across The South Atlantic Region*. Geological Society, London, Special Publications, **294**, 153–172.

PETITGIRARD, S., VAUCHEZ, A. ET AL. 2009. Conflicting structural and geochronological data from the Ibituruna quartz-syenite (SE Brazil): Effect of protracted 'hot' orogeny and slow cooling rate? *Tectonophysics*, doi: 10.1016/j.tecto.2009.02.039.

PINHO-TAVARES, S., CASTAÑEDA, C. & PEDROSA-SOARES, A. C. 2006. O feldspato industrial de Coronel Murta, MG, e a perspectiva de aplicações à indústria cerâmica e vidreira. *Revista Brasileira de Geociências*, **36** (1, supplement), 200–206.

PINTO, C. P. 2008. *Folha Jequitinhonha*. Programa Geologia do Brasil, CPRM–Serviço Geológico do Brasil, Rio de Janeiro.

PINTO, C. P. & PEDROSA-SOARES, 2001. Brazilian Gem Provinces. *The Australian Gemmologist*, **21**, 12–16.

PINTO, C. P., DRUMOND, J. B. & FÉBOLI, W. L. (coord.) 2001. *Projeto Leste, Etapas 1 E 2*. CODEMIG, Belo Horizonte.

POWELL, R. & HOLLAND, T. 2006. *Course Notes for THERMOCALC Short Course*. São Paulo, Brazil. http://www.metamorph.geo.uni-mainz.de/thermocalc/documentation.

PREINFALK, C., KOSTITSYN, Y. & MORTEANI, G. 2002. The pegmatites of the Nova Era-Itabira–Ferros pegmatite district and the emerald mineralization of Capoeirana and Belmont (Minas Gerais, Brazil): geochemistry and Rb–Sr dating. *Journal of South American Earth Sciences*, **14**, 867–887.

QUEIROGA, G., PEDROSA-SOARES, A. C. ET AL. 2007. Age of the Ribeirão da Folha ophiolite, Araçuaí Orogen: The U–Pb zircon dating of a plagiogranite. *Geonomos*, **15**, 61–65.

QUEIROGA, G., PEDROSA-SOARES, A. C. ET AL. 2009. *Folha Nova Venécia*. Programa Geologia do Brasil, CPRM–Serviço Geológico do Brasil, Rio de Janeiro.

RIBEIRO-ALTHOFF, A. M., CHEILETZ, A., GIULIANI, G., FÉRAULT, G., BARBOSA-CAMACHO, G. & ZIMMERMANN, J. 1997. Evidences of two periods (2 Ga and 650–500 Ma) of emerald formation in Brazil by K–Ar and Ar–Ar dating. *International Geology Review*, **39**, 924–937.

ROGERS, J. W. & SANTOSH, M. 2004. *Continents and Supercontinents*. Oxford University Press.

ROMEIRO, J. C. & PEDROSA-SOARES, A. C. 2005. Controle do minério de espodumênio em pegmatitos da Mina da Cachoeira, Araçuaí, MG. *Geonomos*, **13**, 75–85.

RONCATO, J. 2009. *As suítes graníticas tipo-S do norte do Espírito Santo na região das folhas Ecoporanga, Mantena, Montanha e Nova Venécia*. MSc thesis, Instituto de Geociências, Universidade Federal de Minas Gerais, Belo Horizonte.

RONCATO, J., PEDROSA-SOARES, A. C. ET AL. 2009. *Folha Montanha*. Programa Geologia do Brasil, CPRM–Serviço Geológico do Brasil, Rio de Janeiro.

SÁ, J. H. S. 1977. *Pegmatitos litiníferos da região de Itinga–Araçuaí, Minas Gerais*. PhD Thesis, Instituto de Geociências, Universidade de São Paulo.

SAMPAIO, A. R., MARTINS, A. M. ET AL. 2004. Projeto Extremo Sul da Bahia: Geologia e Recursos Minerais. *In*: *Série Arquivos Abertos da Companhia Bahiana de Pesquisa Mineral*, Salvador, **19**.

SCHMIDT-THOMÉ, R. & WEBER-DIEFENBACH, K. 1987. Evidence for frozen-in magma mixing in Brasiliano calc-alkaline intrusions: The Santa Angélica pluton, southern Espírito Santo. *Revista Brasileira de Geociências*, **17**, 498–506.

SCHOLZ, R. 2006. *Mineralogia fosfática do Distrito Pegmatítico de Conselheiro Pena, Minas Gerais*. PhD thesis, Instituto de Geociências, Universidade Federal de Minas Gerais, Belo Horizonte.

SCHOLZ, R., KARFUNKEL, J., BERMANEC, V., DA COSTA, G. M., HORN, A. H., SOUZA, L. A. & BILAL, E. 2008. Amblygonite-montebrasite from Divino das Laranjeiras-Mendes Pimentel pegmatite swarm, Minas Gerais, Brasil. *II. Mineralogy. Romanian Journal of Mineral Deposits*, **83**, 135–139.

SILVA, J. M., LIMA, M., VERONESE, V. F., RIBEIRO-JUNIOR, R. & SIGA-JÚNIOR, O. 1987. *Folha SE.24 Rio Doce, Levantamento De Recursos Naturais, Projeto Radambrasil*. IBGE, Rio de Janeiro.

SILVA, L. C., ARMSTRONG, R. ET AL. 2002. Reavaliação da evolução geológica em terrenos pré-cambrianos brasileiros com base em novos dados U–Pb SHRIMP, parte II: Orógeno Araçuaí, Cinturão Móvel Mineiro e Cráton São Francisco Meridional. *Revista Brasileira de Geociências*, **32**, 513–528.

SILVA, L. C., MCNAUGHTON, N., ARMSTRONG, R., HARTMANN, L. & FLETCHER, I. 2005. The Neoproterozoic Mantiqueira Province and its African connections. *Precambrian Research*, **136**, 203–240.

SILVA, L. C., PINTO, C. P., GOMES, A. C., PAES, V. C. & CHEMALE, F. 2007. Granitogenesis at the northern tip of the Araçuaí Orogen, SE Brazil: LA–ICP–MS U–Pb zircon geochronology, and tectonic significance. *In*: *Simpósio de Geologia do Sudeste*. Diamantina, Anais, 20–21.

SÖLLNER, F., LAMMERER, B. & WEBER-DIEFENBACH, K. 1991. *Die Krustenentwicklung in Der Küstenregion Nördlich Von Rio de Janeiro, Brasilien*. Münchener Geowissenschaftliche Hefte 11, München, Friedrich Pfeil Verlag, **4**.

SÖLLNER, H. S., LAMMERER, B. & WIEDEMANN-LEONARDOS, C. 2000. *Dating the Araçuaí–Ribeira Mobile Belt of Brazil*. Sonderheft, Zeitschrift f. Angewandte Geologie, SH 1, 245–255.

STEIGER, R. & JÄGER, E. 1977. Subcommision on geochronology convention on the use of decay constants in geo- and cosmochronology. *Earth Planetary Science Letters*, **36**, 359–362.

VAUCHEZ, A., EGYDIO-SILVA, M., BABINSKI, M., TOMMASI, A., UHLEIN, A. & LIU, D. 2007. Deformation of a pervasively molten middle crust: insights from the Neoproterozoic Ribeira–Araçuaí orogen (SE Brazil). *Terra Nova*, **19**, 278–286.

VIEIRA, V. S. 2007. *Significado do Grupo Rio Doce no Contexto do Orógeno Araçuaí*. PhD thesis, Instituto de Geociências, Universidade Federal de Minas Gerais, Belo Horizonte.

WHITTINGTON, A. G., CONNELLY, J., PEDROSA-SOARES, A. C., MARSHAK, S. & ALKMIM, F. F. 2001. Collapse and melting in a confined orogenic belt: preliminary results from the Neoproterozoic Aracuai belt of eastern Brazil. *In*: *AGU Fall Meeting, Abstract T32B–089. American Geophysical Union*, **82**, 1181–1182.

WIEDEMANN, C. 1993. The evolution of the early Paleozoic, late to post-collisional magmatic arc of the Coastal mobile belt in the State of Espírito Santo, eastern Brazil. *Anais Academia Brasileira de Ciências*, **65**, 163–181.

WIEDEMANN, C., MENDES, J. C. & LUDKA, I. P. 1995. Contamination of mantle magmas by crustal contributions: evidence from the Brasiliano Mobile Belt in the State of Espírito Santo, Brazil. *Anais Academia Brasileira de Ciências*, **67**, 279–292.

WIEDEMANN, C. M., MEDEIROS, S. R., MENDES, J. C., LUDKA, I. P. & MOURA, J. C. 2002. Architecture of late orogenic plutons in the Araçuaí–Ribeira folded belt, southeast Brazil. *Gondwana Research*, **5**, 381–399.

Tourmaline nodules: products of devolatilization within the final evolutionary stage of granitic melt?

DRAŽEN BALEN[1]* & IGOR BROSKA[2]

[1]*University of Zagreb, Faculty of Science, Horvatovac 95, Zagreb, Croatia*

[2]*Slovak Academy of Sciences, Geological Institute, Dúbravská cesta 9, Bratislava, Slovakia*

**Corresponding author (e-mail: drbalen@geol.pmf.hr)*

Abstract: The origin of tourmaline nodules, and of their peculiar textures found in peripheral parts of the Moslavačka Gora (Croatia) Cretaceous peraluminous granite are connected with the separation of a late-stage boron-rich volatile fluid phase that exsolved from the crystallizing magma. Based on field, mineralogical and textural observations, tourmaline nodules were formed during the final stage of granite evolution when undersaturated granite magma intruded to shallow crustal horizons, become saturated and exsolved a fluid phase from residual melt as buoyant bubbles, or pockets. Calculated P–T conditions at emplacement level are *c.* 720 °C, 70–270 MPa, and water content in the melt up to 4.2 wt%.

Two distinct occurrence types of tourmalines have been distinguished: disseminated and nodular tourmalines. Disseminated tourmaline, crystallized during magmatic stage, is typical schorl while nodular tourmaline composition is shifted toward dravite. The increase of dravite in nodular tourmaline is attributed to mixing of the fluid phase from the residual melt with fluid from the wall rocks.

The pressure decrease and related cooling at shallow crustal levels can be considered as a major factor controlling fluid behaviour, formation of a volatile phase, and the crystallization path in the Moslavačka Gora granite body.

Tourmaline nodules are typically spherical bodies found in some evolved granitic rocks of varying age, origin and occurrence. Usually they are 1 to 10 cm in diameter and consist of a core of tourmaline + quartz (± feldspar) surrounded by a quartz + feldspar rim, often called a halo or bleached zone. Although the host rocks of tourmaline nodules may differ in age, origin and occurrence, the nodules themselves look similar and quite unique throughout a wide span of time and space.

A review of older reports and localities can be found in Didier (1973). Recent papers describe tourmaline nodules from the Palaeoproterozoic Scrubber granite, Australia (Shewfelt 2005; Shewfelt *et al.* 2005), an S-type granite from the Neoproterozoic Cape Granite Suite, South Africa (Rozendaal & Bruwer 1995), the 470 Ma old leucocratic Dalmatian granite in Antarctica (Grantham *et al.* 1991), Variscan leucocratic granites (Jiang *et al.* 2003; Buriánek & Novák 2004, 2007; Marschall & Ludwig 2006), as well as from paragneiss from the Central Bohemian Pluton (Nemec 1975), a mid-Cretaceous (100 Ma) leucocratic granite from the Seagull batholith, Yukon Territory in North America (Sinclair & Richardson 1992; Samson & Sinclair 1992), the Cretaceous Erongo granite in Namibia (Trumbull *et al.* 2008), a Tertiary leucocratic metagranite from Menderes Massif, Turkey (Bozkurt 2004), the Miocene Manaslu leucogranites (Himalaya, Le Fort 1991) and the late Miocene (*c.* 8 Ma) Capo Bianco aplite at Elba Island (Dini *et al.* 2002, 2007; Dini 2005; Perugini & Poli 2007).

As can be seen from the literature overview, the tourmaline nodule occurrences around the world are described by numerous authors as a distinctive and common feature of mainly leucocratic granite rocks. However, this world widespread texture is often described in terms of 'bizarre', 'conspicuous', 'geological curiosity', 'spherical entities' and under different and vague names like clots, clusters, spots, coca(r)des, ovoids and orbic(u)les (Didier 1973; LeFort 1991; Shewfelt 2005). In spite of their widespread occurrence, these nodules still constitute a peculiar texture of granite, and the physical and chemical parameters of their origin are not well-known. There are at least four main hypotheses for the origin of the nodules in granitic rocks:

(1) The nodules result from post-magmatic replacement related to metasomatic and hydrothermal alteration of previously crystallized granite by externally derived, boron-rich fluids, accompanied by pegmatite injection and percolating along micro fractures and diffusing along grain boundaries (Nemec 1975; Rozendaal & Bruwer 1995);

From: Sial, A. N., Bettencourt, J. S., De Campos, C. P. & Ferreira, V. P. (eds) *Granite-Related Ore Deposits*. Geological Society, London, Special Publications, **350**, 53–68.
DOI: 10.1144/SP350.4 0305-8719/11/$15.00

(2) The nodules are magmatic–hydrothermal features related to the exsolution, separation and entrapment of immiscible aqueous boron-rich fluids from coexisting granitic magma (Sinclair & Richardson 1992; Samson & Sinclair 1992; Shewfelt 2005; Dini *et al.* 2007; Trumbull *et al.* 2008);
(3) The nodules may represent pelitic xenoliths that have been replaced by boron-rich fluids (LeFort 1991).
(4) The nodules may result from crystallization of a boron-rich granitoid magmatic mass (Perugini & Poli 2007).

The aim of this paper is to describe a new occurrence of tourmaline nodules in granite from the Srednja Rijeka (Moslavačka Gora hill, northern Croatia) locality including field observations, plus data on nodule morphologies, nodule petrography, mineral and whole rock chemistry of nodule core, nodule rim (halo) and the host granite. The proposed scenario of origin may put constraints on Alpine granite evolution for the granites of the Pannonian Basin.

Geological setting of the Moslavačka Gora hill

The hill of Moslavačka Gora in Croatia, located between the Sava and Drava rivers, about 50 km E–SE of Zagreb, is built of the Moslavačka Gora crystalline complex and Tertiary and Quaternary sediments of the southern Pannonian Basin (Fig. 1). The Moslavačka Gora crystalline complex is located within the southwestern part of the Pannonian Basin. Generally, the basement of this Tertiary basin is formed by several crustal blocks, and their present day arrangement is the result of large-scale tectonic movements (e.g. Csontos 1995; Fodor *et al.* 1999; Csontos & Vörös 2004).

One view of the geological setting of Moslavačka Gora is that it is in the southern part of the Tisia Unit, which is interpreted as a tectonic fragment broken off from the southern margin of the European plate during the Middle Jurassic (e.g. Csontos 1995; Pamić & Jurković 2002; Pamić *et al.* 2002; Csontos & Vörös 2004 and references therein). This unit is inferred to have moved by

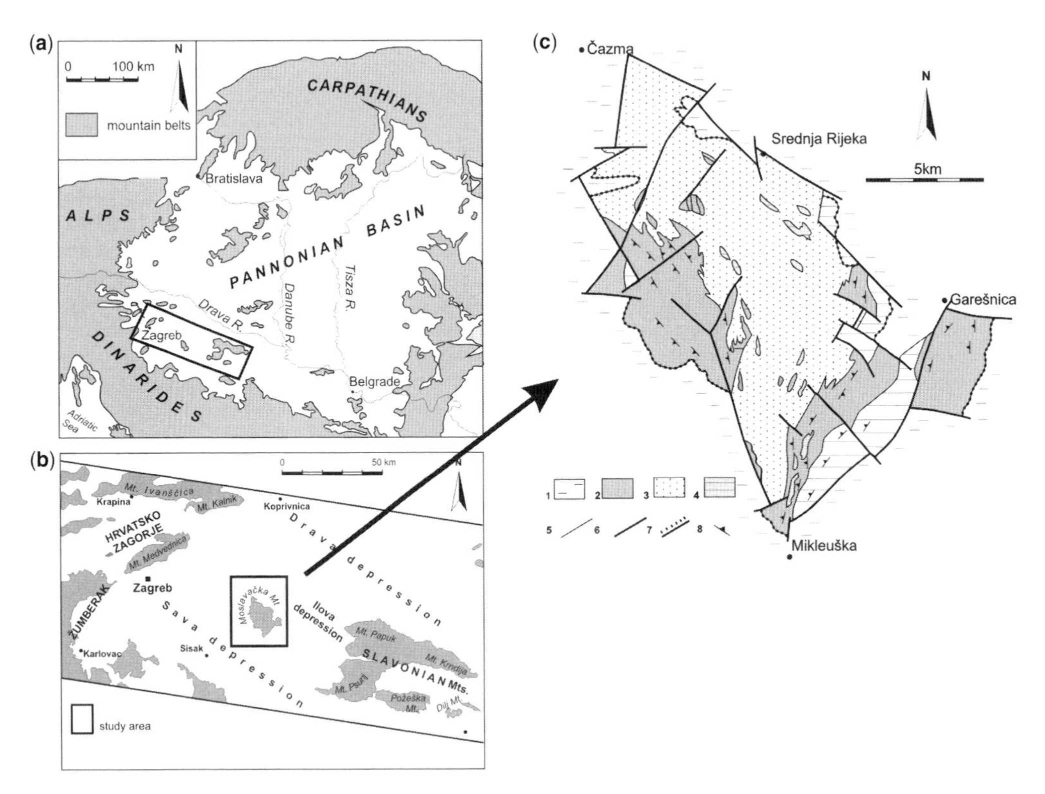

Fig. 1. (**a**) Sketch map showing location of the study area inside the Pannonian Basin; (**b**) sketch map of northern Croatia with position of Moslavačka Gora and (**c**) simplified geological map of the Moslavačka Gora modified after Korolija & Crnko (1985), Pamić (1990), and Pamić *et al.* (2002). (1) Tertiary and Quaternary sedimentary rocks of the Pannonian Basin; (2) migmatite; (3) granite; (4) metamorphic rocks (amphibolite facies); (5) contact line; (6) normal fault; (7) unconformable contact line; (8) foliation.

horizontal block displacements and settled after complex drifting and rotation during Mesozoic and Cenozoic times in its present-day tectonic position. The Tisia Unit is surrounded by the regional-scale tectonic zones, most of them representing oceanic sutures (Schmid *et al.* 2008).

An alternative view of the geological setting of the Moslavačka Gora crystalline complex, based mainly on results of age dating (Pamić 1990; Balen *et al.* 2001, 2003; Starijaš *et al.* 2006) which gave Alpine (Cretaceous) formation ages for granite and medium-grade metamorphic rocks, is that it was part of a belt originally proposed by Pamić *et al.* (1984) and Pamić (1990) to lie within the Moslavačka Gora – Prosara – Motajica – Cer – Bukulja zone. Later Neubauer (2002) recognized that part of this area was an east–west orogen zone with magmatism at *c.* 80 Ma, and Schmid *et al.* (2008) included the area in their so-called Sava zone that is, the suture between the Dinarides and the Tisia terrane.

Field, petrographic and chemical studies reveal the existence of numerous varieties of granitic rocks (andalusite- and tourmaline-bearing granite, leucogranite, biotite granite, monzogranite, granodiorite). Most of the granites occur together with medium-grade metamorphic rocks (amphibolite, marble, metapelite, gneiss, migmatite) and often contain metapelitic xenoliths (biotite + quartz + feldspar ± sillimanite ± andalusite ± garnet schists and/or hornfelses). The granitic rocks cover an area of about 110 km^2 and represent one of the major surface exposures of crystalline basement within the Tertiary sediments in the Pannonian Basin (for a basic geological map and review see Korolija & Crnko (1985) and Pamić (1990), respectively). The geochemical data (Pamić *et al.* 1984; Pamić 1990) and zircon typology analysis (Starijaš *et al.* 2005) further confirm the presence of several groups of granitoids in the crystalline core of Moslavačka Gora.

The isotopic age of the Srednja Rijeka granite is constrained by the data of Balen *et al* (2001) to be 74 ± 1.0 Ma (muscovite Ar–Ar) and 77 ± 33 Ma (garnet–whole rock Sm–Nd isochron). Also Palinkaš *et al.* (2000) obtained Ar–Ar age on muscovite from the Srednja Rijeka pegmatite with a plateau age of 73.2 ± 0.8 Ma, and considered it to be the age of muscovite crystallization and/or cooling. All measured ages have been obtained from leucogranites that penetrate main granite body. A field chronological relation can be established between the main body of two-mica granite and the leucogranites. The intrusion of the two-mica granite precedes the intrusion of the leucogranite, which is often coarser grained and form pegmatites and aplites. Overall age dating of monazites (Starijaš *et al.* 2006) from various types of granites and metamorphic rocks in the area confirms these Cretaceous ages.

Analytical techniques

We have analysed *c.* 20 hand-specimen samples. Selected rock samples (4 granite, 1 nodule core and 1 nodule halo) were analysed in ACME Analytical Laboratories Ltd., Vancouver (Canada). The air-dried samples were sieved to pass 0.125 mm stainless-steel screen and were analysed for 49 elements by the ICP-MS (Inductively coupled plasma-mass spectrometry) and ICP-ES (Inductively coupled plasma-emission spectrometry) analytical procedures. The following elements were analysed: Si, Ti, Al, Fe, Mn, Mg, Ca, Na, K, P, Cr, As, B, Ba, Be, Co, Cs, Cu, Ga, Hf, Mo, Nb, Ni, Pb, Rb, Sn, Sr, Ta, Th, U, V, W, Y, Zn, Zr, La, Ce, Pr, Nd, Sm, Eu, Gd, Tb, Dy, Ho, Er, Tm, Yb and Lu. Sample preparation included splitting of 0.2 g sample for $LiBO_2/Li_2B_4O_7$ fusion decomposition for ICP-ES (macro elements) and 0.2 g sample for ICP-MS (micro elements and REE (rare earth elements)). An additional 0.25 g sample split was analysed for boron, after fusion digestion (Na_2O_2) to 100 ml for ICP-MS. Analysis by ICP-MS used the method of internal standardization to correct for matrix and drift effects. Natural rocks of known composition and pure quartz reagent (blank) were used as reference standards. The analytical accuracy was controlled using the geological standard materials DS7, C3 and SO-18 which represent similar materials. The reference materials were certified in-house by comparative analysis with CANMET (Canada Centre for Mineral and Energy Technology) Certified Reference Materials.

Detection limits for major element oxides and trace elements are given in Table 1 (column d.l.). The analytical accuracy (Table 1, column d.a.) proved to be in the range of ±10% of the certified values for most elements except As (11%) and Mo (12%). The precision of the analyses was determined by analysing 1 duplicate split taken from a sample in each batch of 33 samples and statistically expressed as the variation coefficient (%). The laboratory errors (Table 1, column v.c.) were 89% for W, 28% for Sn, 18% for As and V, 15% for U. For the rest, the coefficient values were below 10% which is considered very reliable.

Compositions and backscattered electron (BSE) images of minerals present in the granites (tourmaline, micas, etc.) were obtained on a CAMECA SX 100 at the Geological Survey of the Slovak Republic in Bratislava. The operating conditions for electron microprobe (EMP) analysis were 15 kV accelerating voltage and 20 nA beam current. Counting time ranged from 10 to 40 s for each element.

Table 1. *Selected whole-rock analyses of host granite (analyses 1–4), leucocratic halo (5) and core (6) of selected tourmaline nodule hosted in granite no. 4.*

	Host granite				Nodule				
Sample	1	2	3	4	5 halo	6 core	d.l.	v.c.	d.a.
Major elements (wt%)							(wt%)	(%)	(%)
SiO_2	72.78	72.33	73.99	73.90	74.29	72.81	0.01	0	1
TiO_2	0.25	0.27	0.20	0.14	0.12	0.38	0.01	2	0
Al_2O_3	14.67	14.44	13.69	14.20	14.63	13.20	0.01	0	1
Fe_2O_3	1.63	1.76	1.43	0.93	0.18	3.58	0.04	2	1
MnO	0.02	0.03	0.02	0.01	0.01	0.05	0.01	0	0
MgO	0.64	0.65	0.40	0.25	0.07	1.64	0.01	0	0
CaO	1.22	1.03	0.60	0.48	0.69	1.84	0.01	1	1
Na_2O	3.85	3.42	3.07	2.71	3.38	0.92	0.01	1	1
K_2O	3.65	5.06	5.37	6.01	5.95	0.08	0.01	0	1
P_2O_5	0.16	0.24	0.18	0.28	0.24	1.30	0.01	3	4
Cr_2O_3	0.00	0.00	0.00	0.01	0.00	0.00	0.02	0	0
LOI	1.00	0.70	1.00	1.10	0.30	4.20	0.01	0	0
Total	99.87	99.93	99.95	100.02	99.86	100.00			
Trace elements (ppm)							(ppm)	(%)	(%)
As	11.1	5.0	22.7	1.3	1.3	0.6	0.5	18	11
B	n.a.	n.a.	n.a.	11	12	9915	3	0	1
Ba	453	495	415	332	428	5	1	6	4
Be	<1	<1	2	5	<1	<1	1	0	0
Co	2.1	2.2	1.0	1.3	0.5	7.7	0.2	3	2
Cs	14.0	17.2	21.4	26.2	12.2	0.3	0.1	3	7
Cu	1.9	2.2	2.8	4.3	3.5	0.4	0.1	9	1
Ga	17.1	15.9	14.6	16.2	13.4	40.3	0.5	7	2
Hf	4.0	3.4	3.0	2.7	1.8	7.1	0.1	6	2
Mo	0.2	0.2	0.7	0.5	0.1	<0.1	0.1	0	12
Nb	14.0	14.2	12.3	9.0	4.9	2.7	0.1	2	0
Ni	5.1	8.3	2.7	1.6	0.5	0.3	0.1	0	7
Pb	6.0	5.4	10.6	12.9	4.5	1.2	0.1	4	0
Rb	183.8	215.0	247.2	276.9	256.9	4.1	0.1	4	4
Sn	11	12	9	9	4	5	1	28	0
Sr	161	129	90	69	112	25	0.5	3	2
Ta	2.0	1.5	1.7	1.6	1.4	1.4	0.1	0	5
Th	8.8	9.3	7.9	6.8	5.6	13.7	0.2	8	2
U	3.9	4.4	4.6	3.7	2.6	10.9	0.1	15	4
V	25	21	11	6	5	55	2	18	3
W	1.4	1.6	1.5	1.3	0.5	0.2	0.5	89	3
Y	19.4	22.4	20.2	22.7	15.2	47.0	0.1	7	6
Zn	36	38	31	16	6	1	1	0	1
Zr	137.5	107.3	77.9	70.2	47.9	226.9	0.1	3	1
La	26.4	19.8	14.2	11.7	10.3	31.8	0.1	9	2
Ce	52.4	44.0	28.3	26.9	22.0	69.9	0.1	7	5
Pr	5.77	5.07	3.67	3.07	2.38	7.57	0.02	6	2
Nd	20.3	18.8	14.2	11.8	8.4	28.2	0.3	3	1
Sm	3.79	3.98	3.13	3.30	2.20	6.90	0.05	9	5
Eu	0.64	0.59	0.48	0.43	0.56	0.60	0.02	5	4
Gd	3.26	3.74	3.19	3.11	2.01	6.85	0.05	9	1
Tb	0.54	0.66	0.64	0.72	0.42	1.31	0.01	8	6
Dy	3.39	4.15	3.77	4.19	2.52	8.19	0.05	6	5
Ho	0.63	0.72	0.75	0.79	0.50	1.48	0.02	4	3
Er	1.90	1.99	2.26	1.96	1.40	4.36	0.03	4	5
Tm	0.25	0.31	0.34	0.31	0.22	0.64	0.01	5	0
Yb	1.84	1.98	2.07	1.82	1.41	4.67	0.05	7	3
Lu	0.27	0.28	0.29	0.24	0.21	0.59	0.01	5	0
(La/Yb)N	9.67	6.74	4.62	4.33	4.92	4.59			
Eu/Eu*	0.56	0.47	0.46	0.41	0.81	0.27			
Σ REE	121.38	106.07	77.29	70.34	54.53	173.06			

Abbreviations: LOI, loss of ignition; d.l., detection limit; v.c., variation coefficient; d.a., data accuracy; n.a., not analysed.

Standards included wollastonite (Ca, Si), TiO_2 (Ti), Al_2O_3 (Al), chromium (Cr), fayalite (Fe), rhodonite (Mn), forsterite (Mg), LiF (F), nickel (Ni), vanadium (V), willemite (Zn), NaCl (Cl), $SrTiO_3$ (Sr), barite (Ba), orthoclase (K) and albite (Na).

Analytical accuracy is variable depending on elemental concentration in the analysed mineral. Detection limits for major elements of rock forming minerals are 0.01–0.02 (in wt%). The matrix effects were corrected by the conventional ZAF (atomic number-absorption-fluorescence) method. Main tourmaline elements show high accuracy ranging from $\pm 0.5\%$ for Si, Al to $\pm 1-2$ wt% for Fe, Mg, Na and $\pm 3-5$ wt% for Ti, Ca. Standard deviation of K for tourmaline is *c.* ± 20 wt% but for biotite there is ± 1 wt%. The rest of trace elements in tourmaline show relatively high uncertainty.

Crystal–chemical formulae of tourmaline were calculated based on the general formula $XY_3Z_6T_6O_{18}(BO_3)_3V_3W$, where X = Ca, Na, K, vacancies (□); Y = Al, Ti, Cr, Mg, Mn, Fe^{2+}; Z = Al; T = Si, Al; B = B; V + W = OH + F + Cl = 4, normalized to 31 (O, OH, F) atoms per formula unit (apfu). The Z-site was considered to be fully occupied by Al, Fe_{tot} as FeO. Alkali deficiency has been calculated on ideal stoichiometry on X site.

The cation assignment of major rock minerals is calculated on the basis of 8 O for plagioclase and 22 O for micas.

Tourmaline nodules

Sample location

Tourmaline nodule bearing granites are located at the northern flanks of Moslavačka Gora near the town Čazma. There is a recently reopened quarry in Srednja Rijeka with fresh outcrops that allowed us to make good observations on the distribution of various types of granites, metapelite and gneiss xenoliths, rare K-feldspar and biotite megacrysts up to 10 cm in size and tourmaline nodule occurrences.

Structural and textural features

Tourmaline is found in Srednja Rijeka granites as scattered (disseminated) interstitial crystals and as a part of nodules. The tourmaline nodules are typically spherical, consisting of 1–2 mm up to 10–20 mm long crystal needles, but may be flattened as well. They are scattered through the two-mica granite (Fig. 2a), which is also rich in metapelitic xenoliths (biotite + quartz + feldspar + sillimanite ± andalusite schist). The nodules are ovoidal (Fig. 2b), are not spatially associated with tourmaline microveins or fracture fill, and they are without visible connection to pegmatite and aplite veins and dikes. The nodules are often found in inhomogeneous randomly distributed swarms aligned along streamlines which are defined by alignment of magmatic minerals (Fig. 2c). Voids and/or miarolitic cavities are occasionally present inside the nodule cores. The erosion of the porphyric granite by weathering isolates both the K-feldspar megacrysts and the tourmaline nodules and reveals the relatively spherical shape of the latter (Fig. 2d).

Textural features, including their often rounded shape, their physical distinctness from the groundmass, the occurrence of included magmatic phenocrysts inside the nodules, slight differences in grain size and texture, lack of (micro)vein network connecting the nodules are important in constraining the origin of the tourmaline nodules. The same features have been already noticed at Capo Bianco aplite (Elba Island; Dini *et al.* 2007; Perugini & Poli 2007).

Petrographic characterization of the host rock and nodule

We have investigated samples of granites hosting tourmaline nodules, including both fine-grained, equigranular granite and porphyric, two-mica granite. The latter has a weak foliation and biotite predominates over muscovite. The major mineral assemblage of the host granite is quartz, plagioclase (albite to oligoclase), biotite, muscovite and K-feldspar. Accessory phases include zircon, apatite, monazite, xenotime and opaque mineral(s) – Figure 3a. *Quartz* is about 1 mm in size forming interlocking polycrystalline aggregates that display a mosaic texture. *Albite* is tabular, similar to quartz in size, and is randomly distributed in the rock. Often secondary tiny apatite is present as inclusions in plagioclase. Alteration of albite to white mica and kaolinite is common (Fig. 3b, c). *Biotite* grains define a weak foliation throughout the rock. Individual flakes are 1–2 mm in length, but may form large clots, up to 5 mm in length, together with subordinate muscovite. Locally the biotite grains are very elongated (Fig. 3d). *Muscovite* flakes occur also as separate grains and inclusions in feldspars. *K-feldspar* is similar to plagioclase and quartz in size, around 1 mm, and is tabular in form, randomly distributed through granite, intergrowing with albite and quartz. White mica flakes are often included in K-feldspar.

A leucocratic biotite-free rim (halo) surrounds each tourmaline nodule core and is considered to be an integral part of the nodule. It consists of quartz, feldspar and muscovite, with generally the same textural features as in the host granite but

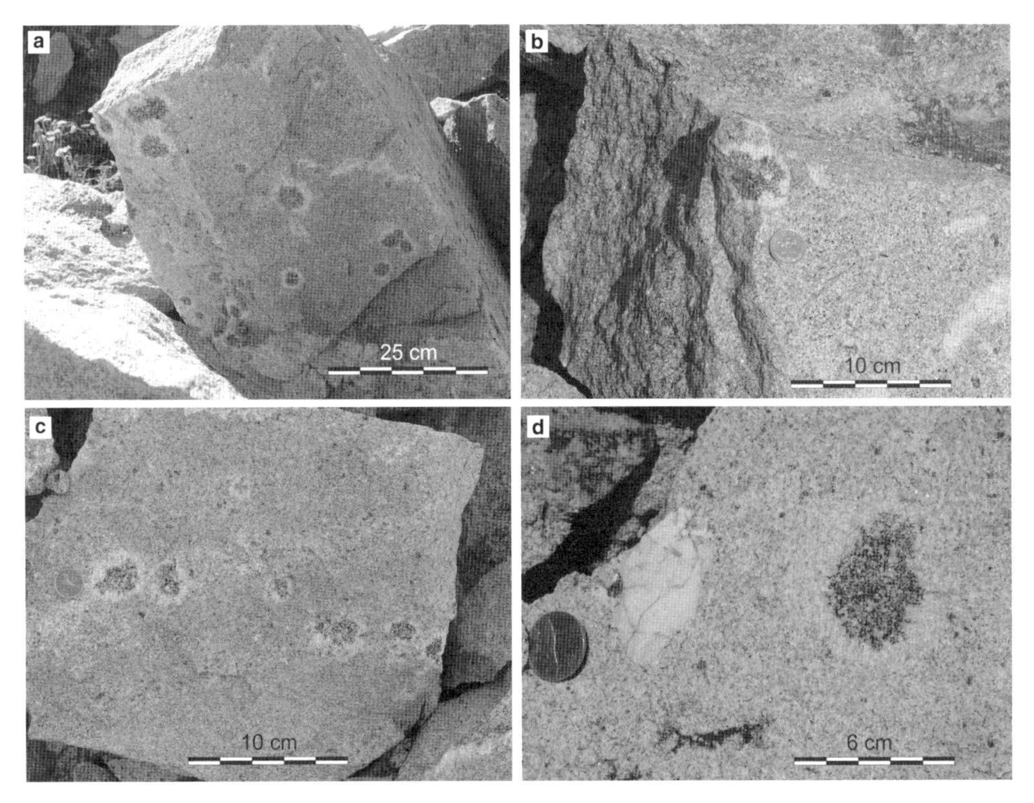

Fig. 2. (**a**) Typical outcrop of granite with tourmaline nodule with the characteristic leucocratic rim (halo); (**b**) tourmaline nodule with typical spherical shape; (**c**) a tourmaline nodule 'train'; (**d**) tourmaline nodule and K-feldspar megacryst.

with slightly different grain size. Usually leucocratic minerals in the nodule rim (halo) grow from the edge of the nodule halo/core toward to the core. Inside the nodule core the feldspar and quartz occasionally form euhedral crystals. Biotite flakes are very rare in the halo.

Tourmaline rich nodule cores consist of tourmaline + quartz + albite + K-feldspar ± muscovite, with very fine grained tourmaline and interstitial microgranular quartz. Internal arrangement of tourmaline into the nodule shows random (without preferred orientation) distribution of grains (Fig. 4a, b). Small tourmaline patches that look like separate grains in fact belong to the same grain which may be recognized with the same extinction. Anhedral to subhedral *tourmaline*, commonly 1–2 mm in size, exhibits brown to light greenish-brownish pleochroism, but is locally blue, mainly in the centres of nodules. It also exhibits variable zoning from almost homogenous to patchy zoning (Fig. 4c, d). Tourmaline is typically interstitial between grains of quartz and feldspars. Irregular tourmaline crystals are embayed by quartz and feldspars, and generally display regular grain contacts. Throughout the core *quartz* crystals form interlocking polycrystalline aggregates that display mosaic texture. The nodules often enclose larger grains of quartz and feldspars, comparable in size and shape with those set in the main groundmass. *Albite* is tabular and is partly replaced by tourmaline near the nodule margins.

Irregular vugs or miarolitic cavities are locally present in the rock as important evidence for a late, but trapped, volatile phase. The high volatile activity during nodule origin seems to be the general feature for the Moslavačka Gora granite.

The alkali feldspars, especially albite, commonly contain newly-formed apatite crystals. Such secondary apatite is very small, typically less than 2 μm in size, apatites within K-feldspar crystals form much larger grains. The crystallization of secondary apatite was enhanced by the high F activity in the late-stage corrosive fluid. The formation of pure end-member albite and muscovite accompanied this late-stage process.

At the Srednja Rijeka locality, beside tourmaline nodule bearing granite, leucogranite dikes occur. Garašić *et al.* (2007) found that mineral assemblage

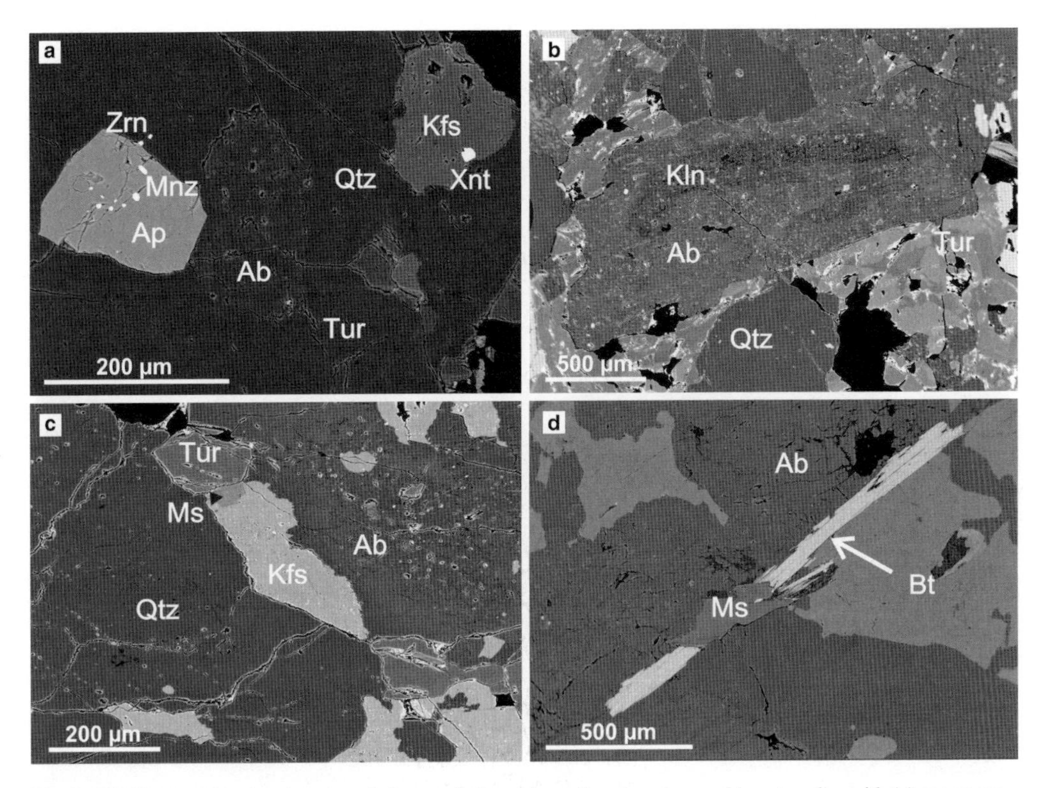

Fig. 3. BSE images showing the textural characteristics and granite mineral assemblage together with (**a**) accessory phases include zircon, apatite, monazite and xenotime; (**b**) & (**c**) alteration of albite is common forming white mica and kaolinite; (**d**) rapidly crystallized elongated Fe-rich biotite. Mineral abbreviations are after Kretz (1983), Xnt – xenotime.

of leucogranite is characterized by subequal proportions of quartz, K-feldspar (microcline and orthoclase) and plagioclase (Ab_{93-95}) with variable contents of muscovite, biotite, garnet, andalusite and tourmaline. Black elongated, several mm to cm long tourmaline grains are randomly distributed (i.e. disseminated) through rock and represent a distinct occurrence type of tourmaline at this locality.

Geochemistry

Whole-rock chemistry of the host rock and nodule

Selected chemical analyses (Table 1) representing host granite (analyses 1–4), tourmaline nodule halo (an. 5) and core (an. 6). Selected nodule is typical of a nodule found inside granite represented with analysis no. 4.

Granite analyses indicate 72.33–73.99 wt% SiO_2, 13.69–14.67 wt% Al_2O_3, low concentrations of Fe_2O_3 (0.93–1.76 wt%), MgO (0.25–0.65 wt%), CaO (0.48–1.22 wt%) and TiO_2 (0.14–0.27 wt%), and relatively high K_2O (3.65–6.01 wt%) and Na_2O (2.71–3.85 wt%). Strong peraluminosity [ASI (alumina saturation index) = 1.1–1.2] and high SiO_2 content are coupled with low concentrations of ferromagnesian elements together with high LIL (large ion lithophile) trace elements Ba (332–495 ppm), Sr (69–161 ppm), Cs (14–26 ppm), Rb (184–277 ppm), Zr (70–138 ppm). The host granite is characterized by moderately flat and fractionated REE chondrite normalized patterns (Fig. 5 – after Boynton 1984), slight enrichment in the light REE (LREE) and strongly negative Eu anomaly [$(La/Yb)_N = 4.33–9.67$; low $\Sigma REE = 70.34–121.38$ ppm; $Eu/Eu^* = 0.41–0.56$; where $Eu/Eu^* = Eu_N/\sqrt{(Sm_N * Gd_N)}$]. The REE pattern (Fig. 5) is typical for crustal granites produced by partial melting of metapelitic rocks.

The nodule halo has low Fe_2O_3 (0.18 wt%) and MgO (0.07 wt%), while K_2O (5.95 wt%), Na_2O (3.38 wt%) and CaO (0.69 wt%) together with SiO_2 (74.29 wt%) and Al_2O_3 (14.63 wt%) correspond well with values in the host granites. Values for trace elements Ba (428 ppm), Sr (112 ppm), Cs (12 ppm), Rb (257 ppm) are high and also close to

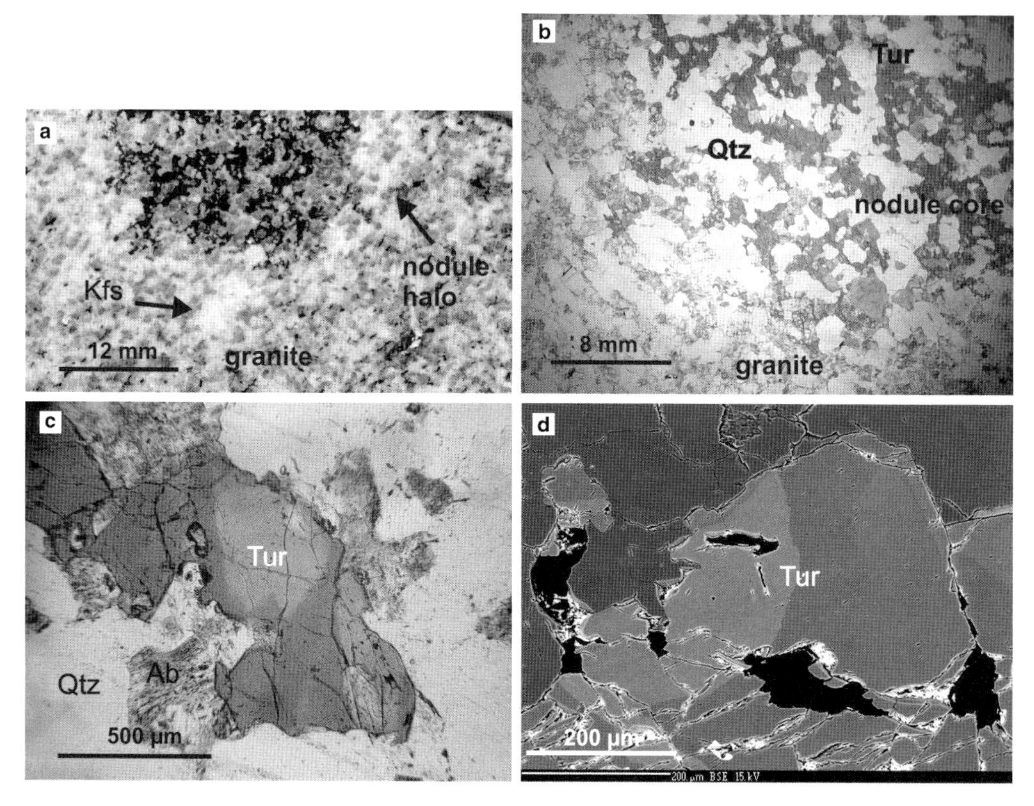

Fig. 4. (**a**) Internal arrangement of anhedral to subhedral tourmaline into the larger nodule show random distribution of grains, also visible larger K-feldspar captured inside halo of nodule; (**b**) random (optically non-oriented) distribution of tourmaline grains inside nodule core (photo obtained with Petroscope – microdiascope for thin sections, plane polarized light, parallel polars); (**c**) tourmaline shows chemical zoning and pleochroism (plane polarized light, parallel polars); (**d**) tourmaline zoning as visible in BSE images. Mineral abbreviations are after Kretz (1983).

the granite values. Europium anomaly is not very distinct and ΣREE is low (54.53 ppm).

The tourmaline nodule cores have relative high content of Fe_2O_3 (3.58 wt%), MgO (1.64 wt%) and CaO (1.84 wt%) while Na_2O and K_2O are low (0.92 and 0.08 wt%, respectively). High loss of ignition (LOI) (4.2 wt%) is a consequence of large quantity of tourmaline in assemblage accompanied

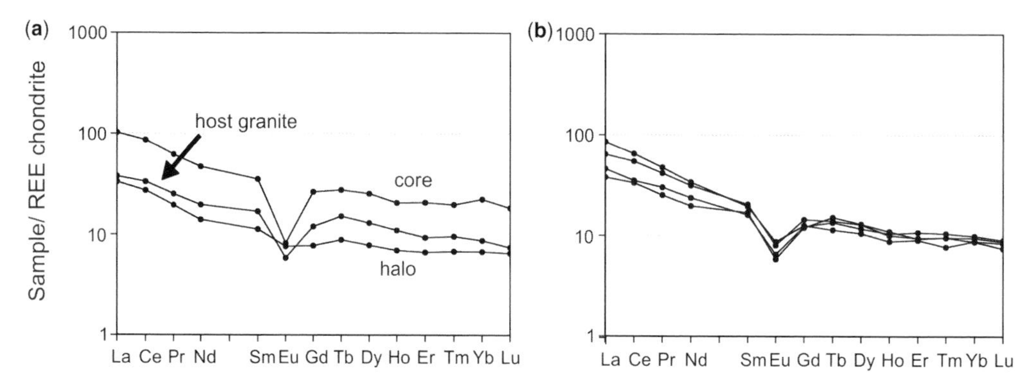

Fig. 5. Chondrite-normalized rare elements patterns (normalization after Boynton 1984) for (**a**) one selected host granite (an. 4), tourmaline nodule halo (an. 5) and core (an. 6); (**b**) REE variations in host granites (analyses 1–4).

with alteration processes on feldspar and fluids in mineral structure (mica) and/or trapped volatile phase. The LIL trace elements Ba (5 ppm), Sr (25 ppm), Cs (0.3 ppm), Rb (4.1 ppm) are low and significantly differ from the granite and halo values. Europium anomaly in the nodule core is a strong negative with value $Eu/Eu^* = 0.27$ and ΣREE is high (173.06 ppm).

The boron content of the host rock is low in the tourmaline-free portions (11–12 ppm) and reaches high values (*c.* 9900 ppm) in the tourmaline nodule core (Table 1).

The difference between core, rim and host rock is visible on the chondrite-normalized REE patterns (Boynton 1984) – Figure 5. The LREE distribution in all samples is higher than heavy REE (HREE) with pronounced Eu anomaly in host granite and nodule core (Fig. 5a, b). The core in comparison to the halo or host granite shows a larger negative Eu anomaly; on the other hand, in the feldspar-rich halo Eu anomaly is weak that is, not so pronounced (Fig. 5a). That feature of halo is related to high feldspar content accompanied with high concentration of Sr and Ba, elements isomorphous with Eu. Also halo shows low ΣREE because of a decrease in the amount of mafic minerals.

The core and halo in the chondrite normalized diagrams show slight tetrad effect ($T_{1,3} > 1.1$; according to Irber 1999). Such trends indicate suppression of the control of charge and radius on element behaviour under water-rich conditions in the presence of boron and fluorine-rich fluids.

Mineral chemistry

Compositional variation in tourmalines

Representative compositions of both disseminated and nodule tourmalines are shown in Table 2. Crystals of disseminated tourmaline are schorl, while nodular tourmaline range from schorl to dravite, depending on the level of MgO present. MgO generally rise toward the rims of nodular tourmaline and weak zoning can be traced on backscatter electron (BSE) images (Fig. 4d; Table 2) and $^{X}\square/(^{X}\square + Na)$ v. $Fe/(Fe + Mg)$ diagram for tourmalines (Fig. 6).

For disseminated type of tourmaline Garašić *et al.* (2007) found that it is schorl-foitite showing chemical zonation with Na^{+} and F^{-} contents increasing from core (0.488; 0.287 apfu) to rim (0.623; 0.389 apfu), respectively.

The homovalent *dravite* substitution expressed by exchange vector $FeMg_{-1}$ is the most widespread in the investigated samples. Electron microprobe analysis (EMPA) compositions of disseminated tourmaline (schorl) reveal higher $Fe/(Fe + Mg) = 0.75–0.85$ than in nodular schorl-dravite crystals ($Fe/(Fe + Mg) = 0.4–0.6$). The T-site in both tourmalines is occupied by Si and Al, the Z-site is fully occupied by Al. The nodular tourmaline in comparison to the disseminated shows slight enrichment in Ca and Ti in the Y-site. Alkali-deficiency in both tourmaline types is significant $^{X}\square = 0.4$ and following heterovalent *foitite* exchange vector can explain the substitution: $^{X}\square + {}^{Y}Al = {}^{X}Na + {}^{Y}Mg$ (Fig. 7a). On the other, hand the *uvite* substitution $^{X}Ca + {}^{Y}Mg = {}^{X}Na + {}^{Y}Al$ is characteristic for the nodular tourmaline (Fig. 7b).

Chemical compositions of the selected major rock forming minerals in the nodule and host granite are shown in Table 3. Beside tourmaline and quartz, mineral assemblage of nodule core comprises plagioclase An_{13-17} while in the nodule halo plagioclase composition is An_{11-21}. Variations in chemical composition of host rock micas (biotite and muscovite) are shown in Table 3. Low Ti and typical K concentration of the biotites indicate that these are only partly altered to chlorite rich in water, and their low sum probably reflects increasing of ferric component in biotite due to oxidic regime.

Discussion

Melt production and setting

Tourmaline nodules have been recently described in evolved granitic rocks of various age, origin and occurrence around the world. They are commonly interpreted, in spite of some models that derive them from hydrothermal fluids (e.g. Rozendaal & Bruwer 1995), to be the result of the final stage of granitic melt crystallization, in settings where late-magmatic volatile-rich melts did not manage to escape the magmatic system (e.g. Sinclair & Richardson 1992; Shewfelt 2005; Dini *et al.* 2007; Perugini & Poli 2007; Trumbull *et al.* 2008).

In the case of Moslavačka Gora potential field evidence for the generation of early melt by melting of crust in a collisional environment is the presence of metapelitic xenoliths, whose strong depletion in muscovite argues for a restitic nature. Muscovite dehydration melting at low a_{H_2O} of (meta)pelite is a plausible mechanism for production of melt (Patiño Douce & Harris 1998; Patiño Douce 1999; Dini *et al.* 2002). Garašić *et al.* (2007) have already related the formation of Moslavačka Gora leucogranite to melting with a K-feldspar rich residue and muscovite dehydration melting at low a_{H_2O}, and assigned the process to the melting of continental crust in a collisional environment.

The Moslavačka Gora two-mica granite, which hosted the tourmaline nodules, solidified in a shallow plutonic environment, as can be inferred from the occurrence of magmatic andalusite in the

Table 2. *Selected microprobe analyses of disseminated tourmaline and of tourmaline from the nodule assemblage. The cation assignment is calculated on the basis of 31 anions (O, OH, F).*

Location	Disseminated tourmaline					Nodular tourmaline				
	1	2	3	4	5	core 6	7	rim 8	rim 9	10
SiO_2	34.39	34.59	35.08	35.59	35.96	35.78	36.63	36.46	36.42	36.45
TiO_2	0.57	0.73	0.54	0.50	0.58	0.79	0.01	0.88	0.68	1.06
Al_2O_3	34.77	32.78	33.01	32.55	33.24	33.84	32.47	33.96	34.52	33.24
Cr_2O_3	0.01	0.91	0.00	0.00	0.00	0.00	0.03	0.04	0.00	0.01
FeO	12.96	12.71	12.81	12.84	12.59	10.38	11.46	7.70	7.38	8.06
MgO	1.30	1.65	1.74	1.87	1.98	2.98	3.48	4.84	4.85	5.26
CaO	0.18	0.18	0.17	0.16	0.15	0.30	0.24	0.40	0.30	0.45
MnO	0.31	0.21	0.15	0.30	0.22	0.08	0.15	0.07	0.04	0.10
Na_2O	1.67	1.90	1.72	1.92	2.02	1.64	2.00	1.79	1.98	1.91
K_2O	0.03	0.06	0.03	0.05	0.05	0.04	0.01	0.05	0.04	0.03
F	0.06	0.34	0.27	0.45	0.14	0.00	0.00	0.06	0.00	0.12
Cl	0.01	0.00	0.00	0.00	0.00	0.00	0.01	0.00	0.00	0.00
H_2O*	3.55	3.40	3.42	3.36	3.54	3.61	3.61	3.65	3.69	3.62
B_2O_3*	10.38	10.30	10.30	10.35	10.46	10.48	10.49	10.66	10.68	10.66
Total	100.38	100.03	99.49	100.25	100.93	99.92	100.59	100.55	100.59	100.96
O = F	0.02	0.14	0.11	0.19	0.06	0.00	0.00	0.02	0.00	0.05
Total	100.35	99.89	99.38	100.06	100.87	99.92	100.59	100.53	100.59	100.91
T position										
Si	5.756	5.836	5.921	5.975	5.977	5.935	6.072	5.946	5.925	5.942
Al	0.244	0.164	0.079	0.025	0.023	0.065	0.000	0.054	0.075	0.058
B	3.000	3.000	3.000	3.000	3.000	3.000	3.000	3.000	3.000	3.000
Z position										
Al	6.000	6.000	6.000	6.000	6.000	6.000	6.000	6.000	6.000	6.000
Y position										
Al	0.616	0.354	0.488	0.415	0.490	0.551	0.342	0.474	0.543	0.330
Ti	0.071	0.093	0.068	0.064	0.073	0.098	0.001	0.108	0.083	0.130
Cr	0.002	0.121	0.000	0.000	0.000	0.000	0.003	0.005	0.000	0.001
Mg	0.323	0.415	0.437	0.467	0.491	0.738	0.860	1.176	1.177	1.278
Mn	0.044	0.030	0.021	0.043	0.032	0.011	0.021	0.010	0.006	0.013
Fe_{2+}	1.814	1.793	1.808	1.802	1.750	1.440	1.588	1.050	1.004	1.098
Y sum	3.000	3.000	3.000	3.000	2.836	2.837	2.816	2.823	2.813	2.850
X position										
Ca	0.032	0.032	0.030	0.029	0.027	0.054	0.043	0.070	0.053	0.079
Na	0.543	0.620	0.564	0.625	0.651	0.529	0.643	0.565	0.626	0.603
K	0.006	0.013	0.007	0.010	0.010	0.008	0.002	0.010	0.009	0.007
□	0.420	0.335	0.399	0.336	0.312	0.409	0.312	0.355	0.313	0.312
W position										
OH	3.967	3.821	3.855	3.758	3.926	4.000	3.997	3.972	4.000	3.935
F	0.029	0.179	0.145	0.242	0.074	0.000	0.000	0.028	0.000	0.064
Cl	0.003	0.000	0.000	0.000	0.000	0.000	0.003	0.000	0.000	0.001

*H_2O and B_2O_3 are calculated value for ideal stoichiometry giving for B = 3 and W = 4. In total 28 analyses have been performed for nodular tourmaline and 35 for disseminated ones

assemblage and presence of miarolitic cavities (Tućan 1904; Pamić 1990). Conditions calculated for Alpine (Cretaceous) metamorphic rocks associated with granite intrusion are temperatures of 550–650 °C and pressures up to 400 MPa (Balen 1999). Shallow crystallization of the granite melt is also indicated by the low zircon saturation temperature (*c.* 730 °C; using equation from Watson & Harrison 1983); this relatively low temperature for zircon saturation is a maximum temperature since part of the zircon is inherited. REE thermometry (*c.* 720 °C) derived from monazite saturation experiments (Montel 1993) supports this conclusion. The calculated water content of the melt may reach value as high as 4.2 wt% using Montel's (1993) equation. From these temperatures, and using the *P–T*

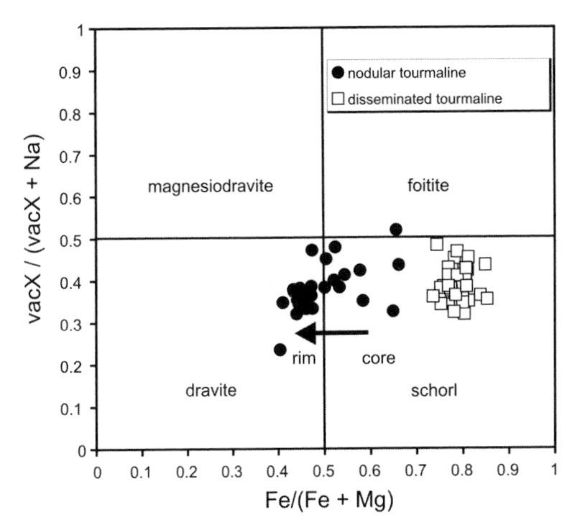

Fig. 6. Quadrilateral $^{X}\square/(^{X}\square + Na)$ v. Fe/(Fe + Mg) diagram for tourmalines occurring in the Moslavačka Gora granite. Arrow shows apparent increase of Mg content from core to rim in nodular tourmaline.

diagram of andalusite stability after Clarke *et al.* (2005), we estimate the pressure was in the range of 70–270 MPa (pressure from the water-saturated peraluminous granite solidus and And = Sil reaction curve, respectively), which corresponds to an average depth of 5–6 km.

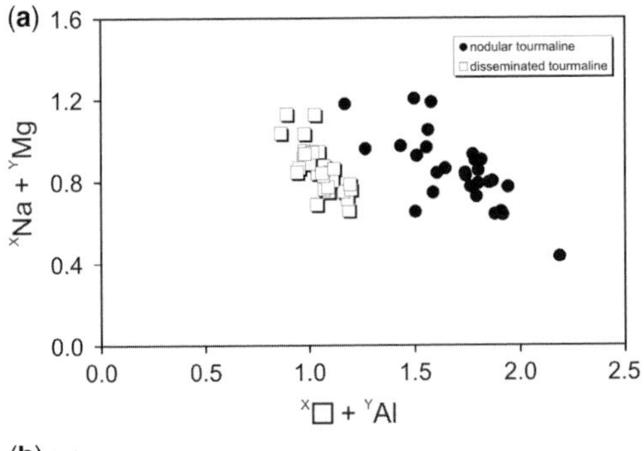

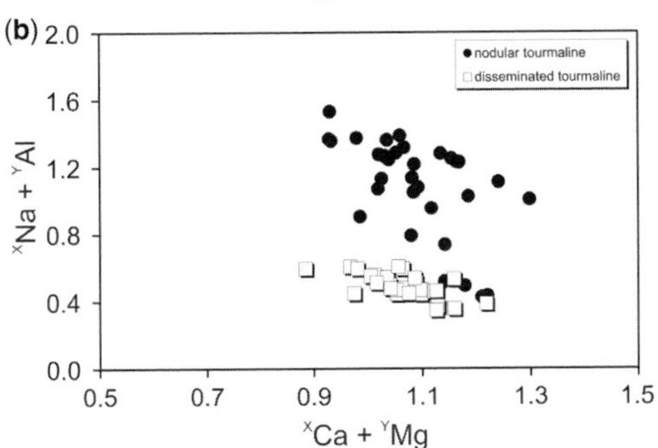

Fig. 7. Composition differences and substitutions of disseminated and nodular tourmalines in the Moslavačka Gora granites: (**a**) $^{X}Na + {}^{Y}Mg$ v. $^{X}\square + {}^{Y}Al$; (**b**) $^{X}Na + {}^{Y}Al$ v. $^{X}Ca + {}^{Y}Mg$.

Formation of tourmaline nodules

The concentration of boron in the low-grade metapelites is generally sufficient to explain concentrations if the granite melt formed by partial melting of the metapelite (Wilke *et al.* 2002). Extraction of low fractions of partial melt during muscovite breakdown in the region of anatexis can result in an initially high boron and water content. However, that melt would not be tourmaline-saturated as indicated by experimental data, and conditions for crystallization of tourmaline would be reached only with a further enrichment, after separation of a boron-rich volatile phase (e.g. London *et al.* 1996, London 1997, 1999; Dini *et al.* 2007).

Magmatic differentiation leads to liquid immiscibility between high-silica and hydrous melts in the roof zone, with partitioning of B, Na and Fe to hydrous melt. Liquid immiscibility in the evolved melt could be mechanism which produces spherical segregations (Trumbull *et al.* 2008). Veksler & Thomas (2002) and Veksler *et al.* (2002) experimentally confirmed the immiscibility of aluminosilicate and water-rich melts with extreme boron enrichment (5 wt%; Thomas *et al.* 2003). Veksler (2004) noted that the more water-rich depolymerized melt are strongly enriched in B, Na, Fe. Therefore, liquid immiscibility concentrates the elements necessary for formation of the tourmaline nodules in this water-rich, highly mobile melt phase, which can percolate through crystal mush and coalesce in discrete bodies (Trumbull *et al.* 2008). The aluminium necessary for tourmaline growth is hypothesized to come from replacement of feldspar, a reaction observed inside tourmaline nodules. The massive precipitation of tourmaline nodules would largely deplete the residual melt in boron.

The study of the generation and crystallization conditions of peraluminous leucocratic granitic magma (London 1992; Scaillet *et al.* 1995) suggests that the initial magma is generated from metasedimentary sources at low a_{H_2O}, and low to medium pressures and temperatures. Following the hypothesis proposed by Candela (1991, 1994) there is an exsolution of the volatile phase within the crystallization interval or at the crystal-melt interface of the magma chamber. As crystallization continues, and as the concentration of the volatile phase in the crystallization interval increases, the rise of pockets of fluid (bubbles ± crystals ± melt) upward is possible. The crystallization of volatile-free minerals from the host melt increases the amount of this fluid phase. As the volume of crystals increases in the magma and the viscosity of the residual melt rises, any aqueous fluid that has not previously escaped

Table 3. *Selected mineral analyses for the host granite assemblage. The cation assignment is calculated on the basis of 8 O for plagioclase and 22 O for micas.*

Mineral		Plagioclase					Biotite		Muscovite		
		Nod. core Core	Nod. core rim	Halo core	Halo rim		Host	Rock	Host	Rock	
SiO_2		65.00	65.73	63.14	65.93		36.51	35.53	47.50	48.05	47.22
TiO_2							0.07	3.22	0.00	0.03	0.03
Al_2O_3		22.59	22.16	23.24	21.57		20.63	18.79	37.22	39.37	38.49
FeO		0.00	0.01	0.00	0.02		17.50	19.75	0.98	0.29	0.30
MnO							0.32	0.32	0.07	0.02	0.00
MgO							10.36	7.57	0.60	0.18	0.20
CaO		3.48	2.78	4.38	2.33		0.04	0.00	0.02	0.25	0.21
Na_2O		9.54	10.08	9.25	10.67		0.00	0.00	0.14	1.92	1.56
K_2O		0.16	0.28	0.23	0.15		9.42	9.65	10.49	6.74	7.59
Total		100.78	101.05	100.23	100.65		94.84	94.83	97.04	96.85	95.60
	Si	2.841	2.865	2.788	2.883	Si IV	5.502	5.453	6.168	6.121	6.123
	Al	1.164	1.138	1.209	1.112	Al IV	2.498	2.547	1.832	1.879	1.877
	Ca	0.163	0.130	0.207	0.109	Al VI	1.165	0.852	3.864	4.032	4.006
	Na	0.809	0.852	0.792	0.905	Ti VI	0.008	0.372	0.000	0.003	0.003
	K	0.009	0.016	0.013	0.008	Fe^{+2}	2.205	2.535	0.106	0.031	0.033
	Total	4.986	5.000	5.010	5.017	Mn^{+2}	0.041	0.042	0.008	0.002	0.000
						Mg	2.327	1.732	0.116	0.034	0.039
	An	16.6	13.0	20.5	10.7	Ca	0.006	0.000	0.003	0.034	0.029
	Ab	82.5	85.4	78.2	88.5	Na	0.000	0.000	0.035	0.474	0.392
	Or	0.9	1.6	1.3	0.8	K	1.811	1.889	1.738	1.095	1.256
						Total	15.563	15.422	13.870	13.705	13.758

becomes trapped within the granite body. However, it appears that this process of increasing the amount of fluid by early crystallization of anhydrous minerals will not be significant in the case of Moslavačka Gora granite. The rarity of the K-feldspar megacrysts occurring in the same fine-grained granite argues against the exsolution of large volume of fluids by extensive early crystallization. Igneous microstructures of the megacrysts such as crystal shape, simple twinning, zonal growth (compositional zoning) are consistent with a free growth of megacrysts from melt. Thus, the megacrysts formed early, though K-feldspar is commonly among the last minerals crystallizing in granitic magmas.

Field and textural features in granites from Moslavačka Gora that are rich in tourmaline nodule occurrences suggest the nodules formed above the solidus of the granite. The spherical shape of tourmaline nodules, their physical distinctness from the groundmass, the occurrence of included phenocrysts from the surrounding magma, the concentration of nodules within the granite at the top (peripheral or roof zone) of the body and the absence of micro-vein network connecting the nodules imply the separation and entrapment of water- and boron-rich fluid forming pockets in a crystal poor magma.

Decompression during emplacement at shallow levels further promotes exsolution of the fluid phase from the melt and its mixing with fluids from the roof rock. The physical model (Shinohara & Hedenquist 1997) for the exsolution of volatiles from a convecting magma shows that at low degrees of crystallization, individual fluid bubbles formed in the magma will buoyantly rise, coalesce at the top of the magma body and tend to accrete into a sphere to decrease surface tension. An important factor in controlling the system kinetics will be the viscosity of the melt and the ability of volatiles to change it (Dingwell 1999).

Boron does not enter the structures of most common granite-forming silicate minerals (quartz, K-feldspar, oligoclase-albite and muscovite). Moreover, boron has a strong affinity with the aqueous phase. It has been reported that boron forms hydrated borate clusters in hydrous melts (Dingwell *et al.* 1996). At low pressure, dissolution of high quantities of H_2O in a silicate melt requires the presence of fluxing volatiles in extraordinarily high amounts. In such cases the initially homogenous melt unmixes into two phases, a 'normal' melt and highly mobile flux rich in aluminosilicate(s) (Thomas *et al.* 2005). Such a boron-rich evolved melt will be depolymerized and will have lower density, viscosity, and liquidus and solidus temperatures (Kubiš & Broska 2005). Those characteristics will promote concentration of boron in isolated areas, that is, bubbles or pockets.

We propose that tourmaline nodules grew from bubbles of such immiscible volatile-rich melt, where they did not manage to escape from the coexisting host granite magma. The authors have the opinion that this scenario is consistent with the observed characteristics of the tourmaline nodules at this locality.

Differences between tourmaline occurrences

The two contrasting tourmaline occurrences in Moslavačka Gora granite, scattered schorl in granite and dravite/schorl in nodules, indicate the separate origin of these phases. Primary tourmaline in granites including that in nodules is typical schorl in composition (Fig. 6). The dravite in nodular tourmaline is attributed to interaction with a fluid derived from the wall rock environment. Wallrock and granite interaction is indicated also by the metapelitic xenoliths in the granite body. The moderate Mg composition in tourmaline shows that the granite environment was affected by wall-rock fluids.

Boron is concentrated in tourmaline, which is the principal host mineral for that element. Tourmaline crystallized in nodule cores has relatively high dravite (Mg) content, which presumably resulted from the mixing of juvenile fluids and volatiles coming from the wall rocks to magma chamber (Kubiš & Broska 2005). The contribution of wall rock fluids at shallow crustal level increased the amount of volatiles. Nodular tourmaline is formed during the final stage of supersolidus crystallization or near solidus crystallization. Textures and chemical composition suggest crystallization in a quasi-closed system where tourmaline crystallized as the last mineral. In contrast, disseminated tourmaline from leucogranite is schorl-foitite and can be considered as a typical magmatic product.

Following the models of Dini *et al.* (2002, 2007) and Dini (2005), developed at Elba Island, the peculiar texture and distribution of tourmaline

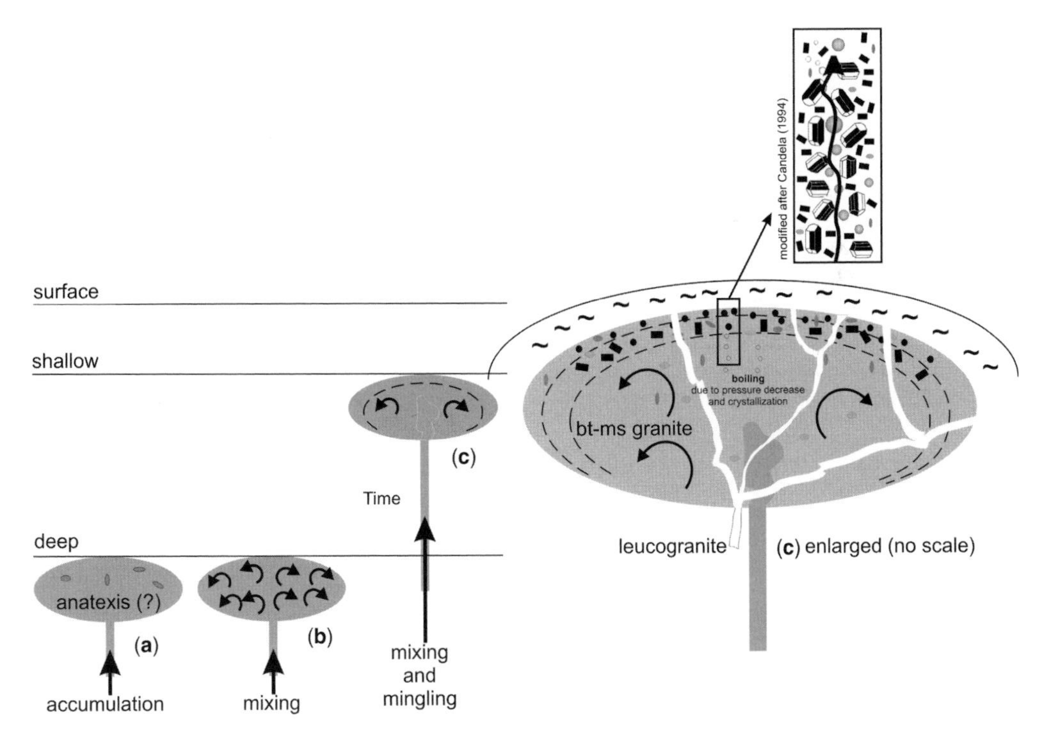

Fig. 8. Proposed scenario for Cretaceous magmatism and origin of tourmaline nodules; (**a**) partial melting and accumulation of granite melt; (**b**) mixing and mingling; (**c**) emplacement of biotite–muscovite granite at shallow crustal level, pressure drop, crystallization and cooling accompanied by boiling. This leads to separation and mixing of volatiles and formation of fluid pockets and bubbles enriched in boron, as the bubbles crystallize to form tourmaline nodules. The inset figure shows the bubble ‘escape route’ in peripheral parts of granite body, and the enlarged sketch shows the relationship between the tourmaline nodule host granite and leucogranites with disseminated tourmaline.

nodules are best explained in the frame of the emplacement setting of this intrusion.

In the case of magma that reaches shallow levels rapidly, the resulting decompression and cooling leads to the separation of boron-rich melt as distinct volatile-rich bubbles $\pm$ crystals. If these bubbles are unable to escape the system owing to impermeable host and wallrock, they cause the formation of tourmaline nodules in nearly-crystallized granite (Fig. 8).

The Moslavačka Gora granite that hosts the tourmaline nodules can thus be regarded as an occurrence of magma that escaped from the plutonic levels and stalled in a low pressure setting. Here, the separation of a boron-rich fluid phase gave way to the formation of tourmaline nodules in a solidifying magma which produced the host two-mica granite.

Conclusion

Based on field, mineralogical and textural observations of two mica Moslavačka Gora granite, it can be concluded that the tourmaline nodules are formed from a late-stage boron-rich fluid phase that separated from the crystallizing magma. When undersaturated granite magmas are intruded into shallower crustal horizons they become saturated and exsolve a fluid phase at relatively high temperatures. This volatile phase is exsolved as buoyant bubbles, or pockets, from which the tourmaline nodules crystallized. There is a significant influx of fluids derived from the wallrock and mixing with the fluid phase from the residual granitic melt. This has resulted in increased dravite content in nodular tourmaline in comparison to the schorl compositions in typical disseminated magmatic tourmaline. The pressure drop and related cooling is a major factor controlling fluid behaviour and crystallization path in the Moslavačka Gora granite body.

Authors are grateful to R. L. Helz for language improvement and comments on an early manuscript draft. A. Dini and an anonymous reviewer provided thoughtful reviews that were a great help in clarifying the paper. Careful editorial work by corresponding editor A. Sial and volume editor C. De Campos is very much appreciated. This research was supported by Croatian Ministry of Science, Education and Sports grant 119-1191155-1156 and by the Slovak Research and Development Agency under the contract No. APVV-0557-06.

References

BALEN, D. 1999. *Metamorphic reactions in amphibole-bearing rocks of Moslavačka Gora*. PhD thesis, University of Zagreb.

BALEN, D., SCHUSTER, R. & GARAŠIĆ, V. 2001. A new contribution to the geochronology of Mt. Moslavačka Gora (Croatia). *In*: ÁDAM, A., SZARKA, L. & SZENDRŐI, J. (eds) *PANCARDI 2001*. Geodetic and Geophysical Research Institute of the Hungarian Academy of Science, DP-2, 2–3.

BALEN, D., SCHUSTER, R., GARAŠIĆ, V. & MAJER, V. 2003. The Kamenjača olivine gabbro from Moslavačka Gora (South Tisia, Croatia). *Rad Hrvatske akademije znanosti i umjetnosti Zagreb*, **486**, 57–76.

BOYNTON, W. V. 1984. Geochemistry of the rare earth elements: meteorite studies. *In*: HENDERSON, P. (ed.) *Rare Earth Element Geochemistry*. Elsevier, Amsterdam, 63–114.

BOZKURT, E. 2004. Granitoid rocks of the southern Menderes Massif (southwestern Turkey): field evidence for Tertiary magmatism in an extensional shear zone. *International Journal of Earth Sciences*, **93**, 52–71.

BURIÁNEK, D. & NOVÁK, M. 2004. Morphological and compositional evolution of tourmaline from nodular granite at Lavičky near Velké Meziříčí, Moldanubicum, Czech Republic. *Journal of the Czech Geological Society*, **49**, 81–90.

BURIÁNEK, D. & NOVÁK, M. 2007. Compositional evolution and substitutions in disseminated and nodular tourmaline from leucocratic granites: examples from the Bohemian Massif, Czech Republic. *Lithos*, **95**, 148–164.

CANDELA, P. A. 1991. Physics of aqueous phase evolution in plutonic environments. *American Mineralogist*, **76**, 1081–1091.

CANDELA, P. A. 1994. Combined chemical and physical model for plutonic devolatilization: A non-Rayleigh fractionation algorithm. *Geochimica et Cosmochimica Acta*, **58**, 2157–2167.

CLARKE, D. B., DORAIS, M. *ET AL.* 2005. Occurrence and origin of andalusite in peraluminous felsic igneous rocks. *Journal of Petrology*, **46**, 441–472.

CSONTOS, L. 1995. Tertiary tectonic evolution of the Intra-Carpathian area: a review. *Acta Vulcanologica*, **7**, 1–13.

CSONTOS, L. & VÖRÖS, A. 2004. Mesozoic plate tectonic reconstruction of the Carpathian region. *Palaeogeography Palaeoclimatology Palaeoecology*, **210**, 1–56.

DIDIER, J. 1973. Mineral nodules. *In*: DIDIER, J. (ed.) *Granites and Their Enclaves. The Bearing of Enclaves on the Origin of Granites*. Elsevier, Amsterdam, Developments in Petrology, **3**, 357–368.

DINGWELL, D. B. 1999. *Granitic Melt Viscosities*. Geological Society, London, Special Publications, **168**, 27–38.

DINGWELL, D. B., PICHAVANT, M. & HOLTZ, F. 1996. Experimental studies of boron in granitic melts. *In*: GREW, E. S. & ANOVITZ, L. (eds) *Boron: Mineralogy, Petrology, and Geochemistry in the Earth's Crust*. Mineralogical Society of America, Reviews in Mineralogy, **33**, 331–385.

DINI, A. 2005. The boron (F-Li) rich Capo Bianco aplite (Elba Island, Italy): a snapshot of fluid separation processes during subvolcanic emplacement of a pegmatite-like magma. *In*: *Crystallization Processes in Granitic Pegmatites, International Meeting Elba*

Island. http://www.minsocam.org/MSA/Special/Pig/PIG_articles/Elba%20Abstracts%206%20Dini.pdf.

Dini, A., Innocenti, F., Rocchi, S., Tonarini, S. & Westerman, D. S. 2002. The magmatic evolution of the late Miocene laccolith–pluton–dyke granitic complex of Elba Island, Italy. *Geological Magazine*, **139**, 257–279.

Dini, A., Corretti, A., Innocenti, F., Rocchi, S. & Westerman, D. S. 2007. Sooty sweat stains or tourmaline spots? The Argonauts at Elba Island (Tuscany) and the spread of Greek trading in the Mediterranean Sea. *In*: Piccardi, L. & Masse, W. B. (eds) *Myth and Geology*. Geological Society, London, Special Publications, **273**, 227–243.

Irber, W. 1999. The lanthanide tetrad effect and its correlation with K/Rb, Eu/Eu*, Sr/Eu, Y/Ho, and Zr/Hf of evolving peraluminous granite suites. *Geochimica et Cosmochimica Acta*, **63**, 489–507.

Fodor, L., Csontos, L., Bada, G., Györfi, I. & Benkovics, L. 1999. Tertiary tectonic evolution of the Pannonian Basin system and neighbouring orogens: a new synthesis of palaeostress data. *In*: Durand, D., Jolivet, L., Horváth, F. & Séranne, M. (eds) *The Mediterranean Basins: Tertiary Extension Within the Alpine Orogen*. Geological Society, London, Special Publications, **156**, 295–334.

Garašić, V., Kršinić, A., Schuster, R. & Vrkljan, M. 2007. Leucogranite from Srednja Rijeka (Moslavačka Gora, Croatia). 8th Workshop on Alpine Geological Studies, Akademie der Naturwissenschaften, Schweizerischer Nationalfonds zur Foerderung der Wissenschaftlichen Forschung, 21–21.

Grantham, G. H., Moyes, A. B. & Hunter, D. R. 1991. The age, petrogenesis and emplacement of the Dalmatian Granite, H.U. Sverdrupfjella, Dronning Maud Land, Antarctica. *Antarctic Science*, **3**, 197–204.

Jiang, S.-Y., Yang, J. H., Novák, M. & Selway, J. B. 2003. Chemical and boron isotopic compositions of tourmaline from the Lavičky leucogranite, Czech Republic. *Geochemical Journal*, **37**, 545–556.

Korolija, B. & Crnko, J. 1985. Basic Geological Map Of Yugoslavia In Scale 1:100000, L 33–82 Sheet Bjelovar. Geološki zavod Zagreb, Savezni geološki zavod Beograd.

Kretz, R. 1983. Symbols for rock-forming minerals. *American Mineralogist*, **68**, 277–279.

Kubiš, M. & Broska, I. 2005. The role of boron and flourine in evolved granitic rock systems (on the example of the Hnilec area, Western Carpathians). *Geologica Carpathica*, **56**, 193–204.

Lefort, P. 1991. Enclaves of the Miocene Himalayan leucogranites. *In*: Didier, J. & Barbarin, B. (eds) *Enclaves And Granite Petrology*. Elsevier, Amsterdam, Developments in Petrology, **13**, 35–47

London, D. 1992. The application of experimental petrology to the genesis and crystallization of granitic pegmatites. *Canadian Mineralogist*, **30**, 499–540.

London, D. 1997. Estimating abundances of volatile and other mobile components in evolved silicic melts through mineral-melt equilibria. *Journal of Petrology*, **38**, 1691–1706.

London, D. 1999. Stability of tourmaline in peraluminous granite systems: the boron cycle from anatexis to hydrothermal aureoles. *European Journal of Mineralogy*, **11**, 253–262.

London, D., Morgan, G. B., VI. & Wolf, M. B. 1996. Boron in granitic rocks and their contact aureoles. *In*: Grew, E. S. & Anovitz, L. (eds) *Boron: Mineralogy, Petrology, and Geochemistry in the Earth's Crust*. Mineralogical Society of America, Reviews in Mineralogy, **33**, 299–330.

Marschall, H. R. & Ludwig, T. 2006. Re-examination of the boron isotopic composition of tourmaline from the Lavicky granite, Czech Republic, by secondary ion mass spectrometry: back to normal. Critical comment on 'Chemical and boron isotopic compositions of tourmaline from the Lavicky leucogranite, Czech Republic' by S.-Y. Jiang *et al.*, *Geochemical Journal*, **37**, 545–556, 2003. *Geochemical Journal*, **40**, 631–638.

Montel, J. M. 1993. A model for monazite/melt equilibrium and application to the generation of granitic magmas. *Chemical Geology*, **110**, 127–146.

Nemec, D. 1975. Genesis of tourmaline spots in leucocratic granites. *Neues Jahrbuch für Mineralogie Monatshefte*, **7**, 308–317.

Neubauer, F. 2002. Contrasting Late Cretaceous With Neogene Ore Provinces in the Alpine-Balkan-Carpathian-Dinaride Collision Belt. *In*: Blundell, D. J., Neubauer, E. & Von Quadt, A. (eds) *The Timing and Location of Major Ore Deposits in an Evolving Orogen*. Geological Society, London, Special Publications, **204**, 81–102.

Palinkaš, A. L., Balogh, K., Strmić, S., Pamić, J. & Bermanec, V. 2000. Ar/Ar dating and fluid inclusion study of muscovite, from the pegmatite of Srednja Rijeka, within granitoids of Moslavačka gora Mt., North Croatia. *In*: Tomljenović, B., Balen, D. & Saftić, B. (eds) *PANCARDI 2000 Special Issue*, Vijesti HGD, **37**, 95–96.

Pamić, J. 1990. Alpine granites, migmatites and metamorphic rocks from Mt. Moslavačka Gora and the surrounding basement of the Pannonian Basin (Northern Croatia, Yugoslavia). *Rad JAZU Zagreb*, **10**, 7–121.

Pamić, J. & Jurković, I. 2002. Paleozoic tectonostratigraphic units in the northwest and central Dinarides and the adjoining South Tisia. *International Journal of Earth Sciences*, **91**, 538–554.

Pamić, J., Krkalo, E. & Prohić, E. 1984. Granites from the northwestern slopes of Mt. Moslavačka Gora in northern Croatia. *Geologija Ljubljana*, **27**, 201–212.

Pamić, J., Balen, D. & Tibljaš, D. 2002. Petrology and geochemistry of orthoamphibolites from the Variscan metamorphic sequences of the South Tisia in Croatia – an overview with geodynamic implications. *International Journal of Earth Sciences*, **91**, 787–798.

Patiño Douce, A. E. 1999. What do experiments tell us about the elative contributions of crust and mantle to the origin of granitic magmas. *In*: Castro, A., Fernández, C. & Vigneresse, J. L. (eds) *Understanding Granites. Integrating New and Classical Techniques*. Geological Society, London, Special Publications, **68**, 55–75.

Patiño Douce, A. E. & Harris, N. 1998. Experimental Constraints on Himalayan Anatexis. *Journal of Petrology*, **39**, 689–710.

PERUGINI, D. & POLI, G. 2007. Tourmaline nodules from Capo Bianco aplite (Elba Island, Italy): an example of diffusion limited aggregation growth in a magmatic system. *Contributions to Mineralogy and Petrology*, **153**, 493–508.

ROZENDAAL, A. & BRUWER, L. 1995. Tourmaline nodules: indicator of hydrothermal alteration and Sn–Zn–(W) mineralization in the Cape Granite Suite, South Africa. *Journal of African Earth Sciences*, **21**, 141–155.

SAMSON, I. M. & SINCLAIR, W. D. 1992. Magmatic hydrothermal fluids and the origin of quartz-tourmaline orbicles in the Seagull Batholith, Yukon Territory. *Canadian Mineralogist*, **30**, 937–954.

SCAILLET, B., PICHAVANT, M. & ROUX, J. 1995. Experimental crystallization of leucogranite magmas. *Journal of Petrology*, **36**, 663–705.

SCHMID, S. M., BERNOULLI, D. ET AL. 2008. The Alpine-Carpathian-Dinaridic orogenic system: correlation and evolution of tectonic units. *Swiss Journal of Geosciences*, **101**, 139–183.

SHEWFELT, D. 2005. *The nature and origin of Western Australian tourmaline nodules; a petrologic, geochemical and Isotopic study*. MS Thesis, University Saskatchewan.

SHEWFELT, D., ANSDELL, K. & SHEPPARD, S. 2005. The origin of tourmaline nodules in granites; preliminary findings from the Paleoproterozoic Scrubber Granite. *Geological Survey of Western Australia Annual Review*, 59–63.

SHINOHARA, H. & HEDENQUIST, J. W. 1997. Constraints on magma degassing beneath the Far Southeast porphyry Cu–Au deposit, Philippines. *Journal of Petrology*, **38**, 1741–1752.

SINCLAIR, D. W. & RICHARDSON, J. M. 1992. Quartz–tourmaline orbicles in the Seagull Batholith, Yukon Territory. *Canadian Mineralogist*, **30**, 923–935.

STARIJAŠ, B., BALEN, D., TIBLJAŠ, D. & FINGER, F. 2005. Zircon Typology in Crystalline Rocks of Moslavačka Gora (Croatia) – Preliminary Petrogenetic Insight from Transmitted Light (TL) and Scanning Electron Microscopy (SEM). 7th Workshop on Alpine Geological Studies, Croation Geological Survey, 89–90.

STARIJAŠ, B., GERDES, A. ET AL. 2006. Geochronology, metamorphic evolution and geochemistry of granitoids of the Moslavačka Gora Massif (Croatia). *In*: Proceedings XVIIIth Congress of the Carpathian-Balkan Geological Association, 594–597.

THOMAS, R., FÖRSTER, H. J. & HEINRICH, W. 2003. The behavior of boron in a peraluminous granite–pegmatite system and associated hydrothermal solutions: a melt and fluid inclusion study. *Contributions to Mineralogy and Petrology*, **144**, 457–472.

THOMAS, R., FÖRSTER, H-J., RICKERS, K. & WEBSTER, J. D. 2005. Formation of extremely F-rich hydrous melt fractions and hydrothermal fluids during differentiation of highly evolve tin-granite magmas: a melt/fluid-inclusion study. *Contributions to Mineralogy and Petrology*, **148**, 582–601.

TUĆAN, F. 1904. Pegmatite from crystalline rocks of Moslavačka hills (Pegmatit u kristaliničnom kamenju Moslavačke gore). *Rad JAZU Zagreb*, **159**, 166–208.

TRUMBULL, R. B., KRIENITZ, M.-S., GOTTESMANN, B. & WIEDENBECK, M. 2008. Chemical and boron-isotope variations in tourmalines from an S-type granite and its source rocks: the Erongo granite and tourmalinites in the Damara Belt, Namibia. *Contributions to Mineralogy and Petrology*, **155**, 1–18.

VEKSLER, I. V. 2004. Liquid immiscibility and its role at the magmatic hydrothermal transition: a summary of experimental studies. *Chemical Geology*, **210**, 7–31.

VEKSLER, I. V. & THOMAS, R. 2002. An experimental study of B-, P- and Frich synthetic granite pegmatite at 0.1 and 0.2 GPa. *Contributions to Mineralogy and Petrology*, **143**, 673–683.

VEKSLER, I. V., THOMAS, R. & SCHMIDT, C. 2002. Experimental evidence of three coexisting immiscible fluids in synthetic granite pegmatite. *American Mineralogist*, **87**, 775–779.

WATSON, E. B. & HARRISON, T. M. 1983. Zircon saturation revisited: temperature and composition effects in a variety of crustal magma types. *Earth and Planetary Science Letters*, **64**, 295–304.

WILKE, M., NABELEK, P. I. & GLASCOCK, M. D. 2002. B and Li in Proterozoic metapelites from the Black Hills, U.S.A.: Implications for the origin of leucogranitic magmas. *American Mineralogist*, **87**, 491–500.

Geochemical characteristics of Miocene Fe–Cu–Pb–Zn granitoids associated mineralization in the Chichibu skarn deposit (central Japan): evidence for magmatic fluids generation coexisting with granitic melt

D. ISHIYAMA[1]*, M. MIYATA[1], S. SHIBATA[1], H. SATOH[1], T. MIZUTA[1], M. FUKUYAMA[2] & M. OGASAWARA[3]

[1]*Faculty of Engineering and Resource Science, Akita University, Akita, Japan*

[2]*Institute of Earth Sciences, Academia Sinica, 128 Academia Road Sec. 2, Nankang Taipei 115 Taiwan, ROC*

[3]*Geological Survey of Japan, Higashi 1-1-1 Central 7, Tsukuba, Ibaraki 305-8567, Japan*

**Corresponding author (e-mail: ishiyama@galena.mine.akita-u.ac.jp)*

Abstract: In this work we study mechanisms and timing of magmatic fluid generation during magma emplacement. Our focus is the Miocene calc-alkaline granitic rocks from the Chichibu mining area, in Japan. The granitoids consist of northern and southern Bodies and of the Daikoku Altered stocks. Cathodoluminescence observation of quartz phenocrysts from the northern body point towards magmatic resorption, which is thought to be caused by mixing between a more differentiated and a more primitive magma. The coexistence of vapour-rich two-phase and halite-bearing polyphase fluid inclusions in a single quartz crystal from the northern Body supports the possibility of pressure decrease during magma emplacement. The magmatic fluids that originated the Chichibu deposit are thought to have been generated by pressure release, related to magmatic differentiation when the SiO_2-content reaches about 65 wt%. As a result, heavy metals, such as copper, gold and arsenic, coexisting with the silicate melt, were transported into the sedimentary strata through degassing of magmatic fluids. A later major fault system caused the intercalation between heavy-metal-free limestone and orebodies, as a secondary skarn-building process took place in the dominant limestone area.

Many studies on the genesis of skarn deposits have unravelled metal sources and mineralizing fluids generation (Kwak 1986; Fulignati *et al.* 2001; Meinert *et al.* 2003; Baker *et al.* 2004). Kwak (1986) proposed that magmatic fluid is the dominant fluid phase for skarn deposits formation. Fulignati *et al.* (2001) suggested that immiscibility between silicate, hydrosaline and carbonate melts predominated at the magma chamber–carbonate wall rock interface: a hydrosaline fluid is thought to be derived from the magmatic system during skarn formation, especially for the Vesuvius volcanic system. Another model considering no external fluid input during skarn ore-forming processes was proposed by Meinert *et al.* (2003) and Baker *et al.* (2004). Based on these studies, over the past decade discussions on the metal source for ore deposits formation suggests that magmatic activities play an important role supplying metals to porphyry copper and skarn deposits (Candela & Piccoli 1995; Hedenquist & Richards 1998; Meinert *et al.* 2005; Williams-Jones & Heinrich 2005). In addition to these ideas, the mechanisms and timing of fluid generation in a silicate melt are also important to consider for skarn and porphyry copper ore deposit genesis. Models called 'first boiling (pressure decrease)' and 'second boiling (saturation of volatile)' have been proposed for fluid generation in a silicate melt. Harris *et al.* (2003) reinforce the importance of 'second boiling' and 'first boiling' for economic porphyry system formation including porphyry copper deposits.

During the early stage of granitic magma emplacement of Chichibu, the presence of fluids in the silicate melt was coeval with the Chichibu Fe–Cu–Pb–Zn–Au–Ag polymetallic skarn deposit generation as pointed out by Ishihara *et al.* (1987). Mechanisms and timing of fluid generation during magmatic differentiation could be possibly better understood by knowing the, (1) distribution and modes of occurrence of sulphide minerals in granitic rocks; (2) changes in major and trace element concentrations, especially heavy metals (e.g. arsenic, copper and zinc), (3) internal growth textures of quartz crystals, and (4) mode of occurrence of fluid inclusions in granite quartz crystals.

From: Sial, A. N., Bettencourt, J. S., De Campos, C. P. & Ferreira, V. P. (eds) *Granite-Related Ore Deposits*. Geological Society, London, Special Publications, **350**, 69–88.
DOI: 10.1144/SP350.5 0305-8719/11/$15.00

The purpose of this study is to characterize the melt forming Chichibu granite bodies and the coexisting fluid phase based on geological, petrological, mineralogical, geochemical and fluid inclusions studies. Our goal is to unravel the mechanisms and timing of the magmatic fluid generation in the Fe–Cu–Pb–Zn–Au–Ag Chichibu skarn deposit and related granite melt.

Outline of geology and ore deposits

The Chichibu deposit, located 100 km NW of Tokyo, occurs in an area on the east side of an active uplifting region composed of late Cenozoic volcanic and sedimentary rocks (Fig. 1). The Chichibu mining area is located at the northern end of Izu-Ogasawara (Izu-Bonin) volcanic arc that is subducting beneath the east–west trending accretionary complexes parallel to the southwestern extension of Japan. Some igneous activities from the Chichibu mining area on the Izu Peninsula are thought to be related to the subduction of this Izu-Ogasawara volcanic arc along the north trending arc (Takahashi 1989).

An outline of the geology and ore deposits of Chichibu mining area has been presented by Ishiyama (2005). The geology around the Chichibu deposit consists of Palaeozoic–Mesozoic sedimentary strata (Southern Chichibu Terrane composed of pebbly mudstone and sandstone) including Carboniferous–Jurassic chert and limestone as olistoliths and Neogene granitoids (Figs 1 & 2). The age of the sedimentary strata is estimated to be Jurassic (Sakai & Horiguchi 1986). Sedimentary rocks around the Chichibu deposit consist of pebbly mudstone with blocks of chert, limestone and basalt (MITI 1975; Nakano *et al.* 1998).

The Chichibu granitoids are one of the Neogene granitoids (Tanzawa, Tokuwa, Kai-Komagatake, Chichibu and Oohinata granitoids) that are located in an area trending north–south, which is east of the uplifting region (Fig. 1). These granitoids are divided into: (1) a Northern Body; (2) a Southern Body; and (3) a Daikoku Altered stock. (Fig. 2). The Daikoku Altered stock crops out between the Northern and Southern Bodies. The radiometric ages determined by the K–Ar method for biotite in both Northern and Southern Bodies are 6.59 and 5.87 Ma, respectively (Ueno & Shibata 1986). Ueno & Shibata (1986) also reported K–Ar ages

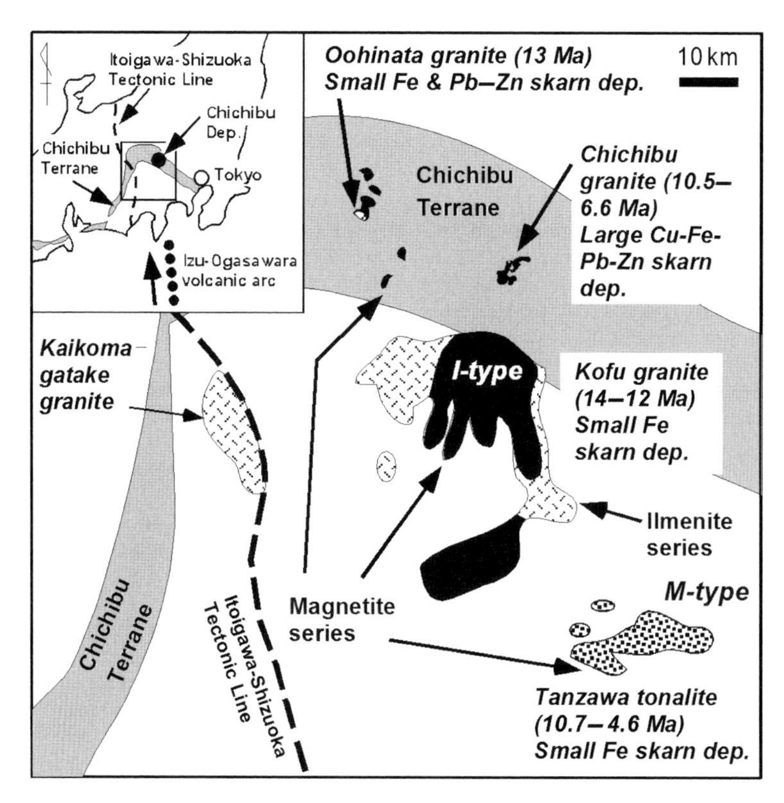

Fig. 1. Location map showing Chichibu mining area.

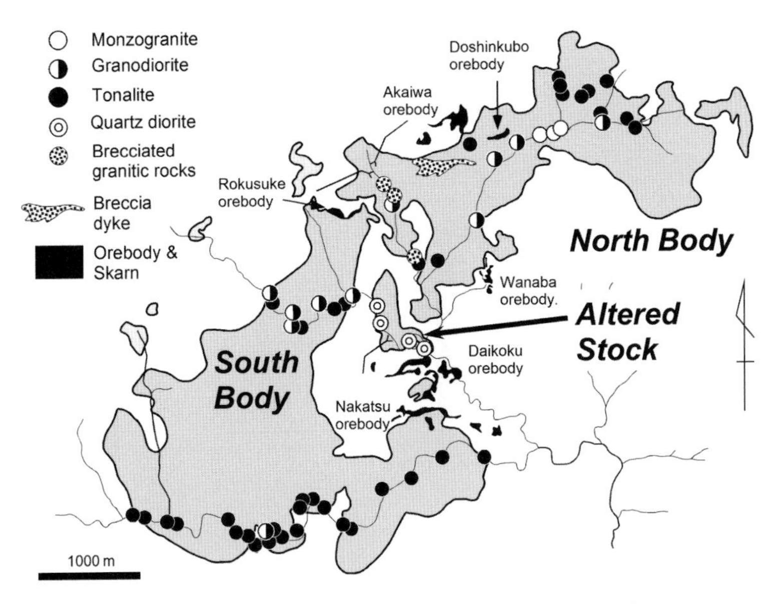

Fig. 2. Map showing distribution of Chichibu granitic rocks and orebodies (Kaneda 1967).

for hornblends from the Southern Body as 10.5 Ma. They suggested that the hornblende may contain excess ^{40}Ar. Saito *et al.* (1996) reported K–Ar ages from 6 biotites, ranging from 5.6 to 6.8 Ma (South and North Bodies), and from 2 hornblende, ranging from 6.0 to 6.1 Ma (South Body). They suggest that the Chichibu granitoids experienced a rapid cooling. Based on the similarity of ages of emplacements (Ueno & Shibata 1986; Saito *et al.* 1996) and short distance between North Body and South Body, the depth of emplacement of North Body and South Body is thought to be similar.

The total metal production of the Chichibu skarn deposit is about 5.8 m.t. with average ore grade of 0.3% Cu, 0.2% Pb, 2.4% Zn, 20 g/t Ag and 1.0 g/t Au (Shimazaki 1975). The Chichibu skarn deposit is a zinc-rich skarn deposit, though some quantities of copper, gold and silver also occur in the deposit. The amounts of metals in the Chichibu deposit are large, though the apparent size of the area for granitoids is small, when compared to the apparent size of areas such as those associated to the Tanzawa, Tokuwa and Kai-Komagatake granitoids. Skarn and hydrothermal ore deposits related to these last granitoids in the same region are rather small (Fig. 1). Orebodies of the Chichibu skarn deposit are mainly distributed in the northwestern part of the Northern Body and Daikoku Altered stock. Major orebodies of the Northern Body are Akaiwa and Dohshinkubo. While the Daikoku orebody is a major orebody associated with the Altered stock, the Southern Body accompanies small skarn orebodies called Nakatsu and Rokusuke (Fig. 2). These ore deposits occur in contact aureole within 300 m from the boundary between the granitic body and contact metamorphosed sedimentary rocks (Kaneda & Watanabe 1961). A vertical zoning according to ore type has been recognized in the Chichibu mining area (Kaneda 1967; Ishihara *et al.* 1987) (Fig. 3). Magnetite ore tends to be dominant at the lower level of ore bodies, while sphalerite and rhodochrosite ores tend to be dominant at the upper level of orebodies. The Dohshinkubo orebody, which occurs in the deeper part of contact aureole of the Northern Body, mainly consists of magnetite ore. The Akaiwa orebody shows vertical zoning, from pyrrhotite-rich ore at the deeper part of a contact aureole of the Northern Body, through sphalerite-rich ore and rhodochrosite-rich ore at the shallower part of the aureole. The Wanaba orebody, which occurs at the lower part of the aureole surrounding the Northern Body, consists of magnetite–sphalerite ore. The Rokusuke orebody occurs in the northern and at the upper part of a contact aureole from the Southern Body and consists of sphalerite ores. The Nakatsu orebody is restricted to a deeper part of the contact aureole, consists of magnetite-rich ore, while the Daikoku orebody shows vertical zoning, from magnetite at the deeper part, through pyrrhotite and sphalerite/rhodochrosite ore in the shallower part of the aureole (Fig. 3) (Kaneda 1967). Large fault systems crosscut the large Dohshinkubo, Akaiwa and Daikoku orebodies, while no fault system have been observed in small orebodies such as Wanaba (Abe *et al.* 1961; Kaneda & Watanabe 1961; Shoji *et al.* 1967).

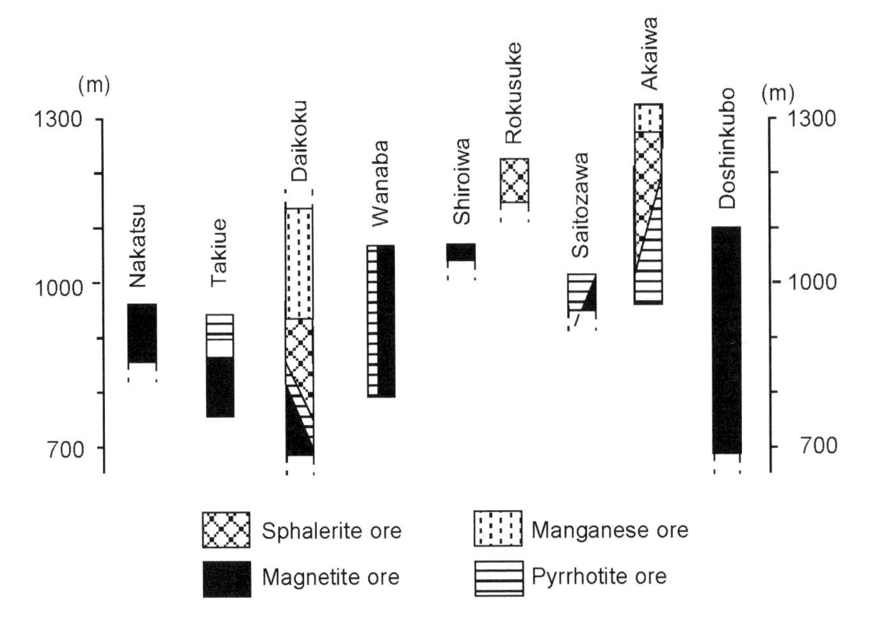

Fig. 3. Schematic diagram showing vertical zoning of type of ores in Chichibu mining area. After Kaneda (1967).

These facts suggest that the formation of large orebodies is structurally controlled by a large fault system, which is thought to be the main path for mineralizing fluids. Funnel shaped breccia dykes also occur near the Akaiwa deposit, in granitic outcrops of the Northern Body (Fig. 2). The maximum size of the biggest east–west trending breccia dyke is about seven hundred metres in length (Abe *et al.* 1961; Kaneda & Watanabe 1961) (Fig. 2). Rock fragments found in these breccia dykes are: (1) granitic rocks from the Northern Body; (2) thermally metamorphosed slate; and (3) chert. These rock fragments are commonly altered and contain chlorite. This is interpreted as an evidence for a low pressure environment at a late stage of the Northern Body emplacement.

Characteristics of Chichibu granitoids

Structure, texture and mineral assemblage of granitic rocks

Rock facies were identified by a method of modal analyses under a microscope. Counting of three thousand points was carried out for each sample to carry out the modal analyses. North Body consists of medium- to fine-grained tonalite, granodiorite, and monzogranite from marginal to central parts (Figs 2, 4 & 5). The tonalite and granodiorite have melanocratic features and relatively holocrystalline texture. Many dark inclusions are found in tonalite at the marginal part of North Body (Fig. 5d). Monzogranite shows leucocratic features and porphyritic texture. Miarolitic cavities and tourmaline veins and veinlets are common at the central part of the monzogranite and some tourmaline veins and veinlets cut the monzogranite containing miarolitic cavities. The cavities are occupied by intergrowth of tourmaline (schorl) and pyrite (Fig. 5). The mode of occurrence of intergrowth of tourmaline and pyrite

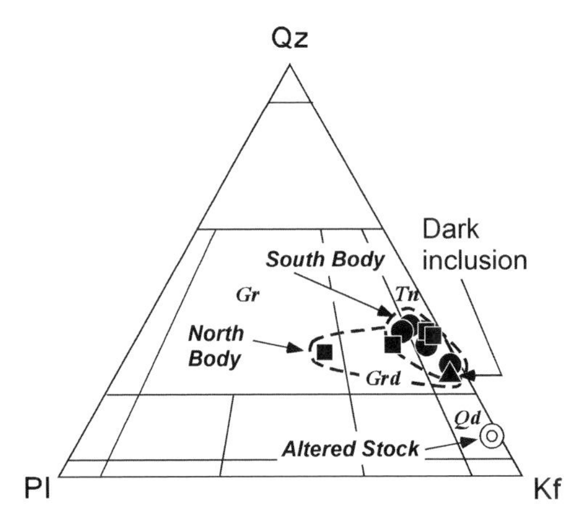

Fig. 4. Diagram showing modal compositions of Chichibu granitic rocks and classification of the granitic rocks (Streckeisen 1976).

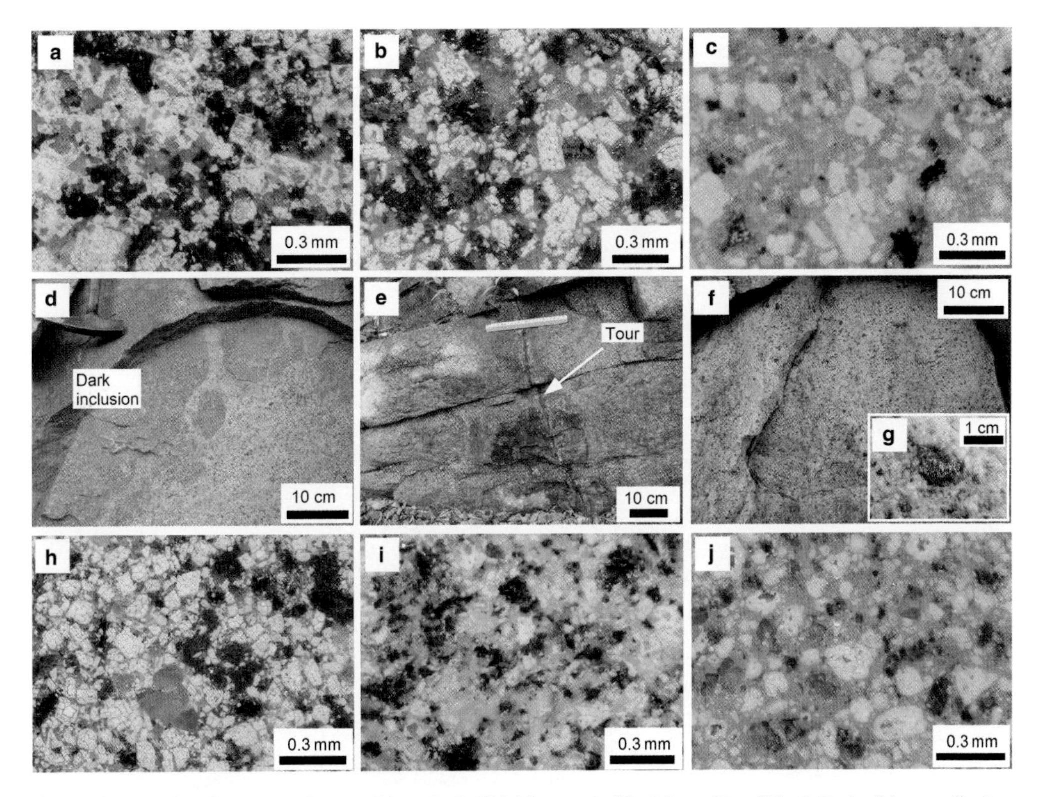

Fig. 5. Photographs of representative granitic rocks in Chichibu granitoids: (**a**) tonalite of North Body, (**b**) granodiorite of North Body, (**c**) monzogranite of North Body, (**d**) dark inclusions in tonalite of North Body, (**e**) a tourmaline vein cutting monzogranite of North Body, (**f**) monzogranite containing miarolitic cavities (brown spots), (**g**) tourmaline and pyrite crystals in miarolitic cavities, (**h**) tonalite of the southeastern part of South Body (magnetite-series), (**i**) tonalite of the southwestern part of South Body (ilmenite-series), (**j**) quartz diorite of Daikoku Altered stock.

suggests that magmatic fluid was present within granitic rocks of North Body at the late stage of emplacement of magma.

The tonalite and granodiorite of North Body consist of large amounts of plagioclase and quartz with lesser amounts of hornblende and biotite and small amounts of K-feldspar, magnetite, ilmenite, pyrite and chalcopyrite. The sizes of plagioclase and quartz crystals range from 1 to 2 mm in diameter. Anhedral K-feldspar occurs between euhedral plagioclase and anhedral quartz crystals. The monzogranite of North Body consist of large amounts of plagioclase and quartz with lesser amounts of K-feldspar and hornblende and small amounts of magnetite, ilmenite, pyrite and chalcopyrite. Groundmass of the monzogranite also consists of anhedral plagioclase, quartz and K-feldspar. The dark inclusions are medium-grained and show more melanocratic features. The dark inclusions are composed of large amounts of euhedral plagioclase and anhederal quartz with small amounts of anhedral K-feldspar, subhedral hornblende and biotite. The tonalite, granodiorite, and monzogranite are classified as magnetite-series granitoids.

South Body mainly consists of medium- to fine-grained tonalite. Rock facies vary from tonalite to medium- to fine-grained granodiorite from marginal to central parts of the northern part of South Body. The tonalite is characterized by holocrystalline texture and shows more leucocratic features compared to the granitic rocks of North Body (Figs 2, 4 & 5). The tonalite and granodiorite of South Body consist of large amounts of plagioclase and quartz with lesser amounts of hornblende and biotite and small amounts of K-feldspar, magnetite, ilmenite, pyrite and chalcopyrite. The size of plagioclase crystals is about 1 mm in diameter. Anhedral K-feldspar occurs between euhedral plagioclase and anhedral quartz crystals. Tonalite in the southern part of South Body contain more biotite compared with the abundance of biotite of granitic rocks in northern part of South Body. Tonalite of the southern part of South Body is divided into magnetite-series and ilmenite-series.

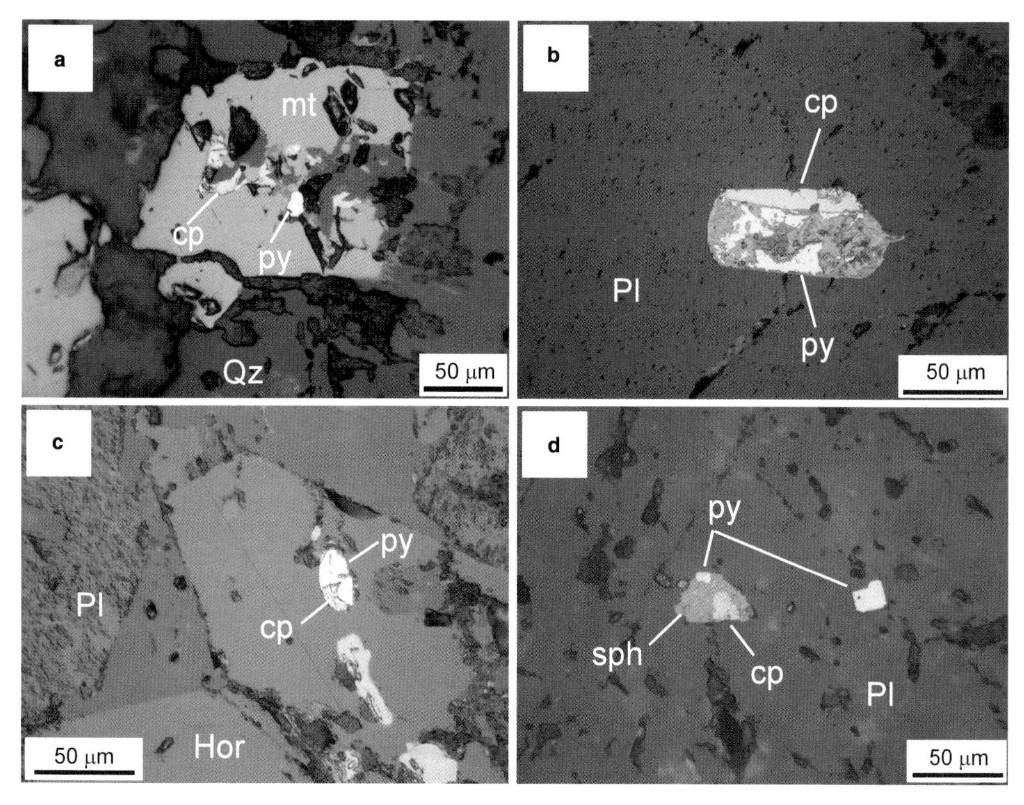

Fig. 6. Photomicrographs showing modes of occurrence of sulphide minerals in granitic rocks: (**a**) and (**b**) opaque minerals in silicate of granitic rocks of North Body, (**c**) opaque minerals in silicate of granitic rocks of South Body, (**d**) opaque minerals in silicate of granitic rocks of Daikoku Altered stock.

Tonalite occurring in the southwestern part of South Body are classified into ilmenite-series granitoid, although most of the granitic rocks of South Body are classified as magnetite-series granitoids. These characteristics of magnetic susceptibility of Chichibu granitic rocks were also described by Ishihara *et al.* (1987).

Daikoku Altered stock is a quartz diorite stock showing a porphyritic structure (Figs 2, 4 & 5). The Altered stock is composed of a large amount of plagioclase with lesser amounts of biotite after hornblende and small amounts of quartz, K-feldspar, magnetite, ilmenite, pyrite and chalcopyrite. Plagioclase phenocryst shows a euhedral shape and quartz phenocrysts have an anhedral shape. Pseudomorph of hornblende replaced by biotite is also common. Groundmass of the Altered stock consists of anhedral quartz and K-feldspar. Daikoku Altered stock is thought to be classified into magnetite-series granitoids because some relatively fresh quartz diorite samples have high magnetic susceptibility. The Altered stock near the Daikoku deposit contains many pyrite grains.

Mode of occurrence of opaque minerals in granitic rocks

Opaque minerals such as oxide and sulphide minerals are observed in Chichibu granitic rocks. Opaque minerals that have been identified are magnetite, ilmenite, pyrite, pyrrhotite, chalcopyrite and sphalerite. Some examples of mode of occurrence of sulphide minerals are pyrite–chalcopyrite–sphalerite intergrowth in fresh plagioclase, magnetite–pyrite–chalcopyrite intergrowth in fresh plagioclase and presence of pyrite and chalcopyrite globules of oval shape in hornblende (Fig. 6). The mode of occurrence of these sulphide minerals is similar to the mode of occurrence of sulphide minerals in latite dykes described by Stavast *et al.* (2006), suggesting magmatic-origin sulphide minerals. The mineral paragenesis and textures described here suggest that sulphide minerals in the Chichibu granitic rocks were formed at higher temperatures that hornblende and plagioclase crystallized in a magma chamber. These sulphide minerals are common in tonalite of the marginal part of North Body and the northern part of South Body, while the amounts of sulphide minerals in granitic rocks are small in monzogranite occurring at the central portion of North Body. The amount of sulphide minerals in the southwestern part of South Body, which is ilmenite-series granitic rocks, is small.

Texture of quartz crystals in granitic rocks determined by the cathodoluminescence technique

Textures of quartz crystals consisting of Chichibu granitic rocks were examined using a scanning electron microscopy with cathodoluminescence (SEM-CL) arrangement (SEM: JEOL JSM-5400LV, CL: Sanyu electron cathodoluminescence detector). Based on the observation of quartz consisting of granitic rocks of North Body, South Body and Daikoku Altered stock using the cathodoluminescence technique, quartz crystals in tonalite and granodiorite of North Body and South Body show an obscure patchy texture and a subtle concentric zoning that dark core is surrounded by light rim (Figs 7 & 8). Some quartz crystals in tonalite containing many dark inclusions at the margin of North Body show an oscillating growth zoning (Figs 7 & 8). Quartz crystals in dark inclusions of North Body have an obscure patchy texture. While quartz crystals in monzogranite of North Body and quartz diorite stock of Daikoku Altered stock are characterized by quartz in oval and rounded shape suggesting dissolution texture. The quartz crystals are surrounded by fine-grained aggregates and radial aggregates of a mixture of quartz, plagioclase and K-feldspar of groundmass (Figs 7 & 8). These quartz crystals are observed as a small-scale porphyritic structure by the naked eye (Fig. 5c, j). Dissolution of quartz crystals is thought to cause the formation of porphyritic structure. Internal textures of rounded and oval-shaped quartz crystals are also similar to the obscure patchy texture observed in quartz crystals of tonalite of North Body and South Body (Fig. 7). The relation of texture shown by the cathodoluminescence studies suggests quartz crystals formed at an earlier stage of emplacement of granitic bodies was corroded at

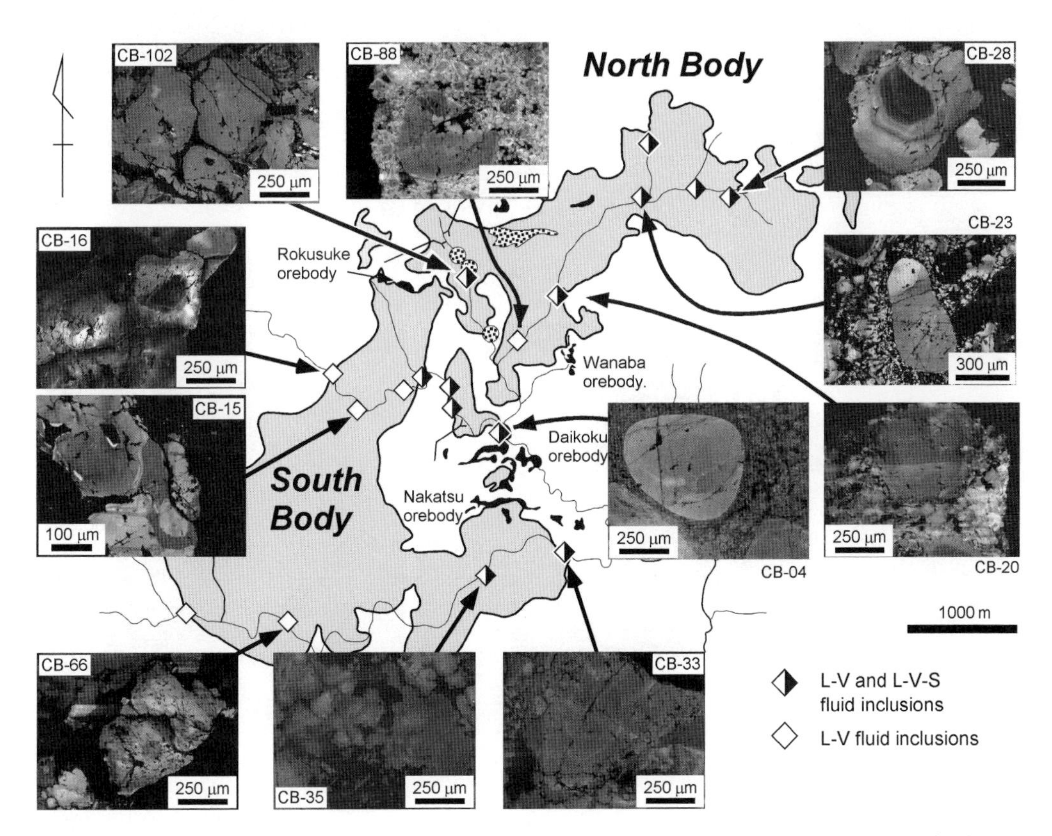

Fig. 7. Diagram showing the distribution of textures of quartz crystals in Chichibu granitic rocks observed by the cathodoluminescence method.

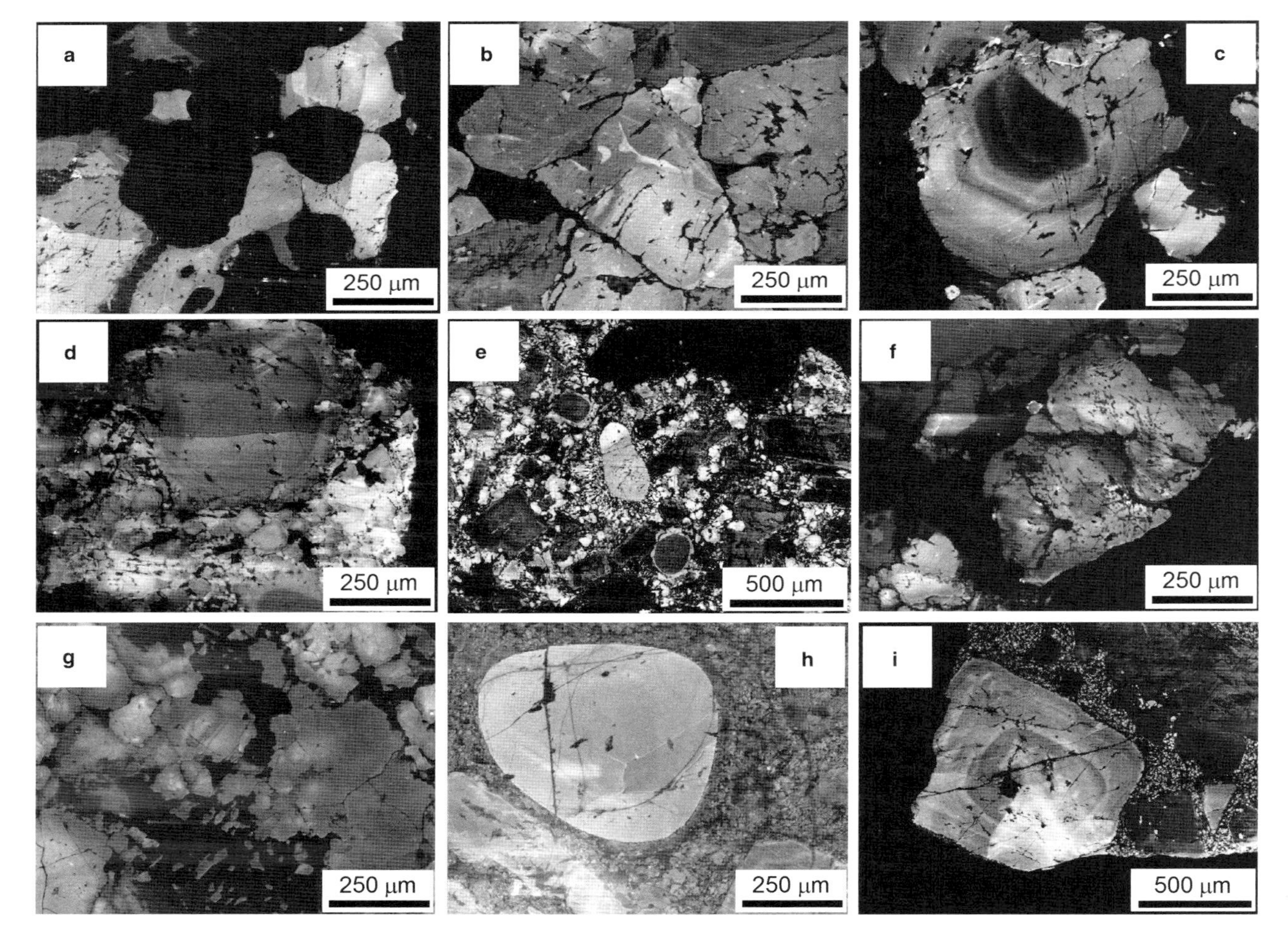

Fig. 8. Cathodoluminescence images showing variation of textures of quartz crystals in Chichibu granitic rocks: (**a**) quartz crystals of dark inclusions in tonalite of North Body (CB-26), (**b**) quartz crystals in tonalite of North Body (CB-26), (**c**) quartz crystals in tonalite of North Body (CB-26), (**d**) quartz crystals in tonalite of North Body (CB-20), (**e**) quartz crystals in monzogranite of North Body (CB-23), (**f**) quartz crystals in tonalite of South Body (CB-66), (**g**) quartz crystals in tonalite of South Body (CB-35), (**h**) quartz crystals in quartz diorite porphyry of Daikoku Altered stock (CB-04), (**i**) quartz crystals in quartz diorite porphyry of Daikoku Altered stock (CB-07).

a later stage of the emplacement. Quartz crystals having an oscillating growth zoning in tonalite at the margin of North Body were formed under the condition at low degree of oversaturation, while rounded-shaped quartz crystals surrounded by radial aggregates of a mixture of quartz and plagioclase in monzogranite at the central part of North Body and Daikoku quartz diorite altered porphyry were formed under the condition at a high degree of oversaturation. The change of texture is drastic according to the process of formation from granodiorite to monzogranite. Given the results obtained by Muller *et al* (2005), Wark *et al.* (2007) and Weibe *et al.* (2007), processes such as mixing of more differentiated magma with more basic magma and a decrease of pressure are considered for mechanisms controlling the change during emplacement of Chichibu granitic magma.

Chemical compositions of Chichibu granitoids

Chemical compositions of Chichibu granitic rocks were examined by X-ray fluorescence (XRF) analyses for major and some minor elements and by inductively coupled plasma mass spectrometry (ICP-MS) and instrumental neutron activation analysis (INAA) analyses for trace elements. The measurements using XRF were carried out by Phillips PW2404 XRF of the Faculty Education and Human Studies, Akita University and VG Elemental PQ-3 ICP-MS of the Faculty of Engineering and Resource Sciences, Akita University. INAA analyses were conducted at the facilities of Kyoto University Nuclear Reactor Institute.

SiO_2 content of the granitic rocks of North Body and South Body range from 60.8 to 66.3 wt% and 61.4 to 65.7 wt%, respectively (Fig. 9, Table 1). These SiO_2 contents are higher than the average SiO_2 content of granitic rocks forming iron and gold skarn deposits (Meinert *et al.* 2005) in the world and are lower than the average SiO_2 content of granitic rocks forming zinc skarn deposits in the world (Meinert *et al.* 2005). The SiO_2 content of dark inclusions in North Body range from 54.0 to 55.6 wt%, which are lower than those of the granitic rocks of North Body (Table 1). The SiO_2 content of the granitic rocks of North body increases from tonalite at the marginal part to monzogranite at the central part of the body. TiO_2, Al_2O_3, total Fe_2O_3, MnO, MgO and CaO content of North Body are decreased and Na_2O and K_2O content are increased with increase in SiO_2 content from marginal to central parts of the body. Monzogranite at the central part of North Body are thought to be more differentiated rock facies in the body. Zoning of the rock facies is not so clear for South Body because of the difficulty in carrying out a geological survey owing to precipitous land features. However,

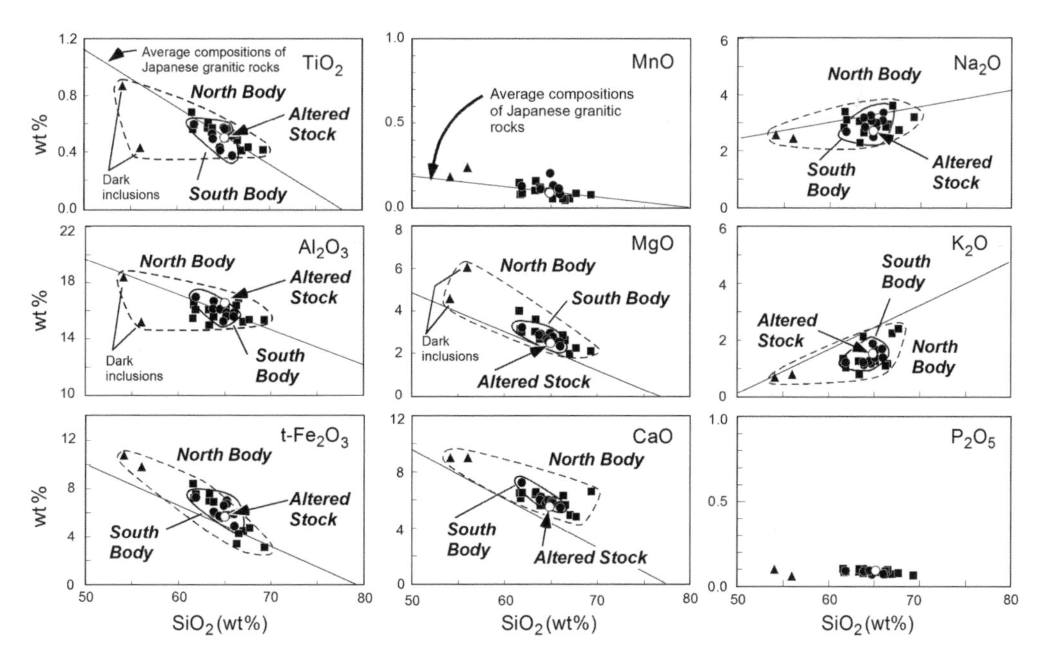

Fig. 9. Harker diagrams showing bulk chemical compositions of major elements of Chichibu granitic rocks. Solid lines in the diagrams show average compositions of Japanese granitic rocks (Aramaki *et al.* 1972).

Table 1. *Chemical compositions of granitic rocks of Chichibu mining area*

	1	2	3	4	5	6	7	8	9	10	11	12	13	14	15
	80CB92	CB26B-1	CB26B-2	80CB83	CB27A	CB28B	CB24	CB23B	CB08E	CB11	CB15	CB61	CB67	CB35	CB07
Granitic body	North	North	North	North	North	North	North	North	South	South	South	South	South	South	Daikoku, Altered, Stock
Rock facies	tonalite	tonalite	dark inclusion	tonalite	tonalite	dark inclusion	tonalite	monzo-granite	tonalite	tonalite	grano-diorite	tonalite	tonalite	tonalite	quartz, diorite, porphyry
wt%															
SiO_2	64.97	62.99	55.61	62.64	61.03	54.03	60.79	66.26	61.41	64.61	64.46	64.39	65.73	63.54	64.13
TiO_2	0.52	0.60	0.43	0.56	0.56	0.87	0.58	0.41	0.60	0.57	0.56	0.41	0.38	0.49	0.49
Al_2O_3	16.09	14.92	15.13	15.31	16.27	18.39	15.82	15.08	16.89	15.79	15.46	16.14	15.83	16.61	16.43
t-Fe_2O_3	3.31	7.53	9.73	6.76	7.10	10.74	7.38	4.42	7.25	6.88	6.59	5.58	4.84	5.93	5.59
MnO	0.05	0.16	0.24	0.12	0.08	0.18	0.08	0.05	0.13	0.11	0.13	0.09	0.09	0.12	0.09
MgO	2.79	3.56	5.99	2.61	3.09	4.56	2.94	1.94	3.22	2.74	2.67	2.54	2.31	2.89	2.40
CaO	6.18	6.51	8.95	5.50	6.03	8.99	6.37	4.85	7.16	5.77	5.62	5.97	5.39	6.18	5.41
Na_2O	2.78	2.26	2.43	2.59	3.36	2.57	3.04	3.57	2.67	2.83	2.95	3.09	3.33	2.84	2.65
K_2O	1.04	0.75	0.75	2.05	1.20	0.65	0.97	2.18	1.18	1.35	1.59	1.16	1.35	1.07	1.50
P_2O_5	0.09	0.10	0.05	0.08	0.08	0.10	0.09	0.07	0.09	0.09	0.08	0.07	0.08	0.10	0.08
LOI	1.89	0.42	0.49	1.76	0.96	0.07	1.40	0.73	0.27	0.47	1.02	0.48	0.39	0.58	1.20
H_2O (−)	0.10	0.18	0.26	0.11	0.14	0.08	0.39	0.42	0.46	0.37	0.13	0.03	0.03	0.01	0.02
Total	99.79	99.98	100.06	100.08	99.91	101.24	99.86	99.98	101.33	101.57	101.27	99.94	99.76	100.36	100.00
ppm															
Sc	20.0	24.3	52.4	20.9	23.2	33.1	18.6	14.5	24.3	22.2	18.7	15.7	13.5	12.5	15.6
Cr*	bd	65.6	129	bd	51.7	29.5	38.1	35.2	44.2	32.7	36.1	48.2	38.6	29.9	39.8

Co	11.3	22.4	34.1	16.3	13.4	28.3	11.1	7.11	20.9	18.9	15.2	14.8	11.9	18.6	7.31
Cu	7.59	49.3	71.0	5.24	14.2	37.9	10.6	6.67	87.4	72.0	27.3	9.81	11.8	160	116
Zn	23.9	66.1	70.8	51.8	33.4	93.7	31.3	21.3	107	196	70.6	43.7	60.8	49.6	149
Ga	15.0	17.1	18.3	13.4	16.0	20.4	14.5	15.8	16.7	18.4	14.3	14.0	12.91	15.5	16.6
As*	na	8.69	13.3	na	9.21	10.5	7.98	10.7	11.8	54.7	5.63	6.43	na	5.16	46.0
Rb	27.4	27.0	24.6	64.1	41.3	18.4	29.7	41.6	38.6	49.5	41.0	30.7	31.1	31.0	43.1
Sr	152	216	212	112	221	220	170	194	226	217	165	147	137	170	199
Y	14.3	13.5	24.6	16.5	20.7	14.8	19.1	13.5	17.4	21.4	21.3	12.4	11.8	7.81	13.0
Nb	2.27	3.21	2.38	3.36	2.27	2.13	1.81	2.35	1.94	2.85	2.26	1.43	1.52	1.59	2.68
Cs	0.71	3.45	2.09	4.69	4.08	3.73	4.11	2.52	2.64	5.53	2.66	2.12	2.70	2.73	1.51
Ba	219	162	172	322	227	199	220	466	271	384	355	295	287	256	238
La	8.29	10.2	10.1	9.75	6.25	5.13	8.02	5.03	10.0	9.38	8.87	7.14	7.28	7.62	10.0
Ce	18.7	23.4	26.4	22.3	15.1	11.7	18.3	11.8	18.3	20.2	19.9	15.1	14.8	15.3	21.6
Pr	2.37	2.40	2.69	2.70	2.37	1.59	2.41	1.54	2.35	2.52	2.60	1.82	1.74	1.79	2.71
Nd	9.68	9.51	10.9	10.5	10.2	7.03	10.5	6.49	9.23	10.6	11.6	7.57	7.09	7.48	10.1
Sm	2.27	2.33	3.07	2.55	3.01	2.07	2.85	1.85	2.48	2.88	3.17	1.42	1.26	1.54	2.53
Eu	0.65	0.89	0.89	0.66	0.94	0.70	0.89	0.52	0.88	0.85	0.78	0.45	0.41	0.54	0.76
Gd	2.85	2.48	3.40	3.13	3.37	2.64	3.61	1.87	2.59	3.38	3.93	1.92	1.34	1.78	2.52
Tb	0.51	0.39	0.63	0.56	0.65	0.40	0.65	0.32	0.51	0.53	0.67	0.30	0.24	0.29	0.41
Dy	3.23	2.46	3.94	3.68	3.71	2.58	4.03	2.01	3.11	3.51	4.11	1.94	1.92	1.85	2.59
Ho	0.70	0.51	0.83	0.78	0.73	0.55	0.85	0.43	0.64	0.74	0.90	0.39	0.40	0.42	0.54
Er	2.01	1.48	2.35	2.16	2.06	1.59	2.63	1.26	1.92	2.21	2.77	1.11	1.12	1.14	1.53
Tm	0.31	0.24	0.36	0.34	0.31	0.25	0.41	0.20	0.30	0.34	0.36	0.14	0.13	0.15	0.19
Yb	1.86	1.40	2.26	2.03	2.00	1.44	2.54	1.34	1.79	2.07	2.48	0.97	0.96	1.12	1.26
Lu	0.29	0.20	0.37	0.34	0.30	0.22	0.36	0.19	0.29	0.36	0.39	0.14	0.06	0.15	0.19
Hf*	bd	0.92	1.28	bd	2.65	2.15	3.21	4.08	2.64	3.45	3.43	2.62	2.67	2.25	3.10
Au*	bd	bd	bd	bd	bd	bd	bd	bd	bd	0.03	0.02	bd	bd	0.01	0.07
Pb	4.14	3.67	3.75	5.37	4.74	5.55	6.18	5.31	19.4	14.0	14.6	6.43	11.9	3.25	27.8
Th	3.07	1.85	3.05	5.93	2.91	1.14	3.00	4.68	2.81	3.11	4.35	2.44	2.46	2.10	4.07
U	0.50	0.32	0.48	1.49	0.70	0.26	0.69	0.96	1.09	0.60	0.83	0.42	0.20	0.43	0.79

*INAA method; n.a., not analysed; bd, below detection limit.

a similar tendency is found in the northern part of South Body, that is, rock facies changing from tonalite to granodiorite. Chemical compositions of major elements of granitic rocks of South Body including magnetite-series granitic rocks of the northern part of the body, ilmenite- and magnetite-series granitic rocks of the southern part of the body are included in the range of chemical compositions of major elements of granitic rocks of North Body and chemical characteristics of major elements of granitic rocks of South Body are similar to those of major elements of granitic rocks of North Body.

SiO_2 content of the granitic rocks of Daikoku Altered stock is 64.1 wt% and the content is classified into the group having high SiO_2 content among granitic rocks in Chichibu mining area. TiO_2, Al_2O_3, total Fe_2O_3, MnO, MgO and CaO contents are decreased with increase in SiO_2 content for the granitic rocks of North Body and South Body. Na_2O and K_2O content of these granitic rocks are slightly increased with increase in SiO_2 content, and those contents are lower than those of the average composition of Japanese granites (Fig. 9). The chemical compositions of Chichibu granitic rocks are shown on a total-FeO v. total-FeO/MgO diagram (Miyashiro 1974) discriminating tholeiite-series and calc-alkaline-series rocks (Fig. 10). Chichibu granitic rocks are classified into calc-alkaline-series rocks. The curved distribution of the data of chemical compositions of Chichibu granitic rocks also suggests other factors are controlling the bulk chemical compositions of Chichibu granitic rocks in addition to magmatic differentiation of calc-alkaline magma during the crystallization of the magma.

Norm compositions of the Chichibu granitoids were estimated on the basis of the chemical compositions of the granitic rocks in this study and Fe^{2+}/Fe^{3+} ratios of granitic rocks of Chichibu granite examined by Ishihara *et al.* (1987). The norm compositions of the Chichibu granitoids are plotted on and around the boundary between quartz and albite at the pressure of 0.5 kbar in the norm Ab–Or–Qz diagram (Tuttle & Bowen 1958). This fact suggests that pressure of the emplacement of the Chichibu granitoids is low (about 0.5 kbar) (Fig. 11) and accords with the occurrence of breccia dykes in granitic rocks of North Body.

Copper, zinc and lead content of the granitic rocks of North Body and South Body determined by ICP-MS and arsenic content of the granitic rocks of North Body and South Body determined by INAA are: Cu, 5.2 to 49.3 ppm for North Body and 9.8 to 160 ppm for South Body; Zn, 21.3 to 66.1 ppm for North Body and 43.7 to 196 ppm for South Body; Pb, 3.7 to 6.2 ppm for North Body and 3.3 to 19.4 ppm for South Body; As, 8.0 to

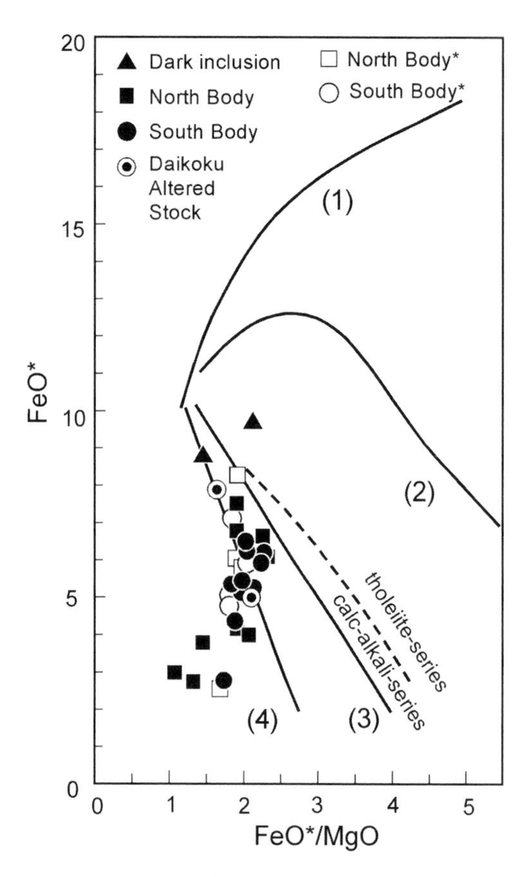

Fig. 10. FeO*–FeO*/MgO diagram (Miyashiro 1974). (1) Skaergaard, (2) Miyake-jima in the Izu-Bonin Arc, (3) Asama volcano, (4) Amagi volcano, belonging to the Izu-Bonin Arc.

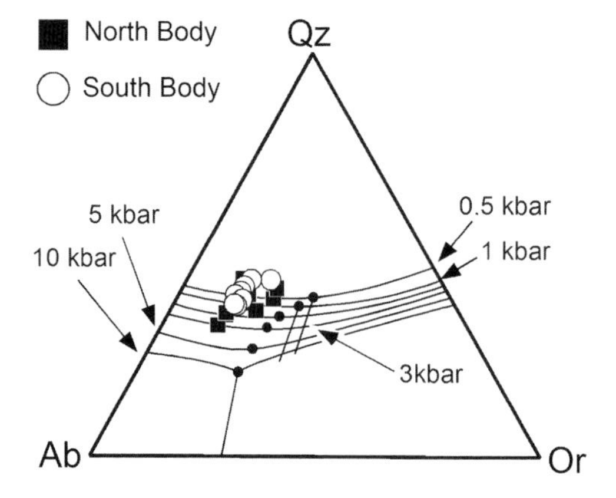

Fig. 11. Ternary diagram in the system $NaAlSi_3O_8$–$KAlSi_3O_8$–SiO_2–H_2O (Tuttle & Bowen 1958) showing schematic composition of Chichibu granitic rocks.

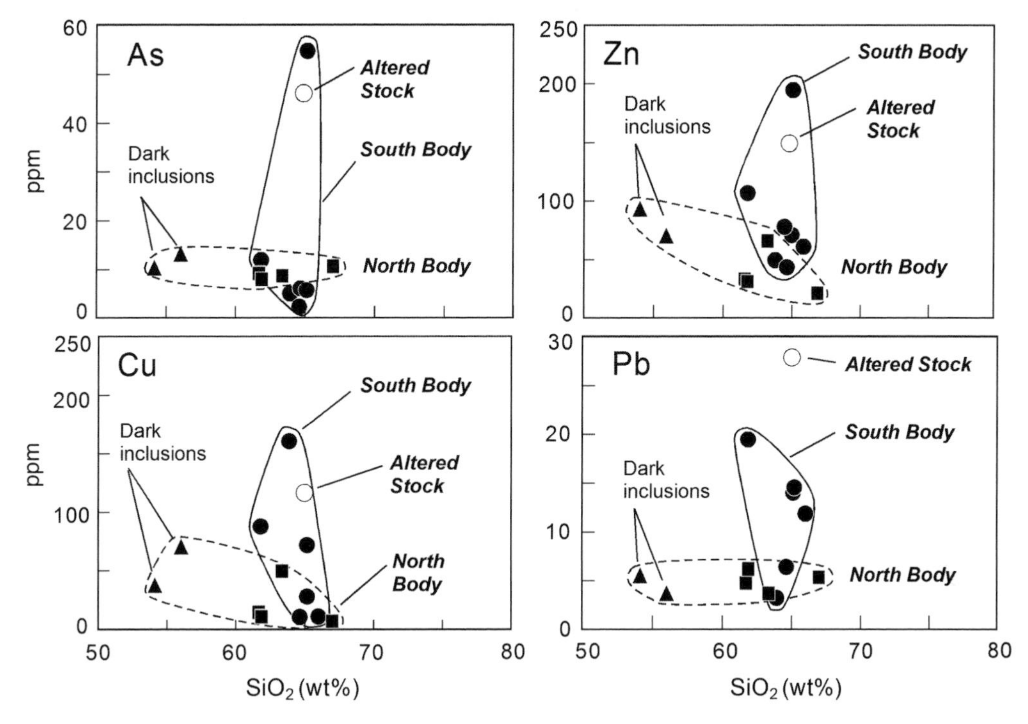

Fig. 12. Diagrams showing the relation between SiO_2 content and As, Cu, Zn and Pb content of granitic rocks of Chichibu mining area.

10.7 ppm for North Body and 5.2 to 54.7 ppm for South Body (Fig. 12, Table 1). Copper, zinc, lead and arsenic content of Daikoku Altered stock are 46, 116, 149 and 27.8 ppm, respectively. Copper content of the granitic rocks of South Body, which has small skarn orebodies (Nakatsu and Rokusuke orebodies), is higher than that of the granitic rocks of North Body, which has some large skarn orebodies in the Chichibu deposit (Akaiwa, Dohshinkubo and Wanaba orebodies). On the other hand, arsenic content of the granitic rocks of South Body is lower than that of the granitic rocks of North Body, which has some large skarn orebodies in the Chichibu deposit. Ishihara *et al.* (1987) reported average copper content of 17.5 ppm for North Body and 82.6 ppm for South Body and average arsenic content of 11.7 ppm for North Body and 4.8 ppm for South Body. The results of this study accord with the results obtained by Ishihara *et al.* (1987). The contents of copper, zinc, lead and arsenic of Daikoku Altered stock are also high compared with the content of these elements of the granitic rocks of North Body. The granitic rocks having around 65 wt% SiO_2 content of South Body tend to have high arsenic, copper, zinc and lead content. Arsenic, copper, zinc and lead contents of the granitic rocks of South Body tend to increase with increase in SiO_2 content during magmatic differentiation and reach maximum content around 65 wt% SiO_2 content and then decrease in the range above 67 wt% of SiO_2 (Fig. 12).

Gold content of granitic rocks of North Body, South Body and Daikoku Altered stock were measured by INAA analyses. Gold content of granitic rocks of North Body were below the detection limit of INAA analyses. Gold content of some granitic rocks of South Body range from 0.01 to 0.03 ppm, although the content of other granitic rocks of South Body were below the detection limit of INAA analyses. The gold content of Daikoku altered stock is 0.07 ppm (Table 1). Ishihara *et al.* (1987) showed that the average gold content of granitic rocks of North Body, South Body and Daikoku Altered stock are 5.6, 11.9 and 146.8 ppb, respectively. Based on the gold contents in this study and gold contents examined by Ishihara *et al.* (1987), gold content of granitic rocks in Chichibu granitic bodies increase in the order of North Body, South Body and Daikoku Altered stock.

Rare earth element (REE) content of Chichibu granite were measured to understand the process of emplacement of magma. Chondrite-normalized REE patterns of granitic rocks of North Body show light REE (LREE)-enriched chondrite-normalized

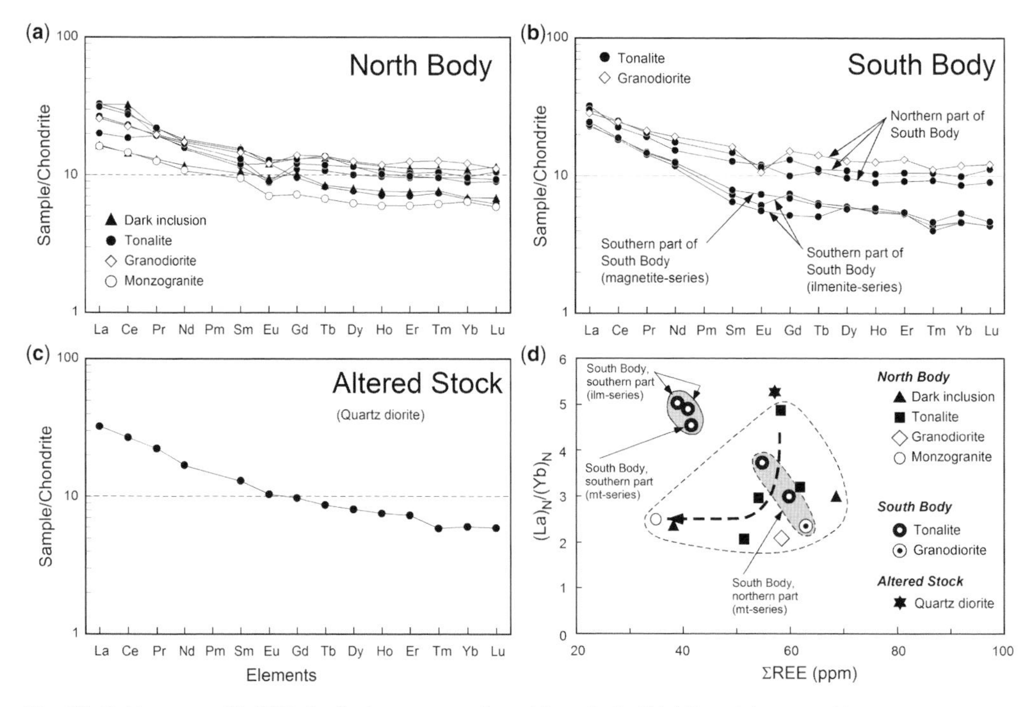

Fig. 13. Spidergrams with REE-distribution patterns of granitic rocks in Chichibu mining area: (**a**) chondrite-normalized REE patterns of granitic rocks of North Body, (**b**) chondrite-normalized REE patterns of granitic rocks of South Body, (**c**) chondrite-normalized REE patterns of granitic rocks of Daikoku Altered stock, (**d**) diagram showing the relation between La_N/Yb_N ratios and total REE contents of Chichibu granitic rocks.

REE patterns (Fig. 13, Table 1). A series of granitic rocks formed according to the process of normal magmatic differentiation generally shows increases of total REE content and La_N/Yb_N ratios. However, the total REE content and La_N/Yb_N ratios of the granitic rocks of North Body decrease according to magmatic differentiation from tonalite to monzogranite (Fig. 13). The magnetite-series granitic rocks of the northern part of South Body differ in several ways from the ilmenite-series granitic rocks of the southwestern part of South Body. Total REE content of the magnetite-series granitic rocks of the northern part of South Body are similar to those of tonalite of North Body, and the total REE content are higher than those of the ilmenite-series granitic rocks of the southwestern part of South Body. The La_N/Yb_N ratios of the magnetite-series granitic rocks of the northern part of South Body are also similar to those of tonalite of North Body, and the ratios are smaller than those of the ilmenite-series granitic rocks of the southern part of South Body (Fig. 13). The chondrite-normalized REE pattern of the least altered quartz diorite porphyry of Daikoku Altered Stock has a high La_N/Yb_N ratio and relatively high total REE content.

Fluid inclusion studies

Various kinds of fluid inclusions occur in quartz crystals of Chichibu granitic rocks. The fluid inclusions are classified into two types: liquid–vapour two-phase fluid inclusions and liquid + vapour + solid-bearing polyphase fluid inclusions. The two-phase fluid inclusions are also divided into vapour-rich and liquid-rich fluid inclusions. Some polyphase fluid inclusions contain opaque minerals such as hematite in addition to NaCl crystal (Fig. 14). The two-phase fluid inclusions are widely distributed in quartz crystals of North Body, South Body and Daikoku Altered stock, while many polyphase fluid inclusions occur in quartz crystals of North Body. The total number of fluid inclusions of granitic rocks of North Body is larger than that of granitic rocks of South Body. The polyphase fluid inclusions account for about 30 and 100% of the total numbers of fluid inclusions in quartz crystals of tonalite at the maginal part and monzogranite at the central part of North Body, respectively, and the number of polyphase fluid inclusions increase from marginal to central parts of North Body. The number of polyphase fluid inclusions of quartz crystals in Daikoku Altered

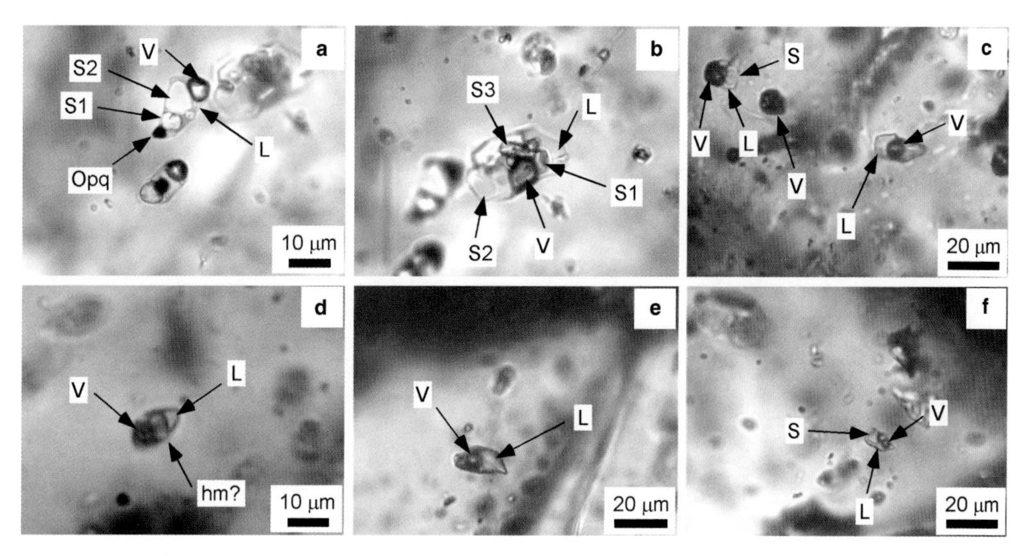

Fig. 14. Photomicrographs of fluid inclusions in quartz crystals in Chichibu granitic rocks: (**a**), (**b**) and (**c**): fluid inclusions in quartz crystals in tonalite of North Body (CB-26), V: vapour, L: liquid, S1 and S2: daughter minerals, Opq: daughter mineral as opaque minerals, (**d**) polyphase fluid inclusions in quartz crystals in monzogranite of North Body (CB-23), hm?: hematite?, (**e**) liquid–vapour two-phase fluid inclusions in quartz crystals in granodiorite of northern part of South Body (CB-15), (**f**) polyphase fluid inclusions in quartz crystals in quartz diorite porphyry of Daikoku Altered stock (CB-07).

stock ranges from 10 to 20% of the total number of fluid inclusions in quartz crystals of the stock. In South Body, the polyphase fluid inclusions in quartz crystals of granitic rocks account for about 10% of the total number of fluid inclusions of quartz crystals, and the abundance of polypahse fluid inclusions for total number of fluid inclusions of granitic rocks of South Body is almost the same for granitic rocks of the northern part (magnetite-series granitic rocks), southwestern part (ilmenite-series granitic rocks) and southeastern part (magnetite-series granitic rocks) of South Body.

Fluid inclusions in quartz crystals of granitic rocks in Chichibu mining area were examined by heating and freezing experiments. The heating and freezing stage used was Linkam TH-600. Accuracy of the measurement is plus/minus 1 degree for the heating experiment and plus/minus 0.2 wt% NaCl equivalent for the freezing experiment. Many fluid inclusions were measured in this study; many of the fluid inclusions were not homogenized by heating up to 600 °C. Final homogenization temperatures of liquid–vapour two-phase fluid inclusions and polyphase fluid inclusions of tonalite of North Body are above 310 °C (mostly above 600 °C) and above 540 °C (mostly above 600 °C), respectively. The dissolution temperatures of daughter crystals in the polyphase fluid inclusions range from 130 to above 600 °C (Figs 15 & 16). The polyphase fluid inclusions in tonalite of North Body are classified into two types: the polyphase fluid inclusions that homogenize by halite disappearance [Th (l-v) < Tm (halite)] and the polyphase fluid inclusions that homogenize by vapour disappearance [Tm (halite) < Th (l-v)]. Polyphase fluid inclusions in monzogranite of North Body homogenize by vapour disappearance [Tm (halite) < Th (l-v)] (Fig. 15). The assemblages of polyphase fluid inclusions in tonalite and monzogranite are different. Homogenization temperatures of liquid–vapour two-phase fluid inclusions and polyphase fluid inclusions of monzogranite of North Body are above 600 °C and above 130 °C (mostly above 460 °C), respectively. Final ice melting temperatures of liquid–vapour fluid inclusions in quartz crystals of tonalite of North Body ranges from −29.7 to −20.7 °C (Fig. 15). These facts suggest that the chemical composition of the liquid–vapour fluid inclusions is a chemical composition such as Na–Ca–Cl–H_2O system.

Final homogenization temperatures of polyphase fluid inclusions of tonalite of the southeastern part of South Body are above 525 °C (Fig. 16). Polyphase fluid inclusions that homogenize by halite disappearance [Th (l-v) < Tm (halite)] and by vapour disappearance [Tm (halite) < Th (l-v)] are recognized in tonalite of South Body. Most of the two-phase fluid inclusions show a mode of occurrence as secondary fluid inclusions aligning in small

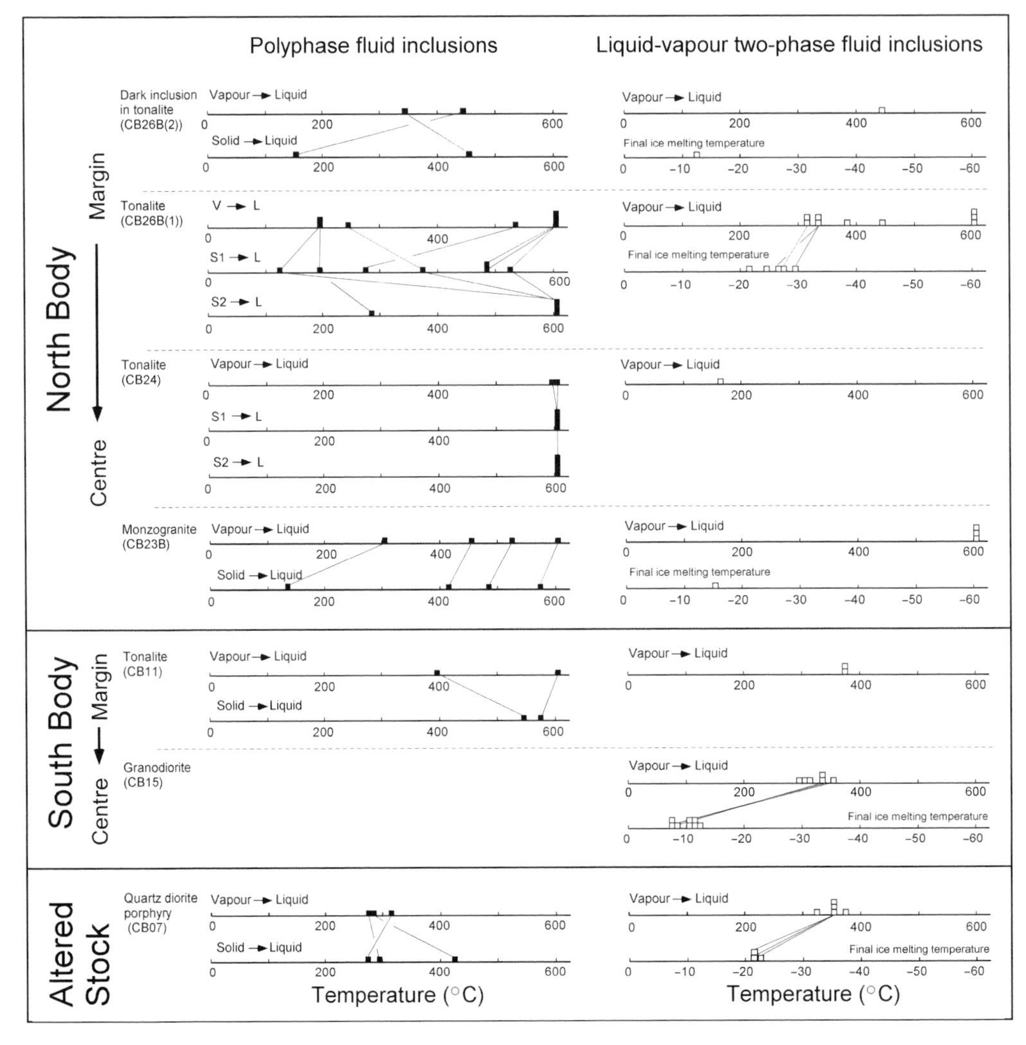

Fig. 15. Diagrams showing homogenization temperatures and final ice melting temperatures of fluid inclusions in quartz of granitic rocks of Chichibu mining area. The lines connecting homogenization temperatures of vapour and solid indicate homogenization temperatures of vapour and solid in a single fluid inclusion. The lines connecting homogenization temperature of vapour and final ice melting temperature also indicate homogenization temperatures of vapour and final ice melting temperature of a single fluid inclusion.

fractures in igneous quartz crystals of the granitic rocks of South Body. The homogenization temperatures of liquid–vapour two-phase fluid inclusions of granodiorite and tonalite of South Body range from 300 to 360 °C (Figs 15 & 16). Final ice melting temperatures of liquid–vapour fluid inclusions in quartz crystals of granodiorite of South Body ranges from −12.4 to −6.5 °C. The temperature range is higher than the temperature of eutectic point of NaCl–H_2O system.

Homogenization temperatures of polyphase fluid inclusions of Daikoku Altered stock range from 300 to 430 °C. Polyphase fluid inclusions that homogenize by halite disappearance [Th (l-v) < Tm (halite)] and by vapour disappearance [Tm (halite) < Th (l-v)] are present in quartz diorite porphyry of Altered stock. In the case of liquid–vapour two phase fluid inclusions in quartz crystals of Daikoku Altered stock, homogenization temperatures of the fluid inclusions in quartz crystals of Daikoku Altered stock range from 280 to 370 °C. Final ice melting temperatures of the two-phase fluid inclusions in the altered stock ranges from −23.5 to −21.4 °C (Figs 15 & 16). The temperature

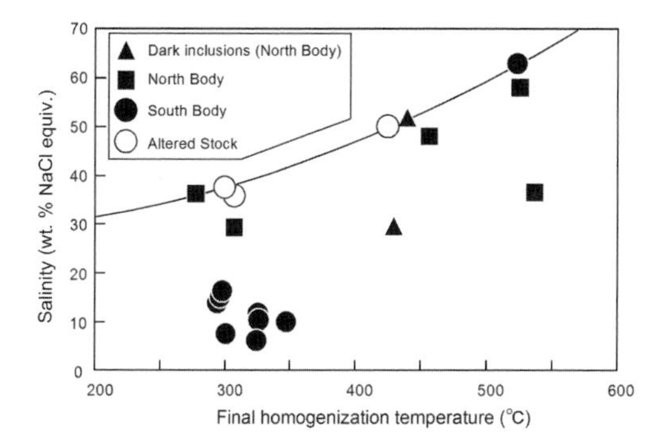

Fig. 16. Diagram showing relationship between homogenization temperatures and salinity of fluid inclusions in quartz of Chichibu granitic rocks.

range is slightly lower than the eutectic temperature of NaCl–H_2O system.

The data of fluid inclusions of North Body, south Body and Daikoku Altered stock suggest that final homogenization temperatures of fluid inclusions in North Body tend to be higher than those of fluid inclusions of South Body and Altered stock.

Discussion

Characteristics of REEs in magmatic fluid at late stage of emplacement of Chichibu granitic magma

Characteristic features of Chichibu granites were described in previous parts. The fact that halite-bearing polyphase fluid inclusions are recognized in North body and Diaikoku Altered stock and the fact that miarolitic cavities and tourmaline veins are present in monzogranite of North Body suggest that magmatic fluid was present at the late stage of emplacement of granitic magma in Chichibu mining area. Fluid inclusions in granitic rocks of South Body are mainly liquid-rich liquid–vapour two-phase fluid inclusions. Most of the two-phase fluid inclusions show a mode of occurrence as secondary fluid inclusions aligning in small fractures in igneous quartz crystals of the granitic rocks. Based on the distribution of polyphase fluid inclusions in North Body and South Body, magmatic fluid is thought to be dominant in granitic melt forming North Body. Quartz crystals in granitic rocks of North Body and South Body show subtle concentric zonal texture and monotonous patchy texture on the basis the texture of quartz crystals in granitic rocks determined by the cathodoluminescence technique. Formation of the zonal texture of quartz crystals of North Body is thought to be formed by change in chemical composition of melt and coexisting fluid during emplacement of the granitic magma of North Body.

The variation in concentrations of REEs in granitic rocks of North Body is different from the variation in concentrations of REEs of granitic rocks solidifying under normal magmatic differentiation free of magmatic fluid (Fig. 13). The concentrations of LREEs decrease from early to late periods (from tonalite to monzogranite) of the emplacement of magma forming North Body. The characteristic tendency can be explained by some processes. One possibility is crystallization and removal of LREE-enriched minerals such as allanite from residual magma in an early period of the emplacement of magma. Another possibility is removal of LREEs from granitic melt by coexisting magmatic fluid. The first possibility can be ruled out because LREE-enriched minerals such as allanite are not common accessory minerals in granitic rocks of North Body. The second possibility is one possible explanation for the variation in REE concentrations of granitic rocks of North Body. If magmatic fluid coexists with melt, LREEs transfer from melt to magmatic fluid because LREEs having a larger ionic radius have incompatible signatures and prefer magmatic fluid compared with melt.

Possibility of first boiling and second boiling

Based on the observation and chemical analyses in this study, (1) coexistence of vapour-rich two-phase fluid inclusions and halite-bearing polyphase fluid inclusions in a single quartz crystal of granitic rock of North Body, (2) presence of breccia dykes and tourmaline veins in North Body and (3) an estimated low pressure (about 0.5 kbars) for the emplacement of granitic rocks of North Body. In addition these facts and estimation, the texture of quartz of oval shape enclosed by radial aggregates of a mixture of fine-grained quartz, plagioclase and K-feldspar of groundmass of the monzogranite of North Body (Fig. 8) is thought to be dissolution texture formed at a certain stage during emplacement of North Body. The processes to form the texture are considered magma mixing and decrease of pressure during the emplacement. Based on the petrological and geochemical data, the process of formation of tonalite to granodiorite of the North Body is controlled by magmatic differentiation. In addition, dark inclusions suggesting magma mixing are not observed in monzogranite. The possibility of magma mixing is not likely for the formation of the quartz showing dissolution texture. The other possibility is dissolution of quartz crystals by pressure decrease during emplacement of North Body. The process of dissolution of quartz crystals is schematically explained on the basis of the phase relation among quartz-albite-K-feldspar

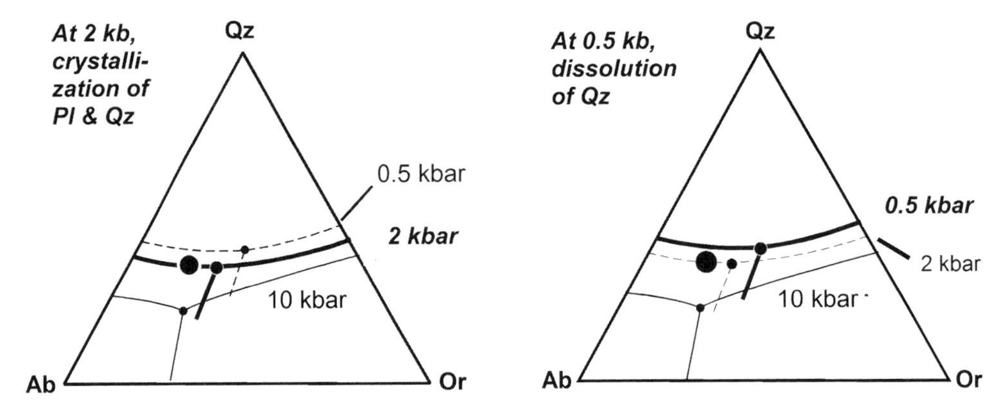

Fig. 17. Diagram showing schematic changes of environments and stability field of Chichibu granitic magma on the ternary diagram in the system $NaAlSi_3O_8-KAlSi_3O_8-SiO_2-H_2O$ (Tuttle & Bowen 1958).

proposed by Tuttle & Bowen (1958) (Fig. 17). If the pressure decreases from 2 to 0.5 kbar, quartz would remain in an albite-stable region and in an unstable environment and it would be dissolved. The decrease in pressure would cause a large temperature decrease and fine-grained needle-like plagioclase crystals surrounding the rounded quartz crystals would be formed under supersaturated conditions. The texture of quartz shown here suggests a phenomenon of pressure decrease during emplacement of granitic magma. The coexistence of vapour-rich two-phase fluid inclusion and halite-bearing polyphase fluid inclusion in a single quartz crystal of granitic rock of North Body accords with low confining pressure based on the phase relation of the $NaCl-H_2O$ system under supercritical conditions (Hedenquist *et al.* 1998). The decrease of confining pressure is also supported by geological evidence that breccia dykes and tourmaline veins occur in monzogranite. The dissolution texture of igneous quartz would be evidence suggesting decrease of pressure.

One of the characteristics of granitic rocks of North Body is presence of sulphide minerals formed under magmatic environments. The amounts of sulphide minerals are relatively high in tonalite of North Body as marginal facies and the amounts decrease from tonalite through granodiorite to monzogranite of North Body. To form sulphide minerals in magmatic conditions, high sulphur fugacity in melt is required (Stavast *et al.* 2006). Some magmas have a high sulphur concentration initially and sulphide minerals can be crystallized in the magma. Another possibility is accumulation of sulphur in volatile phases in magma according to generation of fluid phase by the process of second boiling. It is difficult to determine whether the process of generation of fluid in melt in the early period of emplacement of magma forming North Body was controlled by decrease of pressure (first boiling) according to ascent of magma or by accumulation of volatile phase according to magmatic differentiation (second boiling) because the depth of emplacement of North body is shallow. The possibility of the process of second boiling for the crystallization of sulphide minerals in tonalite would be small because the pressure of emplacement of Chichibu granite is low and pressure is thought to be major factor.

Summary

Based on the study of Chichibu granites, magmatic fluid forming Chichibu deposit would be generated by decrease in pressure when SiO_2 content of the magma exceeds about 65 wt% SiO_2 according to progress of magmatic differentiation. As a result, heavy metals such as copper, gold and arsenic in magmatic fluid coexisting with melt were transported to sedimentary strata intercalating limestone through major fault systems and mineralization as a secondary skarn took place in the limestone dominant area.

Factors for discrimination of ore deposit-related granites from ore deposit-free granites are summarized as follows: (1) presence of miarolitic cavities and/or veins and veinlets composed of minerals such as tourmaline, (2) presence of chemical variation that is decrease–increase–decrease for heavy metals, (3) decrease in concentrations of LREE and total REE concentration according to magmatic differentiation, (4) presence of corroded quartz crystals of oval shape in aggregates consisting of fine-grained plagioclase, quartz and K-feldspar, and (5) presence of halite-bearing fluid inclusions and vapour-rich fluid inclusions in igneous quartz. These factors will provide important information for distinguishing granites related to mineralization from granites free of mineralization.

We would like to acknowledge Chichibu mine, Nitchitsu Co. Ltd. for their cooperation in the research carried out on their property. We would like to thank T. Kawahira, Y. Oyamada and M. Kobayashi, Chichibu mine for provision of information on geology and support of the geological survey. C. De Campos and a reviewer provided thoughtful reviews of an earlier draft of this manuscript.

References

ABE, A., ASANO, K. & KANEDA, M. 1961. Recent prospecting at the Chichibu mine. *Mining Geology*, **11**, 32–36 [in Japanese with English abstract].

ARAMAKI, S., HIRAYAMA, K. & NOZAWA, T. 1972. Chemical composition of Japanese granitoids, Part 2. Variation trends and average composition of 1200 analyses. *Journal of the Geological Society of Japan*, **78**, 39–49.

BAKER, T., ACHTERBERG, E. V., RYAN, C. G. & LANG, J. R. 2004. Composition and evolution of ore fluid in a magmatic-hydrothermal skarn deposit. *Geology*, **32**, 117–120.

CANDELA, P. A. & PICCOLI, P. M. 1995. Model ore-metal partitioning from melts into vapor and vapor/brine mixtures. *In*: THOMPSON, J. F. H. (ed.) *Magma, Fluids, and Ore Deposits*. Mineralogical Association of Canada, Short Course Series, **23**, 101–127.

FULIGNATI, P., KAAMENETSKY, V. S., MARIANELLI, P., SBRANA, A. & MERNAGH, T. P. 2001. Melt inclusion record of immiscibility between silicate, hadrosaline, and carbonate melts: applications to skarn genesis at Mount Vesuvius. *Geology*, **29**, 1043–1046.

HARRIS, A. C., KAMENETSKY, V. S., WHITE, N. C., ACHTERBERGH, E. & RYAN, C. G. 2003. Melt inclusions in veins: linking magmas and porphyry Cu deposits. *Science*, **302**, 2109–2111.

HEDENQUIST, J. W. & RICHARDS, J. P. 1998. The influence of geochemical techniques on the development of genetic models for porphyry copper deposits. *In*: RICHARDS, J. P. & LARSON, P. B. (eds) *Techniques in Hydrothermal Ore Deposits Geology*. Reviews in Economic Geology, Society of Economic Geologists, Littleton, CO, **10**, 235–256.

ISHIHARA, S., TERASHIMA, S. & TSUKIMURA, K. 1987. Spatial distribution of magnetic susceptibility and ore elements, and cause of local reduction on magnetite-series granitoids and related ore deposits at Chichibu, Central Japan. *Mining Geology*, **37**, 15–28.

ISHIYAMA, D. 2005. World Skarn Deposits: Skarns of Japan: 1–10 and 1 Table, in electronic folder '4 Japan' in electronic folder 'Meinert' in CD-ROM supplementary appendix to: Meinert, L. D., Dipple, G. M. & Nicolescu, S. 2005. World Skarn Deposits. *In*: HEDENQUIST, J. W., THOMPSON, J. F. H., GOLDFARB, R. J. & RICHARDS, J. P. (eds) *Economic Geology, 100th Anniversary Volume*. Society of Economic Geologists, Littleton, CO, 299–336.

KANEDA, M. 1967. History of exploration at the Chichibu mine. *Journal of Mining and Metallurgy Institute of Japan*, **83**, 155–157 [in Japanese].

KANEDA, M. & WATANABE, A. 1961. On the geology and prospecting of the Akaiwa and Doshinkubo deposits, Chichibu mine. *Mining Geology*, **11**, 481–490 [in Japanese with English abstract].

KWAK, T. A. P. 1986. Fluid inclusions in skarn (carbonate replacement deposits). *Journal of Metamorphic Geology*, **4**, 363–384.

MEINERT, L. D., HEDENQUIST, J. W., SATOH, H. & MATSUHISA, Y. 2003. Formation of anhydrous and hydrous skarn in Cu–Au ore deposits by magmatic fluids. *Economic Geology*, **98**, 147–156.

MEINERT, L. D., DIPPLE, G. M. & NICOLESCU, S. 2005. World skarn deposits. *In*: HEDENQUIST, J. W., THOMPSON, J. F. H., GOLDFARB, R. J. & RICHARDS, J. P. (eds) *Economic Geology, 100th Anniversary Volume*. Society of Economic Geologists, Littleton, CO, 299–336.

MITI (Ministry of International Trade and Industry). 1975. Reports on the regional geology of the Chichibu district, MITI, Tokyo [in Japanese].

MIYASHIRO, A. 1974. Volcanic rock series in island arcs and active continental margins. *American Journal of Science*, **274**, 321–355.

MULLER, A., BREITER, K., SELTMANN, R. & PECSKAY, Z. 2005. Quartz and feldspar zoning in the eastern Erzgebirge volucano-plutonic complex (Germany, Czech Republic): evidence of multiple magma mixing. *Lithos*, **80**, 201–227.

NAKANO, S., TAKEUCHI, K., KATO, H., HAMASAKI, S., HIROSHIMA, T. & KOMAZAWA, M. 1998. *Geological Map of Japan 1:200 000*. Geological Survey of Japan, Nagano.

SAITO, K., TAKAHASHI, M. & ONOZUKA, N. 1996. A K–Ar investigation of the Chichibu quartz diorite and some discussions on its cooling history. *Journal of Geomagnetism and Geoelectricity*, **48**, 1103–1109.

SAKAI, A. & HORIGUCHI, M. 1986. Chichibu terrane. *In*: KANTO,, OMORI, M., HAYAMA, Y. & HORIGUCHI, M. (eds) *Chapter 1 Paleozoic and Mesozoic Strata*. Regional Geology of Japan, Kyoritsu Shuppan Co Ltd, 12–20 Part 3 [in Japanese].

SHIMAZAKI, H. 1975. The ratios of Cu/Zn–Pb of pyrometasomatic deposits in Japan and their genetical implications. *Economic Geology*, **70**, 717–724.

SHOJI, T., OOTSUKA, M. & IMAI, H. 1967. Consideration on the formation of mineralized faults and non-mineralized farcturs in the Chichibu mine, Saitama Prefecture. *Journal of Mining and Metallurgy Institute of Japan*, **85**, 765–770 [in Japanese with English abstract].

STAVAST, W. J. A., KEITH, J. D., CHRISTIANSEN, E. H., DORAIS, M. J., TINGER, D., LAROCQUE, A. & EVANS, N. 2006. The fate of magmatic sulfides during intrusion or eruption, Bingham and Tintic districts, Utah. *Economic Geology*, **101**, 329–345.

STRECKEISEN, A. L. 1976. To each plutonic rock its proper name. *Earth Science Reviews*, **12**, 1–33.

TAKAHASHI, M. 1989. Neogene granitic magmatism in the South Fossa Magna collision zone, central Japan. *Modern Geology*, **14**, 127–143.

TUTTLE, O. F & BOWEN, N. L. 1958. *Origin of Granite in the Light of Experimental Studies in the System* $NaAlSi_3O_8$–$KAlSi_3O_8$–SiO_2–H_2O. Geological Society of America, Memoir **74**.

Ueno, H. & Shibata, K. 1986. Radiometric ages of quartz diorite bodies related to the Chichibu pyrometasomatic deposits and their relevance to the metallogenic epoch. *Journal of Mineralogy, Petrology and Economic Geology*, **81**, 77–82.

Wark, D. A., Hildreth, W., Spear, F. S., Cherniak, D. J. & Watson, E. B. 2007. Pre-eruption recharge of the Bishop magma system. *Geology*, **35**, 235–238.

Weibe, R. A., Wark, D. A. & Hawkins, D. P. 2007. Insights from quartz cathodoluminescence zoning into crystallization of the Vinalhaven granite, coastal Maine. *Contributions to Mineralogy and Petrology*, **154**, 439–453.

Williams-Jones, A. E. & Heinrich, C. A. 2005. Vapor transport of metals and the formation of magmatic-hydrothermal ore deposits. *Economic Geology*, **100**, 287–1312.

Fractal analysis of the ore-forming process in a skarn deposit: a case study in the Shizishan area, China

QINGFEI WANG[1,2]*, JUN DENG[1,2], HUAN LIU[1,2], LI WAN[1,3] & ZHIJUN ZHANG[1,2]

[1]*State Key Laboratory of Geological Processes and Mineral Resources, China University of Geosciences, Beijing, 100083, China*

[2]*Key Laboratory of Lithosphere Tectonics and Lithoprobing Technology of Ministry of Education, China University of Geosciences, Beijing, 100083, China*

[3]*School of Mathematics and Information Science, Guangzhou University, Guangzhou, Guangdong, 510405, China*

**Corresponding author (e-mail: wqf@cugb.edu.cn)*

Abstract: This paper presents a tool for analysing the element distribution and mineralization intensity. The Hurst exponents and *a-f*(*a*) multifractal spectrum are utilized to analyse the irregular element distribution in Shizishan skarn orefield, China. The Hurst exponents reveal the Cu, Ag, Au and Zn distributions in the skarn-dominated drill cores are persistent and those in marble-dominated drill core are nearly random; the persistence indicates the mineralized segments are repeatedly developed, with accordance to multi-layer structure of the ore-controlling bedding faults and orebodies. The small a_{min} (minimum multifractal singularity) of the Cu, Ag and Au in M_1 reflect bare mineralization. The a_{min} also displays that the mineralization intensities are varied for distinct elements and for different locations, yet the similarity of the distinct ore-forming processes is manifested by constant $a_{min}/f(a_{min})$ ratio. The constant ratio indicates the wider mineralization range denotes a more compact concentration distribution. The compact distributions represent the wide Cu, Ag and Au mineralization in skarns, and the loose distributions reflect the bare Cu, Ag and Au mineralization in marbles. Moreover a_{min} shows a positive correlation with Hurst exponents in the Shizishan skarn orefield. Using fractal analysis the author's show that although the mineralization intensities for different elements and different locations along the Shizishan skarn orefield is not consistent, similar mineralization processes can be correlated to similar fractal exponents.

Skarn deposits occur at the contact between a felsic intrusion and a reactive wallrock, including dolomites, limestones, and clastic sedimentary rocks with calcite cements. Due to the contact metamorphism and water–rock reactions, the content of the wallrock changes by adding SiO_2, H_2O, Fe, and metals. The chemical additions and temperature increase cause the development of a skarn mineral assemblage adjacent to the intrusives by thermal metamorphism, metasomatism, and replacement. In a further location, the metamorphic process causes a complete recrystallization that changes the original carbonate rock into marbles with an interlocking mosaic of calcite, aragonite, and/or dolomite crystals and changes mudstones into hornfels. In addition, the semi-simultaneous multi-stage intrusions of magmatic rocks within a small region often lead to the formation of a series of skarn-type deposits, composing an orefield or an ore cluster area. For example, a couple of orefields, such as the Shizishan orefield and Tongguanshan orefield of the Tongling ore cluster area in the metallogenic belt of the middle and lower reaches of the Yangtze River are mostly composed of the skarn-type ore deposit formed around the granite bodies emplaced during the Yanshanian epoch.

The study of the skarn deposits focuses on the fluid inclusions, mineral assemblage and the metamorphic reaction (van Marcke de Lummen & Verkaeren 1986; Nicolescu & Cornell 1999; Markowski *et al.* 2006). These works explain the ore-forming conditions and mineral formation process. Mass balance also clarifies the mobilization of the elements during the ore-formation. Yet little work has been done on the spatial distribution of ore-forming elements. The intrusive boundaries are often irregular, resulting in a complex orebody geometry and thus a complex spatial element distribution. The quantification of the spatial distribution of ore-forming elements can help when analysing the orebody geometry and provide information for the ore-forming process. The spatial distribution

From: SIAL, A. N., BETTENCOURT, J. S., DE CAMPOS, C. P. & FERREIRA, V. P. (eds) *Granite-Related Ore Deposits*. Geological Society, London, Special Publications, **350**, 89–104.
DOI: 10.1144/SP350.6 0305-8719/11/$15.00

of ore-forming elements can be discussed via studying the concentration curves along several prospect drills.

Along a prospect drill, from top to bottom, element concentrations jump from one location to the next. Whether the concentration variance is totally random or not and if the local concentration fluctuation serves as an indication for the next change or for the mineralization intensity, these are major questions to be answered. If the increase or decrease of the concentrations in different locations is consistent, it is named as persistence; otherwise, it is called anti-persistence. The randomness and persistence can be described by the Hurst exponent in a self-affine domain.

Similar to other physical processes, such as the diffusion limited aggregation (DLA) process, the mineralization process can be considered as the accumulation of anomalous elements with different intensities within a narrow spatial region, which is suitable for the multifractal analysis. The multifractal analysis can describe the characteristics of element concentration levels and their influence range, thus can denote the mineralization features. The multifractal analysis has been widely applied in the description of geological phenomena, including the distribution patterns of mineral deposits (Cheng 2003), fractures (Agterberg *et al.* 1996) and geochemical data (Cheng 2007; Deng *et al.* 2007; Wang *et al.* 2007*a*, 2008).

This paper focuses on the Shizishan orefield in the Tongling ore cluster area as an example, uses the self-affine fractal and multifractal methods to study the elements migration characteristics in the ore-forming process.

Geological setting

The skarn deposit is an important mineralization type in the metallogenic belt of the middle and lower reaches of the Yangtze River, China. The Tongling area in Anhui province is located in the middle segment of the metallogenic belt. The stratigraphic sequence cropping out in the study area ranges from Silurian to Triassic in age. The sequence is approximately 3000 m thick and includes Silurian shallow marine sandstones interbedded with shales, Devonian continental quasimolasse formation and lacustrine sediments, Carboniferous shallow marine carbonates, Permian marine facies alternated by marine-continental facies, and Early to Middle Triassic shallow marine carbonates. Quaternary sediments consist of alluvium in creeks and colluvium along slopes. During the Indosonian and Yanshannian epochs, the strata were deformed due to a pronounced NW compression, resulting in folding and a fault system was formed, in which the bedding faults were pronounced. Intrusive granite bodies emplaced in the caprock around 140 Ma originated broad and intense metasomatism, marble alteration and formation of many skarn deposits. The Tongling ore cluster area mainly comprises five orefields: that is, the Shizishan orefield, Tongguanshan orefield, Fenghuangshan orefield, Xinqiao orefield and Jinlang orefield.

The Shizishan orefield, in the central part of the Tongling area, is situated at the SE limb of the NE-trending Qingshan anticline. The outcropping strata are mainly Low–Middle Triassic thinly bedded carbonate rocks (Fig. 1). The strata were discovered through exploration drilling in the range from Silurian to Triassic. In the caprocks, the Yanshanian intrusions occur as stocks and dykes along faults to form an approximately 3 km long and 1 km wide network system (Deng *et al.* 2004*b*, 2007). Generally most intrusives, such as Baocun, Qingshanjiao, Caoshan, and so on. promote skarn alteration and associated mineralization. Skarns which are alternated with hornfels develop closer to the intrusives. Marble aureoles are widely developed around the skarn zones. Bedding faults in multiple layers develop along the interface between the strata with great lithological differences, from Devonian to Triassic promoting pathways for ore-forming fluids.

The skarn orebody with jagged sharp-cut boundaries develops along the contact zone between the felsic intrusion and sedimentary carbonate. In addition, replacement skarn of both calcic and magnesian types within sedimentary strata along the bedding fault is common and hosts the so-called stratabound copper skarn orebodies. These copper skarn orebodies are multi-layered, with up to 10 layers, within the Carboniferous to Triassic sedimentary rocks (Chang *et al.* 1991; Deng *et al.* 2004*a*, 2006).

The skarn mineralized system is characterized by multi-layered mineralizations (Fig. 2): Dongguashan porphyric type deposit and stratified skarn deposit host in the deep part, Huashupo and Datuanshan interstrata skarn deposits in the middle part, Laoyaling and Xishizishan stratified skarn deposit in the upper part, Dongshizhishan cryptobreccia deposit and Baocun, Baimongshan and Jiguanshan skarn type in the shallow part. From the stratigraphic profiles (Fig. 2), the main orebodies in these deposits are shown as multiple floors within the orefield. While the Dongguashan deposit is a large-scale one, the other skarn-type deposits are mostly intermediate or small.

In general, mineralization in the Shizishan orefield can be divided into an early magnetite–scheelite stage with hydrous silicates for example, epidote, actinolite and phlogopite, and a late sulphide stage from molybdenite to pyrrhotite, pyrite, chalcopyrite, sphalerite and galena. Ore textures change from massive and veinlet in proximal

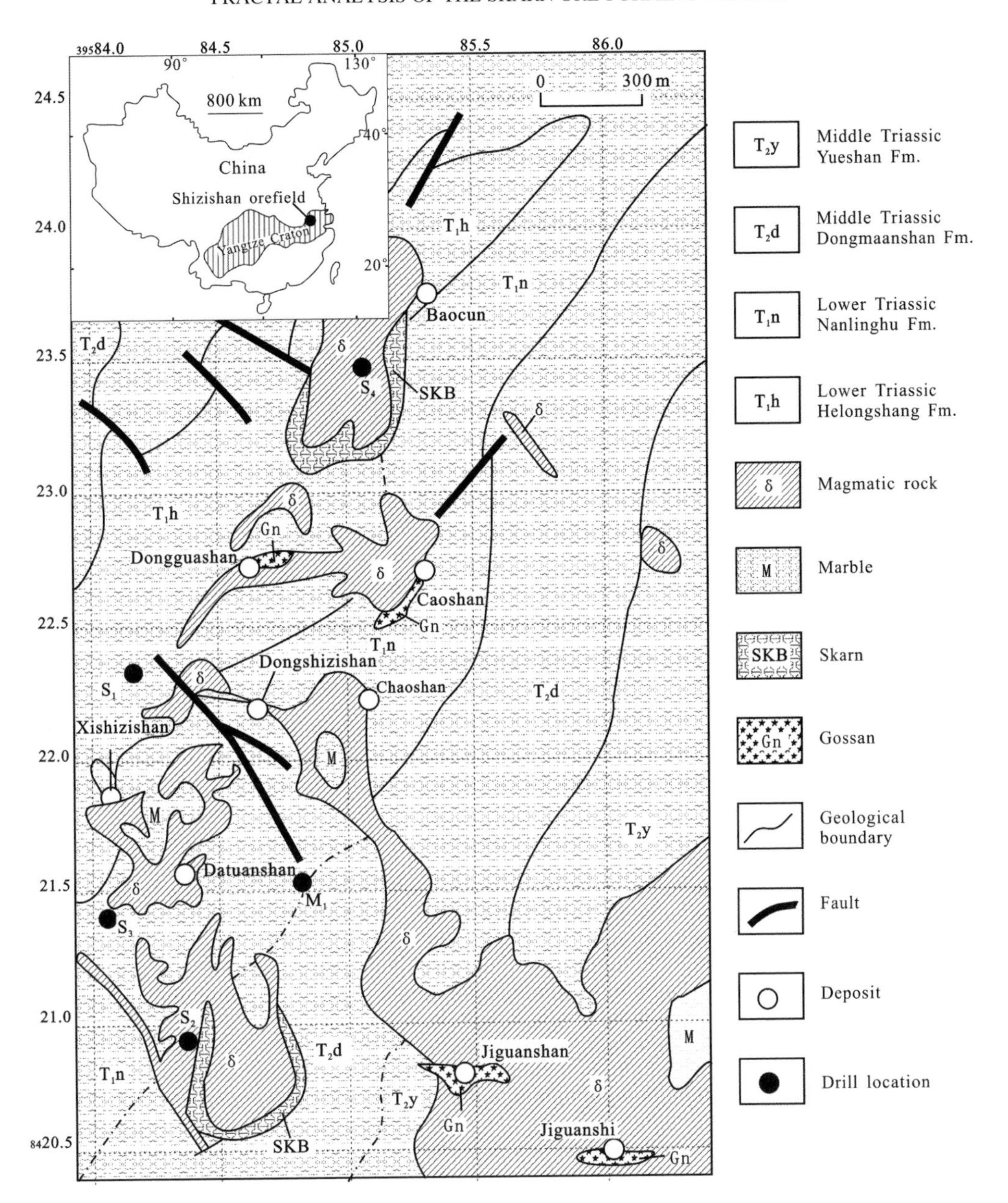

Fig. 1. Geological map of Tongling Shizishan orefield, Anhui province, China.

zones to banded and laminated in distal zones. Sphalerite and galena are generally restricted to the distal zones (Pan & Dong 1999).

Raw data and their distribution

An abundance of prospect drills have been carried out in this area. We select five drill cores, with nearly 1000 m long, from different ore deposits of the Shizishan orefield to study the element distribution along the cores. The strata involved in the drill cores mainly include T_1n (The Lower Triassic Nanlinghu formation), T_1h (the Lower Triassic Helongshan formation), T_1y (the Lower Triassic Yinkeng formation), P_2d (the Upper Permian Dalong formation), P_2l (the Upper Permian Longtan formation), P_1g (the Lower Permian Gufeng formation) and P_1q (the Lower Permian Qixia formation), with additional infiltrations of altered quartz diorite. The lithology in the drills S_1, S_2,

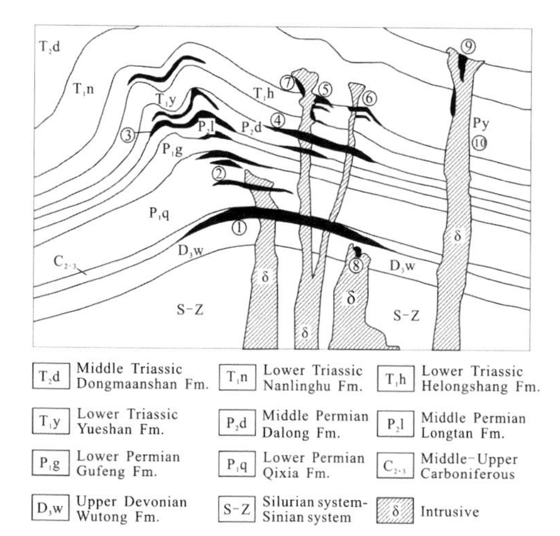

Fig. 2. Metallogenic model of deposit distribution in the Shizishan copper-gold orefield, Tongling area, China (modified after 321 geological team of bureau of geology/mineral resources exploration of Anhui province,1995) (1)Dongguashan Cu (–Au) deposit; (2) Huashupo Cu (–Au) deposit; (3) Laoyaling Cu (–Au) deposit; (4) Datuanshan Cu (–Au) deposit; (5)West Shizishan Cu (–Au) deposit; (6) East Shizishan Cu (–Au) deposit; (7) Hucun Cu (–Au) deposit; (8) The crypto-explosion breccia type Cu (–Au) orebody; (9) Jiguanshan Cu (–Ag) deposit; (10) Caoshan pyrite deposit.

S_3, S_4 are dominated by the skarn, alternated with hornfels, reflecting that these drills are close to the intrusive; M_1 is dominated by marbles.

The drill cores were sampled every 10 m and analysed for Au, Ag, Cu, Pb and Zn. Concentration curves for different elements along the drills show much variation as illustrated in Figure 3. The histograms (Fig. 4a, c) and 'quantile–quantile' plots (Q–Q plots) (Fig. 4b, d) for $\ln_{Cu}$ and $\ln_{Pb}$ in the S_1 drillcore are illustrated in Figure 4, and those in the M_1 drill core are shown in Figure 5. The Q–Q plots for those concentration value distributions are skewed and do not follow a lognormal distribution. They show that the element distributions can be described by fractal tools.

Mathematical models

R/S *analysis*

The Hurst exponent is an important parameter in self-affine fractal. A regular approach to analyse H of a one-dimensional spatial data set is to carry out Fourier transform or the R/S (rescaled-range) analysis. R/S analysis is a valid calculating method for Hurst exponent and widely used in the geosciences (Turcotte 1997; Malamud & Turcotte 1999; Wang *et al.* 2007*b*; Deng *et al.* 2008). The R/S analysis is the one used in this paper.

This method was firstly proposed by Hensy Hurst (Hurst *et al.* 1965). The calculation starts with the whole concentration sequence $\{\xi\}_{i=1}^{N}$ that covers the box with length e and then its mean is calculated over the available data in the range e.

$$(E\xi)_e = \frac{1}{n}\sum_{i=1}^{n}\xi_i \tag{1}$$

Summing up the differences from the mean to get the cumulative total at each data point, $X\,(i, n)$, from the beginning up to any point of the box, then we have:

$$X(i, e) = \sum_{t=1}^{i}[\xi_t - (E\xi)_e] = \sum_{t=1}^{i}\xi_t - i(E\xi)_e, \quad 1 \le i \le n \tag{2}$$

The next step is to find the max $X(i, e)$ representing the maximum of $X\,(i, e)$, min $X\,(i, e)$ representing the minimum of $X\,(i, e)$ for $1 \le i \le e$, and calculate the range $R(e)$:

$$R(e) = \max X(i, e) - \min X(i, e) \tag{3}$$

To be able to compare different phenomena, the range, $R(e)$, is divided by the standard deviation $S(e)$

$$S(e) = \left\{\frac{1}{n}\sum_{i=1}^{n}[\xi_i - (E\xi)_e]^2\right\}^{1/2} \tag{4}$$

and the obtained result is called the rescaled range R/S.

The overall concentration sequence can usually be covered by several non-overlapping boxes with length e. After determining R/S for each box, we can get the mean value of R/S, that is, $E(R/S)_e$. Using successively shorter e to divide the data set into more non-overlapping boxes and finding the $E(R/S)_e$ of these boxes, we can obtain the Hurst exponent, based on the following equation

$$\ln E(R/S)_e = \ln C + H \ln_e \tag{5}$$

where C is a constant.

Multifractal

The one-dimension concentration sequence $\{\xi\}_{i=1}^{N}$ can be subdivided into boxes of the same linear size e. $P_j(e)$ is the distribution probability of mass in the jth box. Thus, the multifractal is described as

$$P_j(e) \propto e^{\alpha_j} \tag{6}$$

$$N_e(\alpha) \propto e^{-f(\alpha)} \tag{7}$$

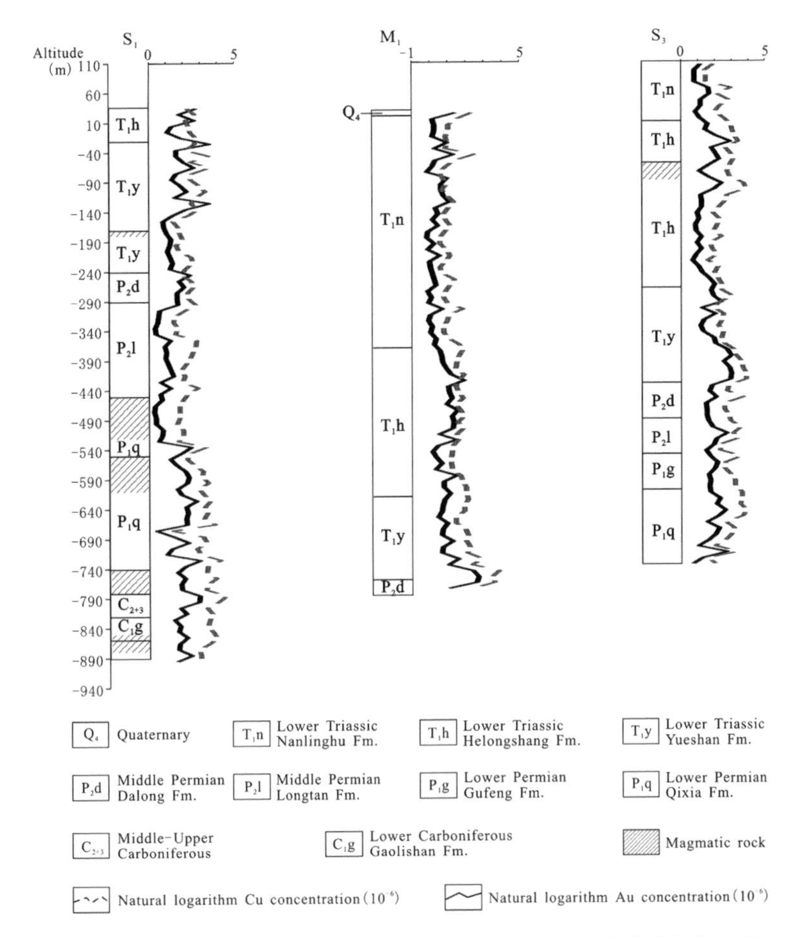

Fig. 3. Element concentration curve along drill cores in the different ore deposits. S_1, S_2, M_1, S_3 and S_4 represent the different drill cores in Dongguashan Cu (–Au) deposit, Changlongshan Cu (–Au) deposit, Datuanshan Cu (–Au) deposit, Huashupo Cu (–Au) deposit and Hucun Cu (–Au) deposit respectively.

where α_j is called the coarse Hölder exponent or singularity. $N_e(\alpha)$ is the number of boxes of size e with the same probability P_j. The values of $f(\alpha)$ could be interpreted loosely as fractal dimensions of subsets of cells with size e having coarse Hölder exponent α and in the limit $e \rightarrow 0$ (Evertsz & Mandelbrot 1992).

There are several methods for implementing multifractal modelling. In this paper, the moment method proposed by Halsey *et al.* (1986) is used. As a first step the fraction of measurement (μ_j) in each box is calculated from:

$$\mu_j = \frac{m_j}{m_T} = \frac{m_j}{\sum_{j=1}^{n(e)} m_j} \tag{8}$$

where, m_j is the sum of the concentrations in box j and m_T is the sum of the total concentrations of the data. The partition function $[M(q, e)]$ is then defined as:

$$M(q, e) = \sum_{j=1}^{n(e)} [M_j(q, e)] \quad q \in \Re \tag{9}$$

where

$$M_j(q, e) = \mu_j^q = \left(\frac{m_j}{\sum_{j=1}^{n(e)} m_j} \right)^q \tag{10}$$

The moment q provides a much more accurate and detailed way for exploring different regions with the singular measure. For $q > 1$, $\mu(q)$ the more singular regions of the measure can be amplified for $q < 1$, it accentuates the less singular regions, and for $q = 1$, the measure $\mu(1)$ replicated the original measure.

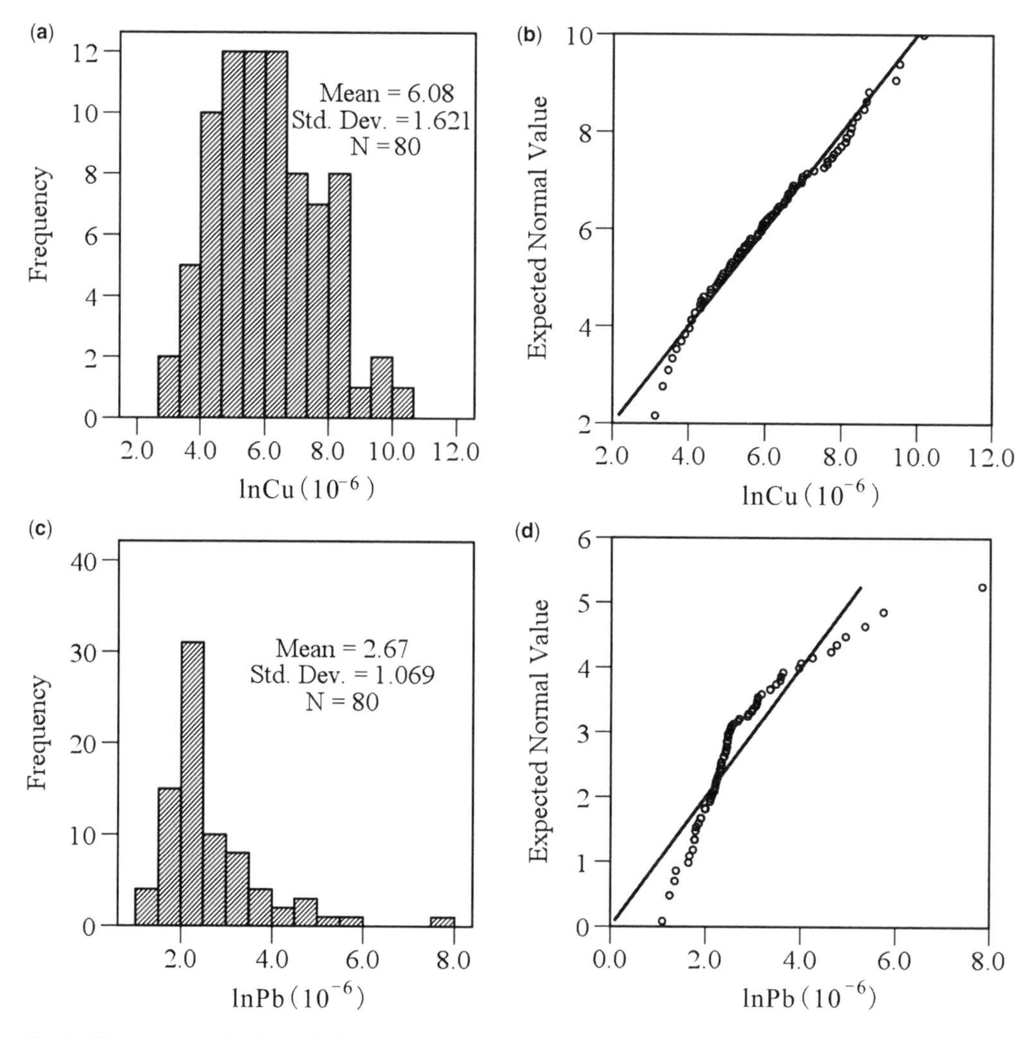

Fig. 4. Histograms and Q–Q plots for $\ln_{Cu}$ and $\ln_{Pb}$ concentrations in the S_1 drill core: (**a**) $\ln_{Cu}$ histogram; (**b**) Q–Q plot for $\ln_{Cu}$; (**c**) $\ln_{Pb}$ histogram; (**d**) Q–Q plot for $\ln_{Pb}$.

The mass exponent $\tau(q)$ for q can be written as (Hentschel & Procaccia 1983):

$$\tau(q) = \lim_{e \to 0} \frac{\log[M(q,e)]}{\log(e)} \quad (11)$$

The function $\alpha(q)$ and $f(\alpha)$ then can be obtained by a Legendre transform as:

$$\alpha(q) = \frac{d\tau(q)}{dq} \quad (12)$$

$$f(\alpha) = q\alpha(q) - \tau(q) = q\frac{d\tau(q)}{dq} - \tau(q) \quad (13)$$

The $\alpha(q)$ and the corresponding $f(\alpha)$ composes a multifractal spectrum with an inverse bell shape. The width of the multifractal spectrum, $\Delta\alpha$, is determined by:

$$\Delta\alpha = \alpha_{max} - \alpha_{min} \quad (14)$$

The height difference $\Delta f(\alpha)$ between the two ends of the multifractal spectrum can be extracted by:

$$\Delta f(\alpha) = f(\alpha_{max}) - f(\alpha_{min}) \quad (15)$$

Calculation process and result analysis

Hurst exponent

Figures 6 and 7 show the calculations of Hurst exponents for Au, Ag, Cu element distribution in drill

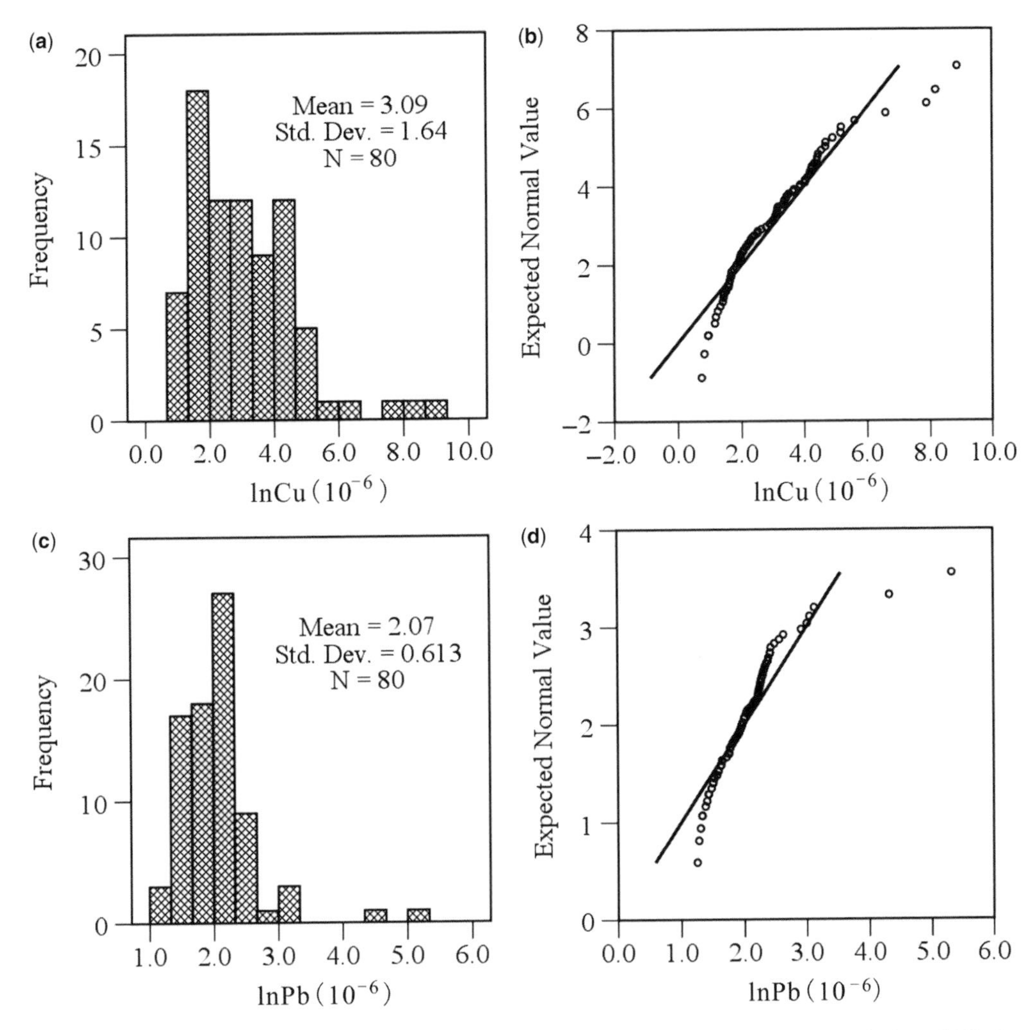

Fig. 5. Histograms and Q–Q plots for $\ln_{Cu}$ and $\ln_{Pb}$ concentrations in the M_1 drill core: (**a**) histogram for $\ln_{Cu}$; (**b**) Q–Q plot for $\ln_{Cu}$; (**c**) histogram for $\ln_{Pb}$; (**d**) Q–Q plot for $\ln_{Pb}$.

core M_1 and drill core S_1 with a good fit. The self-affine parameters of Au, Ag, Cu, Pb, Zn element distribution of the five drill cores are listed in Table 1.

When the Hurst exponent is close to 0.5, there is no correlation between any points and there is 50% probability that the next value may go either up or down. In the case of that the Hurst exponent value ranges from 0 to 0.5, the element concentration distribution shows anti-persistent behaviour, suggesting that any increasing trend in the previous location makes a decreasing trend in the next location more probable, and vice versa. When the Hurst exponent is greater than 0.5, the distribution shows persistent behaviour, meaning a concentration decrease will tend to follow another decrease and the larger the H value, the stronger the trend.

The Hurst exponents of the Au, Ag, Cu and Zn elements in the M_1 drill core are mostly around 0.5, indicating a random distribution; the Hurst exponent for the element Pb is 0.66, greater than that of other elements in the M_1 drill core, showing that the Pb distribution is caused by a process characterized by persistence. The Hurst exponents for Au, Ag, Cu, Zn and Pb in the S_1, S_2, S_3 and S_4 drill cores range from 0.6 to 0.87, are obviously greater than 0.5, representing persistent distributions (Fig. 8). The different elements in a drill core and the same element in different drill cores show

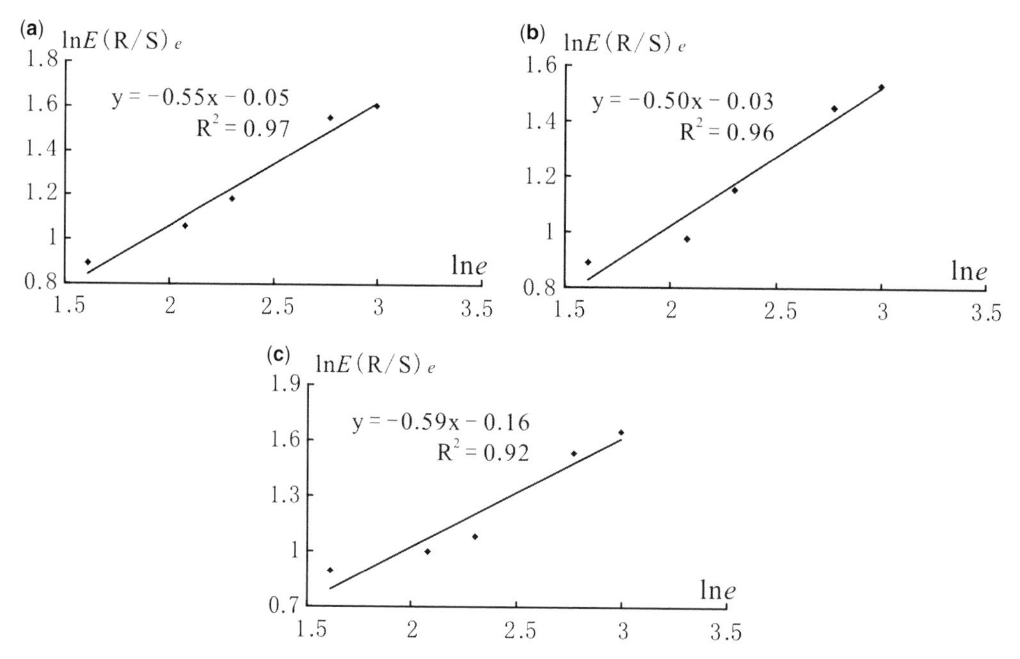

Fig. 6. Calculation diagram of the Hurst exponent by R/S analysis method of the Au, Ag, Cu element distribution in the M_1 drill cores in the Shizishan orefield, Anhui province, China. (**a**) Au; (**b**) Ag;(**c**) Cu.

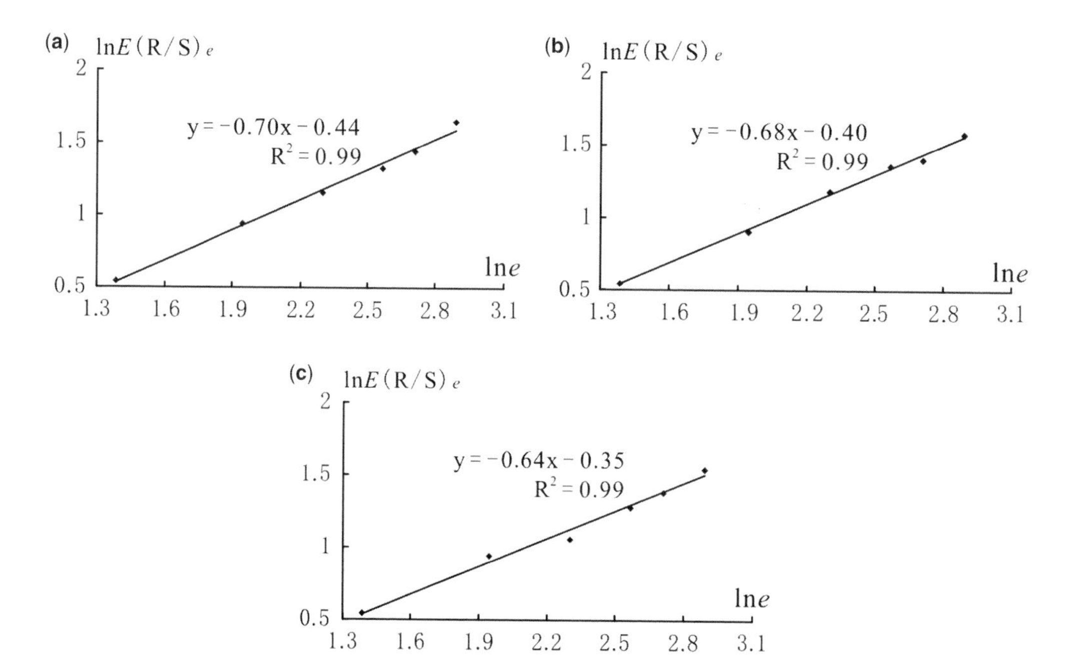

Fig. 7. Calculation diagram of the Hurst exponent by R/S analysis method of the Au, Ag, Cu element distribution in the S_1 drill cores in the Shizishan orefield, Anhui province, China. (**a**) Au; (**b**) Ag; (**c**) Cu.

Table 1. *Hurst exponents of Au, Ag, Cu, Pb and Zn distributions in the Shizishan orefield, Tongling area, China*

Drillcore	Hurst exponent				
	Au	Ag	Cu	Pb	Zn
S_1	0.68	0.64	0.7	0.6	0.63
S_2	0.69	0.71	0.72	0.78	0.71
S_3	0.77	0.8	0.87	0.66	0.62
S_4	0.77	0.84	0.74	0.69	0.71
M_1	0.5	0.59	0.55	0.66	0.53

variation. Since the distributions of most elements in the marble-dominated area are random, it can be identified that the variances of H for Au, Ag, Cu, Zn in the S_1, S_2, S_3 and S_4 drill cores, with respect to those in the M_1 drill core, are caused by the mineralization.

Multifractal spectrum

In the multifractal analysis, $\Delta\alpha$ and $\Delta f(\alpha)$ are often applied. An increase in $\Delta(\alpha)$ means a transition from homogeneous (random, space filling) to heterogeneous (ordered, complex, clustered) pattern. Similarly, a positive $\Delta f(\alpha)$ means the spectrum is right-hooked, indicating that there are more small values within the dataset than great ones; otherwise, they are dominated by greater ones (Wang *et al.* 2008).

Since the ore-forming process can be considered as the enrichment and deposition of the metals in the ores, we choose $q > 0$ and calculate the left part of multifractal spectrum. Therefore, we can obtain a_{min} and $f(a_{min})$ in the case that the moment q is maximal. This can provide more definite information with respect to the implicate meaning of $\Delta\alpha$

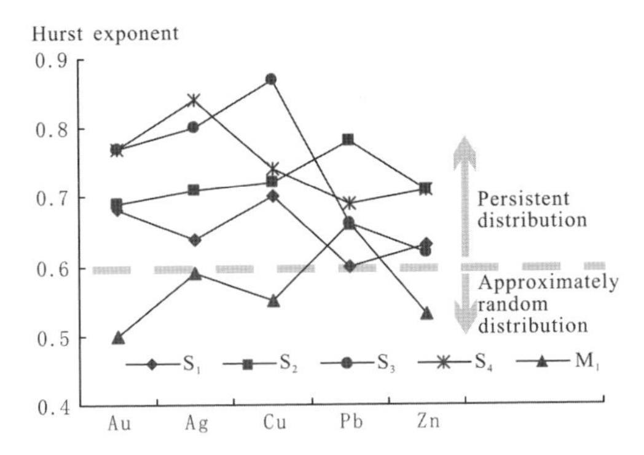

Fig. 8. Comparisons between Hurst exponents for different elements in the studied five drill cores in the Shizishan orefield, Anhui province, China.

and $\Delta f(\alpha)$. The α_{min} denotes the compactness of the concentration spread in the data space. If the concentrations are compact, the concentrations can constitute a subset with a greater a_{min}. Due to the fact that intervals of higher concentrations are generally small. However, with the same moment q, in the sparse dataset, the higher concentrations constitute a subset characterized by a smaller a_{min} for the greater intervals. The corresponding $f(a_{min})$ can describe the spatial influence range of the concentration subset with the same singularity. The lower $f(a_{min})$ means the size of the subsets is smaller, representing the mineralization area is narrower. The great α_{min} or $f(a_{min})$ can be resolved from the compact elevated concentration due to intense mineralization. This may also be induced by the original concentration dataset structure in the strata, where the element concentrations show equally small variance and compact structure. According to the comparison of Hurst exponent between M_1 and the other four drill cores, we can ascertain that the concentration data of the elements with great Hurst exponent is transformed by the mineralization, and therefore their multifractal exponents can reflect the mineralization intensity. The high α_{min} can represent an intense mineralization. In addition, when the mineralization is bare, and elements are only enriched in few segments in space, the concentrations of the dataset are sparse and differentiated to a great extent, meaning a small α_{min}.

In this paper, we set q ranging from 0 to 1.2 with interval 0.2, and calculate the multifractal spectrum. Thus the a_{min} and $f(a_{min})$ are $a(1.2)$ and $f[a(1.2)]$ respectively. The calculation of multifractal spectrum of Cu concentration distribution in S_1 drill is shown in Figure 9. The multifractal parameters of the element distribution of the five drill cores are listed in Table 2.

We overlay the multifractal spectra for Au, Ag, Cu, Pb and Zn in the S_1 drill core and those in the M_1 drill core as shown in Figure 10a, b separately; we also compare the Cu and Pb multifractal spectra from all the five drill cores, which are illustrated in Figure 10c, d. From Figure 10, we can see that the concentration dataset structures of the same elements varied in the different drill cores. The various elements show different enrichment characteristics in the same drill core. It reflects the metallogenic diversity.

We further plot the $\alpha(1.2)$ for Au, Ag, Cu, Pb and Zn in the five drill cores in Figure 11. It shows that $\alpha(1.2)$ of the Au, Ag, Cu and Zn in skarn-dominated drillcores are greater and that of Pb is relatively smaller, indicating that Au, Ag, Cu and Zn are widely enriched in the skarn, and that Pb is only locally accumulated inducing a sparse concentration dataset and bare mineralization in some

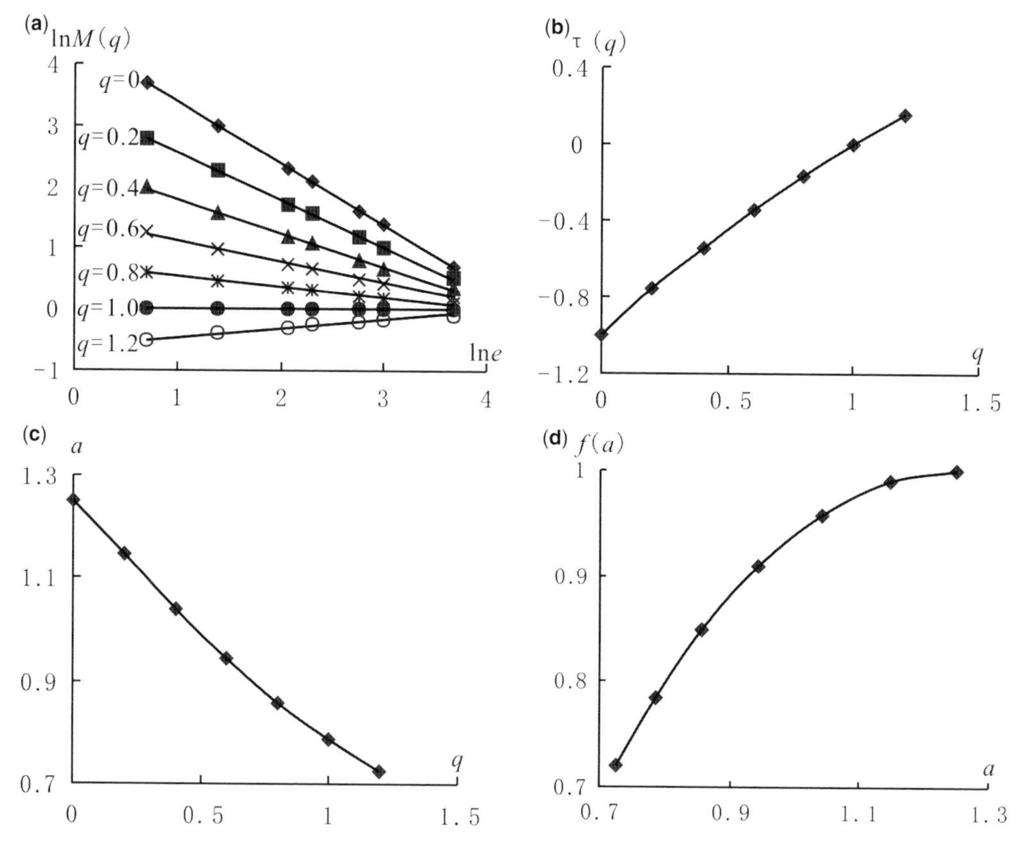

Fig. 9. Multifractal analysis for Cu concentration distribution in S_1 drill in Shizishan orefield, Anhui province, China. (**a**) ln $M_q(e)$ *vs.* ln *e*; (**b**)$\tau(q)$ *vs.* *Q*; (**c**)*a* *vs.* *Q*; (**d**)*f*(*a*) *vs.* *a*.

Table 2. *Multifractal parameters of element distribution in Shizishan orefield, Tongling area, China*

Drill	Parameter	Au	Ag	Cu	Pb	Zn	Drill	Parameter	Au	Ag	Cu	Pb	Zn
S_1	*a*(0)	1.47	1.29	1.25	1.3	1.2	S_2	*a*(0)	1.52	1.47	1.58	1.07	1.1
	f(*a*(0))	1	1	1	1	1		*f*(*a*(0))	1	1	1	1	1
	a(0.4)	1.07	1.03	1.04	1.15	1.09		*a*(0.4)	1.08	1.07	1.09	1.03	1.03
	f(*a*(0.4))	0.92	0.95	0.96	0.96	0.97		*f*(*a*(0.4))	0.91	0.92	0.9	0.99	0.98
	a(0.8)	0.74	0.85	0.86	0.73	0.83		*a*(0.8)	0.7	0.74	0.66	0.94	0.93
	f(*a*(0.8))	0.72	0.84	0.85	0.7	0.81		*f*(*a*(0.8))	0.68	0.73	0.65	0.94	0.93
	a(1.2)	0.52	0.74	0.73	0.33	0.5		*a*(1.2)	0.47	0.57	0.42	0.81	0.82
	f(*a*(1.2))	0.51	0.74	0.72	0.31	0.48		*f*(*a*(1.2))	0.46	0.56	0.41	0.81	0.81
M_1	*a*(0)	1.3	1.26	1.4	1.09	1.44	S_3	*a*(0)	1.28	1.22	1.29	1.26	1.34
	f(*a*(0))	1	1	1	1	1		*f*(*a*(0))	1	1	1	1	1
	a(0.4)	1.14	1.13	1.18	1.04	1.26		*a*(0.4)	1.04	1.04	1.03	1.17	1.12
	f(*a*(0.4))	0.96	0.97	0.94	0.99	0.94		*f*(*a*(0.4))	0.95	0.97	0.95	0.97	0.95
	a(0.8)	0.72	0.76	0.64	0.93	0.54		*a*(0.8)	0.84	0.87	0.85	0.72	0.73
	f(*a*(0.8))	0.7	0.74	0.62	0.92	0.51		*f*(*a*(0.8))	0.83	0.86	0.85	0.69	0.71
	a(1.2)	0.36	0.43	0.34	0.77	0.13		*a*(1.2)	0.7	0.71	0.75	0.24	0.41
	f(*a*(1.2))	0.35	0.42	0.33	0.76	0.12		*f*(*a*(1.2))	0.69	0.7	0.74	0.22	0.4
S_4	*a*(0)	1.24	1.14	1.23	1.41	1.34	S_4	*a*(0.8)	0.86	0.92	0.88	0.75	0.79
	f(*a*(0))	1	1	1	1	1		*f*(*a*(0.8))	0.85	0.92	0.87	0.73	0.78
	a(0.4)	1.04	1.02	1.03	1.08	1.07		*a*(1.2)	0.73	0.86	0.8	0.55	0.63
	f(*a*(0.4))	0.96	0.98	0.96	0.93	0.94		*f*(*a*(1.2))	0.72	0.86	0.8	0.55	0.63

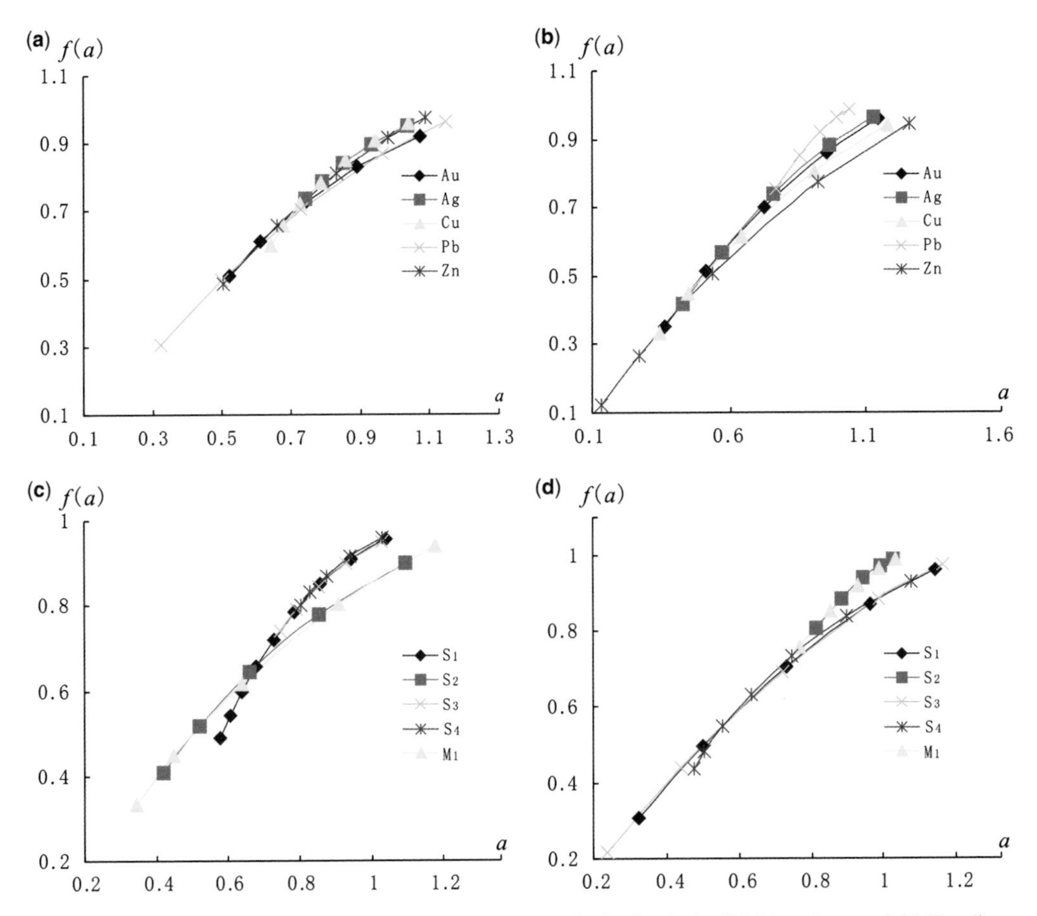

Fig. 10. Overlaying of multifractal spectra of different element distribution in the Shizishan skarn orefield, Tongling area, China: (**a**) Au, Ag, Cu, Pb and Zn distributions in the S_1 drill core; (**b**) Au, Ag, Cu, Pb and Zn distributions in the M_1 drill core; (**c**) Cu distribution in the five drill cores; (**d**) Pb distribution in the five drill cores.

skarn-dominated drill cores. The second greatest $\alpha(1.2)$ of Pb in M_1 represents the intense Pb mineralization occurred in the marble-dominated area (Fig. 11). In Figure 11, the elevated $\alpha(1.2)$ for Au, Ag, Cu, Zn elements in S_1, S_2, S_3 and S_4 drill cores compared to drill core M_1 demonstrate a transit from a local and stochastic mineralization in marbles to a widespread and intense mineralization in skarns.

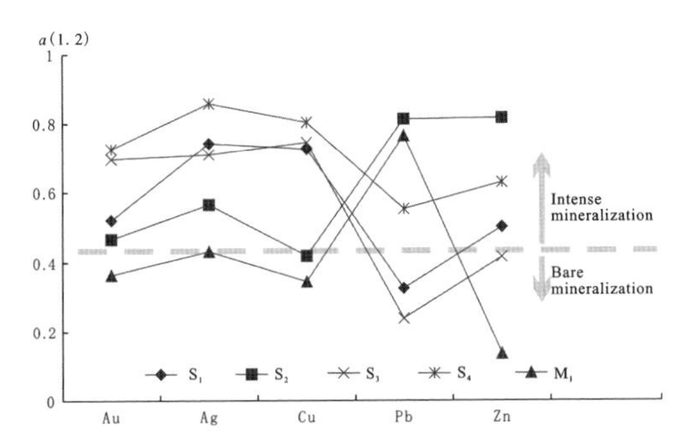

Fig. 11. $a(1.2)$ plots for Au, Ag, Cu, Pb and Zn in the five drill cores in the Shizishan orefield, Anhui province, China.

Moreover, $a(q)$ display linear relationship with $f[a(q)]$ when q is greater than 0.4 (Fig. 12), which has rarely been discussed in the previous multifractal study. The ratio of the different elements in a drill core and that of an element in different drill cores are similar. It indicates that the concentration spread pattern in the dataset is directly related with its size; that is, the wider mineralization range denotes a more compact concentration distribution.

As shown in Figure 13a, five areas are used to cover the $a(1.2)$ range of the different elements in one drill core. The M_1 range is the largest, and the S_4 area is the smallest, indicating the mineralization for different elements show the greatest variance in

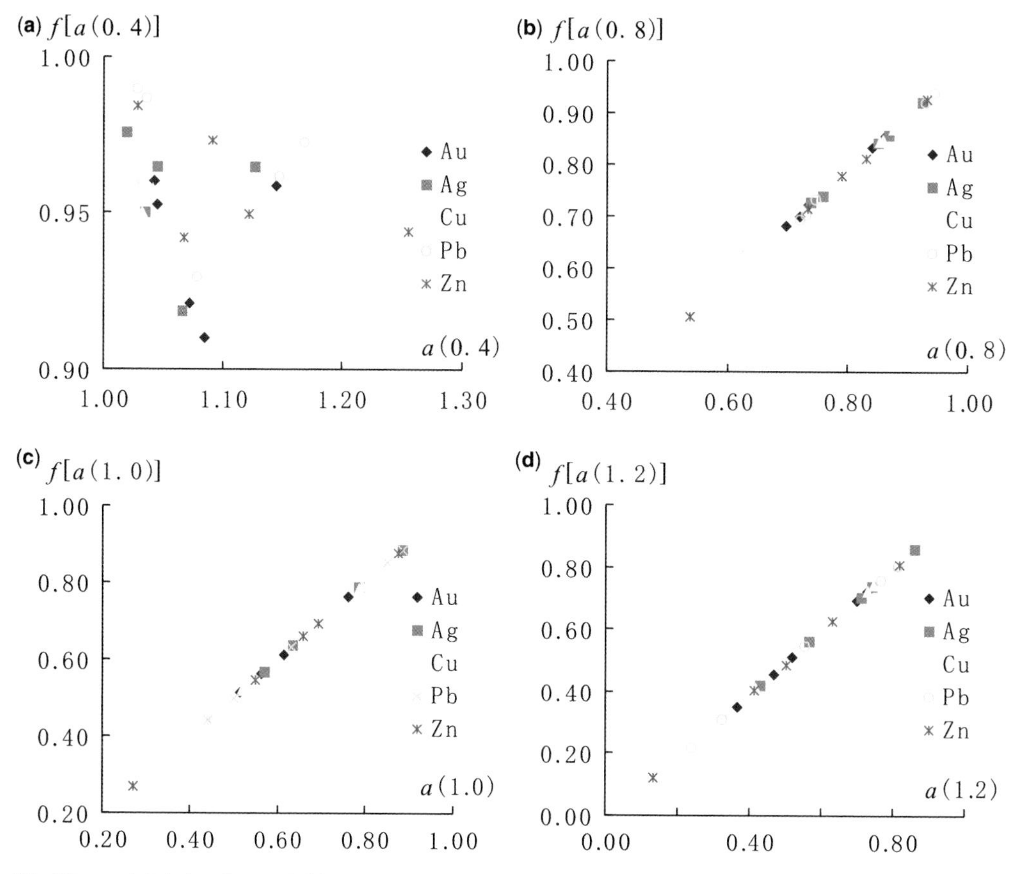

Fig. 12. a and $f(a)$ plots for $q \geq 0.4$ for Au, Ag, Cu, Pb and Zn distribution in Shizishan skarn orefield, Tongling area, China.

the marble-dominated area; and the enrichment patterns of the elements in S_4 display the least discrepancy. Similarly, the ranges covering the same element in the five drill cores are illustrated in Figure 13b. It reveals the Zn range is the largest, while the Au range is the narrowest, meaning that the Zn mineralization shows the greatest variety, while the Au mineralization is relatively common in the orefield.

Discussion

According to the Hurst exponents, the Cu, Ag, Au and Zn distributions in the skarn area show persistent distribution. In the upper part of the orefield where skarns alternated with hornfels develop around the intrusive, the irregular boundaries of the magmatites and discrete skarn development induce the interrupted mineralization. In the middle to the lower part, the multi-layer bedding faults control the mineralization repetition, which causes a great Hurst exponent and persistent distribution. The randomness of element distributions in the marble-dominated area basically reflects the element distribution characteristics in the sedimentary rocks. Therefore it is shown that the mineralization process can increase the persistence of the element distribution.

The α_{min} denotes variations in the enrichment features for different metals in distinct locations. This is partly related to the intrusive system and associated mineralization features. Similarly $a(q)$ to $f[a(q)]$ ratios, when the moment q is greater than 0.4, indicate that the greater singular concentrations take a smaller proportion in the dataset and the lower singular ones take a larger proportion. This rule results from the mineralization structures in the Shizishan orefield. For Cu, Ag, Au and Zn, in the skarn-dominated drills that are closer to the intrusives, the mineralization is characterized by massive and veinlet oretypes, which are widely distributed. This corresponds to compact concentrations and a greater a_{min} and $f(a_{min})$.

Comparatively, the marble-dominated drills, that are far away from the intrusives, are located at the distal part of the orebody, and develop the laminated orebody types; the bare mineralization causes a greater concentration differentiation and a smaller $a_{\min}$ and $f(a_{\min})$. Yet an intense Pb mineralization is also shown in marble-dominated drills, reflecting a metallogenic zoning between Pb and Cu, Ag, Au and Zn. This is consistent with the mineral zoning in the Shizishan orefield. The bare mineralization affects the Hurst exponent of the element distributions to a small extent.

The mineralization can promote a persistent element distribution. The persistence represents repeated concentration increases along the drill core, and the greater persistence means the stronger the trend. The increased element concentration compose a compact concentration subset with a great $\alpha_{\min}$. The stronger trend of repetitions probably results in a more compact subset, that is the larger H could respond to a greater $\alpha_{\min}$. It can be verified that H and $\alpha_{\min}$ of the same element distribution in different drill cores show positive correlation in the Shizishan orefield (Fig. 14).

Resuming all newly identified features means fractal analysis illustrate the main metallogenic changes for the Shizishan skarn orefileds. This is

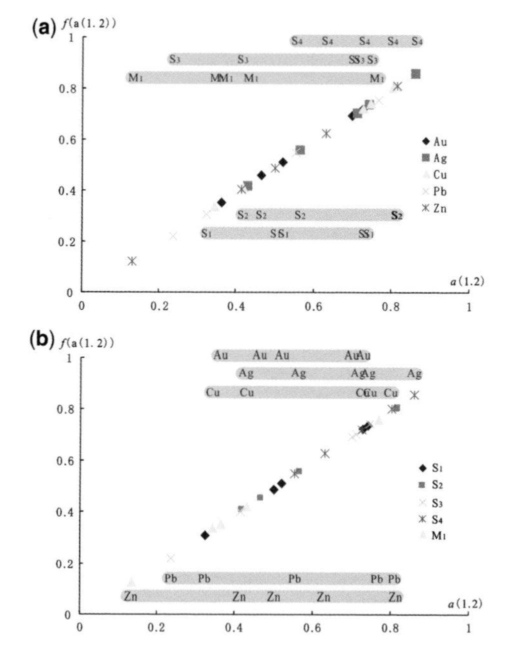

Fig. 13. Multifractal spectra of $a(1.2)$ to $f[a(1.2)]$ for Au, Ag, Cu, Pb and Zn distributions in Shizishan skarn orefield, Tongling area, China.

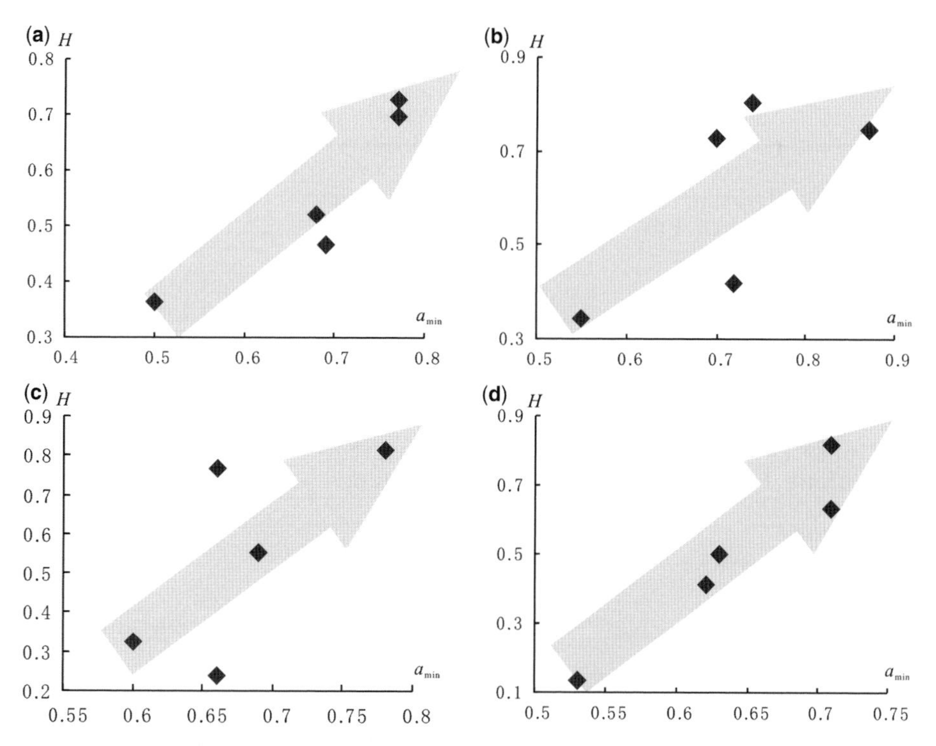

Fig. 14. Plots of H and $a_{\min}$ for element distribution in the Shizishan skarn orefield, Tongling area, China. (**a**) Au; (**b**) Cu; (**c**) Pb; (**d**) Zn.

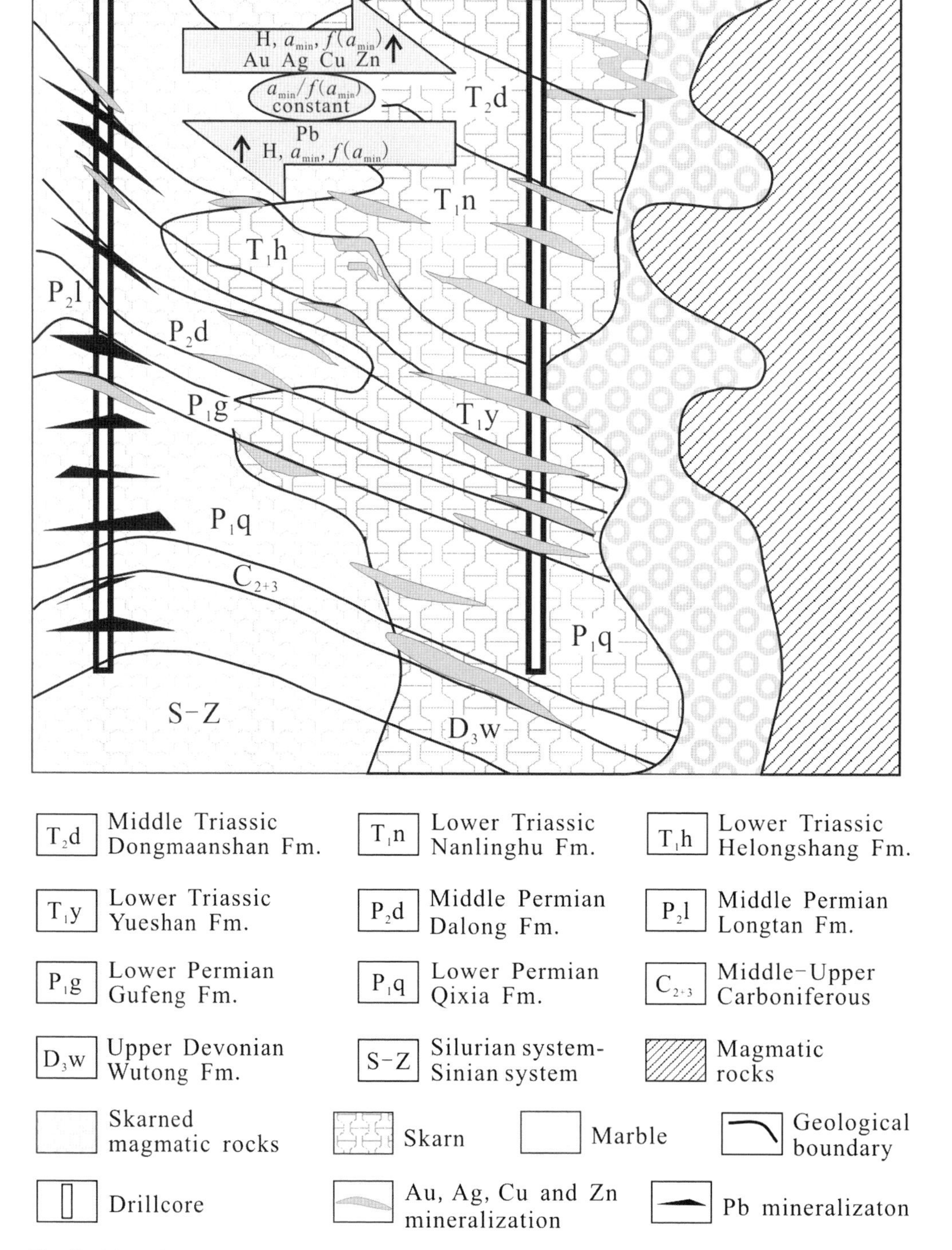

Fig. 15. Schematic model of fractal exponents changes in the skarn deposit, Tongling area, China.

shown in Figure 15. From the skarn-dominated area to marble-dominated area, the mineralization intensity for Au, Ag, Cu and Zn decreases. This is indicated by the decrease of α_{min}, $f(\alpha_{min})$ and by the Hurst exponent of the element distributions. The Pb mineralization becomes prominent with the increase of the three fractal parameters. Although the mineralization intensities vary for distinct elements and for different locations, the similarity between all ore-forming processes is confirmed by similar $a_{min}/f(a_{min})$ exponents.

Conclusion

Selecting five drills in Shizishan skarn orefield in the Tongling ore cluster area, and utilizing the multi-fractal and self-affine fractal tools, the distributions of the ore-forming elements have been analysed to better understand the element transport mechanism during skarn deposit formation.

It is revealed that the ore-forming process is characterized by persistence increase of element distribution. In the marbles, where the original characteristic of the sedimentary rocks is kept relatively unchanged, element concentrations generally show random distributions, although a weak mineralization occurred. On the other hand, in the skarns element, distributions are generally persistent. The persistence represents that the mineralized segments develop repeatedly, according to multi-layer structures of the ore-controlling bedding faults and orebodies.

Hurst exponent and a_{min} generally show positive correlation. Through comparison of the Hurst exponent and a_{min} of different elements in the drill cores, we can distinguish the detailed characteristics of element enrichment. The Cu, Ag, Au and Zn elements in skarn-dominated drill cores and Pb in M_l are widely enriched, displaying greater Hurst exponent and α_{min}; while the Cu, Ag, Au and Zn elements in M_l are only accumulated in few segments. The bare mineralization has little affect on the Hurst exponent, but greatly influences the α_{min}. It induces a great element differentiation and a correspondingly low α_{min}.

The comparison between Hurst exponent and α_{min} indicates metallogenic diversity in the Shizishan orefield. However, constant $a/f(a)$ ratios for different elements in the drills reflect similar transportation process during various elements mineralization, as revealed via multifractal analysis, in which the concentration spread pattern is directly related to the mineralization range. The wider mineralization range denotes a more compact concentration distribution. It is consistent with the interpretation of intense mineralization, which is characterized by the massive and veinlet ores distributed in a large range; and the sparser concentration distribution denotes a bare mineralization where the laminated orebody develops locally, resulting in metal accumulations in a narrow range.

This paper presents a tool for analysing the element distribution and mineralization intensity. Based on the fractal analysis, the ore-forming process in the Shizishan orefield can be further clarified.

We appreciate the valuable and careful comments from editors of *The Geological Society of London* and the two anonymous reviewers. This research is supported by the National Natural Science Foundation of China (No. 40234051), the Program for Changjiang scholars and Innovative Research Team in University (No. IRT0755), the 111 project (No. B07011) and the Special Plans of Science and Technology of Land Resource Department (No. 20010103).

References

Agterberg, F. P., Cheng, Q. M. & Brown, A. 1996. Multifractal modeling of fractures in the Lac Bonnet Batholith, Manitoba. *Computer and Geosciences*, **22**, 497–507.

Chang, Y. F., Liu, X. P. & Wu, Y. C. 1991. *The Cu, Fe Metallogenic Belt in the Middle-Lower Reaches of Yangtze River*. Geological Publishing House, Beijing. [In Chinese with English abstract.]

Cheng, Q. M. 2003. Fractal and multifractal modeling of hydrothermal mineral deposit spectrum, application to gold deposits in Abitibi Area, Ontario, Canada. *Journal of China University of Geosciences*, **14**, 199–206.

Cheng, Q. M. 2007. Mapping singularities with stream sediment geochemical data for prediction of undiscovered mineral deposits in Gejiu, Yunnan Province, China. *Ore Geology Reviews*, **32**, 314–324.

Deng, J., Huang, D. H., Wang, Q. F., Sun, Z. S., Wan, L. & Gao, B. F. 2004*a*. Experimental remolding on the caprock's 3D strain field of the Indosinian-Yanshanian Epoch in Tongling deposit concentrating area. *Science in China, Series D*, **34**, 993–1001.

Deng, J., Huang, D. H. *et al.* 2004*b*. Surplus space method, a new numerical model for prediction of shallow concealed magmatic bodies. *Acta Geologica Sinica* (English edition) , **78**, 358–367.

Deng, J., Wang, Q. F., Huang, D. H., Wan, L., Yang, L. Q. & Gao, B. F. 2006. Transport network and flow mechanism of shallow ore-bearing magma in Tongling ore cluster area. *Science in China Series D*, **49**, 397–407.

Deng, J., Wang, Q. F. *et al.* 2007. Reconstruction of ore controlling structures resulting from magmatic intrusion into the tongling ore cluster area during the Yanshanian Epoch. *Acta Geologica Sinica*, **81**, 287–296.

Deng, J., Wang, Q. F., Wan, L., Yang, L. Q., Zhou, L. & Zhao, J. 2008. The random difference of the trace element distribution in skarn and marbles from Shizishan orefield, Anhui Province, China. *Journal of China University of Geosciences*, **19**, 123–137.

EVERTSZ, C. J. G. & MANDELBROT, B. B. 1992. Multifractal measures. *In*: PEITGEN, H.-O., JUÈRGENS, H. & SAUPE, D. (eds) *Chaos and Fractals*. Springer-Verlag, New York, 849–881.

HALSEY, T. C., JENSEN, M. H., KADANOFF, L. P., PROCACCIA, I. & SHRAIMAN, B. I. 1986. Fractal measures and their singularities, the characterization of strange sets. *Physical Review*, **33**, 1141–1151.

HENTSCHEL, H. G. E. & PROCACCIA, L. 1983. The infinite number of generalized dimensions of fractals and strange attractors. *Physica Series D*, **8**, 435–444.

HURST, H. E., BLACK, R. P. & SIMAIKE, Y. M. 1965. *Long-Term Storage. An Experimental Study*. Constable, London.

MALAMUD, B. D. & TURCOTTE, D. L. 1999. Self-affine time series, I. Generation and analyses. *Advances in Geophysics*, **40**, 1–90.

MARKOWSKI, A., VALLANCE, J., CHIARADIA, M. & FONTBOTÉ, L. 2006. Mineral zoning and gold occurrence in the Fortuna skarn mine, Nambija district, Ecuador. *Mineralium Deposita*, **41**, 301–321.

NICOLESCU, S. & CORNELL, D. H. 1999. P-T conditions during skarn formation in the Ocna de Fier ore district, Romania. *Mineralium Deposita*, **34**, 730–742.

PAN, Y. M. & DONG, P. 1999. The lower Changjiang (Yangzi/Yangtze River) metallogenic belt, east central China: intrusion- and wallrock-hosted Cu–Fe–Au, Mo, Zn, Pb, Ag deposit. *Ore Geology Reviews*, **15**, 177–242.

TURCOTTE, D. L. 1997. *Fractals and Chaos in Geology and Geophysics*. Cambridge University Press, Cambridge.

VAN MARCKE DE LUMMEN, G. & VERKAEREN, J. 1986. Physicochemical study of skarn formation in pelitic rock, Costabonne peak area, eastern Pyrenees, France. *Contributions to Mineralogy and Petrology*, **93**, 77–88.

WANG, Q. F., DENG, J., WAN, L., YANG, L. Q. & LIU, X. F. 2007*a*. Fractal analysis of element distribution in Damoqujia gold deposit, Shandong Province, China. *In*: ZHAO, P. D. (ed.) *12th Conference of the International Association for Mathematical Geology*. Printed by China University of Geosciences Printing House, 262–265.

WANG, Q. F., DENG, J., WAN, L., YANG, L. Q. & GONG, Q. J. 2007*b*. Discussion on the kinetic controlling parameter of the stability of orebody distribution in altered rocks in the Dayingezhuang gold deposit, Shandong. *Acta Petrologica Sininca*, **23**, 590–593. [in Chinese with English abstract.]

WANG, Q. F., DENG, J. ET AL. 2008. Multifractal analysis of the element distribution in skarn-type deposits in Shizishan Orefield in Tongling area, Anhui province, China. *Acta Geologica Sinica*, **82**, 896–905.

Geology, petrology and alteration geochemistry of the Palaeoproterozoic intrusive hosted Älgträsk Au deposit, Northern Sweden

THERESE BEJGARN[1]*, HANS ÅREBÄCK[2], PÄR WEIHED[1] & JUHANI NYLANDER[2]

[1]*Division of Geosciences, Luleå University of Technology, SE-971 87 Luleå, Sweden*

[2]*Boliden Mineral AB, SE-936 81 Boliden, Sweden*

**Corresponding author (e-mail: therese.bejgarn@ltu.se)*

Abstract: The Älgträsk intrusive hosted Au deposit, Skellefte district, northern Sweden, is situated in the oldest, most heterogeneous part of the *c.* 1.89–1.86 Ga Jörn granitoid complex, which intruded a complex volcano–sedimentary succession in an island arc or continental margin arc environment. The Tallberg porphyry Cu deposit, situated only 3 km west of Älgträsk, is associated with quartz feldspar porphyritic dykes. These dykes are suggested to be genetically related to similar porphyry dykes in Älgträsk and the tonalitic host rock in Tallberg. The granodiorite hosting the Älgträsk Au-deposit does not appear to be genetically related to the tonalite or the porphyry dykes.

The mineralization in Älgträsk is structurally controlled and occurs within zones of proximal phyllic/silicic and distal propylitic alteration. It comprises mainly pyrite, chalcopyrite, sphalerite with accessory Te-minerals, gold alloys, and locally abundant arsenopyrite. During hydrothermal alteration an addition of Si, Fe and K together with an increase in Au, Te, Cu, Zn, As occurred. Sodic–calcic and quartz destructive alteration also characterize the deposit. It is suggested that the Älgträsk Au deposit is younger than the Tallberg porphyry deposit but older than the last deformation event in the area, and remobilization of gold during subsequent deformation can be of importance.

The Älgträsk deposit was discovered in the 1980s by Boliden Mineral AB and has over the years been subject to several drilling campaigns, and today the deposit comprises an indicated resource of 2.9 Mt @ 2.6 g/t and additional inferred resources of 1.3 Mt @ 1.8 g/t Au (NewBoliden 2010). The Tallberg porphyry Cu deposit (Weihed 1992*a*, *b*) situated only 3 km west of Älgträsk, and VMS (Volcanogenic massive sulphide) deposits in the Skellefte district south–SW of Älgträsk (Fig. 1) have been thoroughly investigated over the years (Rickard & Zweifel 1975; Lundberg 1980; Claesson 1985; Welin 1987; Wilson *et al.* 1987; Skiöld 1988; Weihed *et al.* 1992, 2002*a*, *b*; Duckworth & Rickard 1993; Claesson & Lundqvist 1995; Allen *et al.* 1996; Bergman Weihed *et al.* 1996; Billström & Weihed 1996; Kathol & Persson 1997; Lundström *et al.* 1997; Bergman Weihed 2001; Bergström 2001; Hannington *et al.* 2003; Barrett *et al.* 2005; Kathol & Weihed 2005; Årebäck *et al.* 2005; Montelius 2005; Skiöld & Rutland 2006; Montelius *et al.* 2007; Schlatter 2007). This is the first publication on the intrusive hosted Älgträsk Au deposit, which is situated in the southern part of the oldest, most heterogeneous, phase of the Jörn Granitoid Complex (JGC, Fig. 1). The Älgträsk area is dominated by a coarse grained quartz porphyritic granitoid, showing mingling relationships with gabbro and tonalite. The area is also crosscut by mafic dykes and felsic quartz porphyritic dykes, similar to the porphyry dykes associated with the Tallberg porphyry Cu deposit (Weihed 1992*b*). The aim of this paper is to discuss the petrogenetic aspects and geochemical characteristics of the different styles of alteration associated with the Älgträsk deposit.

Regional geology

The volcano–sedimentary succession into which the JGC intruded hosts more than 85 economical and sub-economical VMS deposits of similar Palaeoproterozoic age (Claesson 1985; Duckworth & Rickard 1993; Årebäck *et al.* 2005; Barrett *et al.* 2005), among which Renström, Maurliden and Kristineberg are producing mines (Fig. 1). It has been suggested that the Skellefte district constitutes the remnants of an ancient volcanic arc developed in a continental margin or island arc setting on the edge of the Fennoscandian shield during the early Proterozoic (Rickard & Zweifel 1975; Weihed *et al.* 1992; Allen *et al.* 1996). The complex volcano-sedimentary succession comprising the

From: Sial, A. N., Bettencourt, J. S., De Campos, C. P. & Ferreira, V. P. (eds) *Granite-Related Ore Deposits*. Geological Society, London, Special Publications, **350**, 105–132.
DOI: 10.1144/SP350.7 0305-8719/11/$15.00

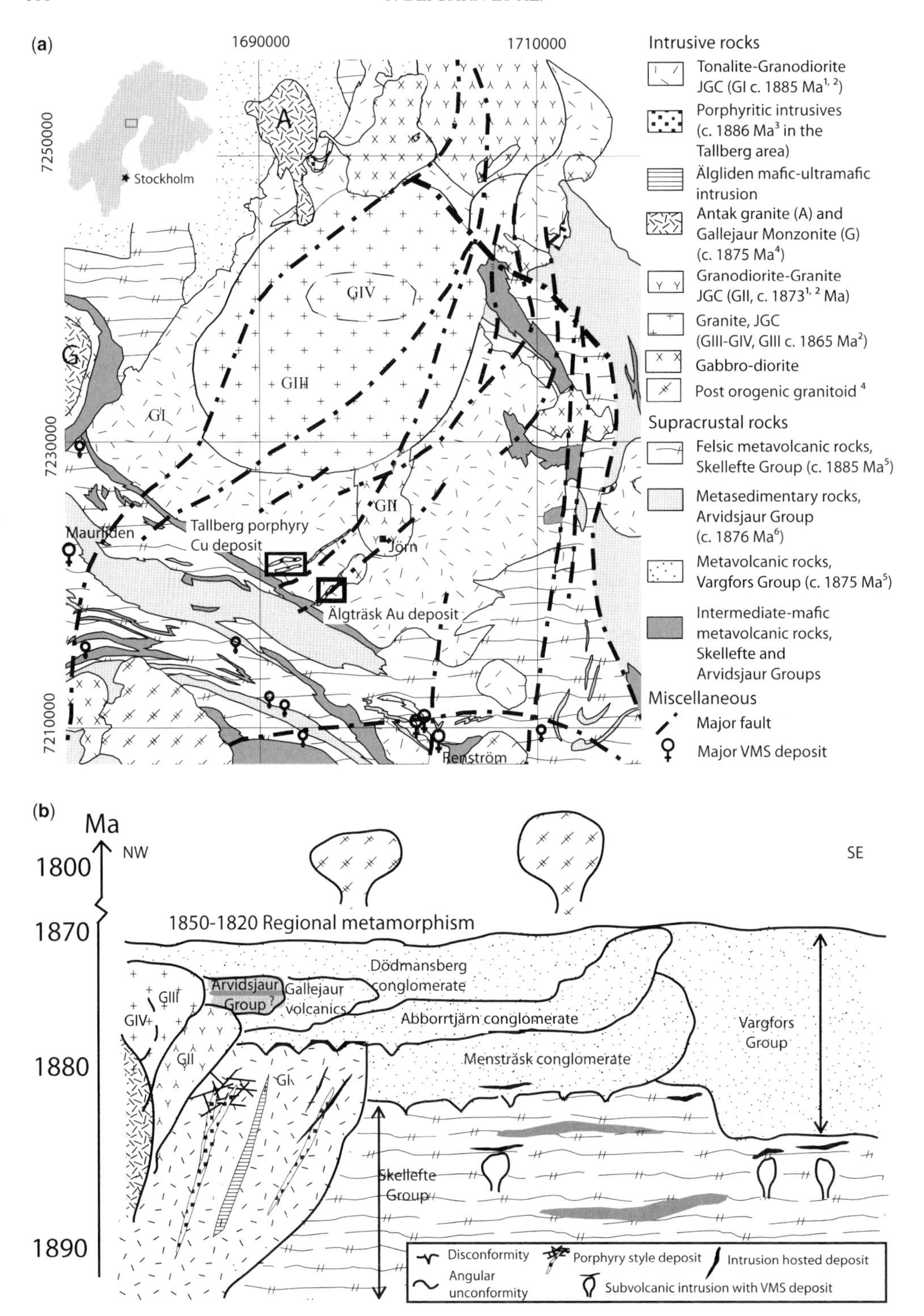

Fig. 1. (**a**) Geological map of the central part of the Skellefte district with major VMS deposits and the study areas indicated. Age references: [1]Wilson *et al.* (1987), [2]Gonzáles-Rondán (2010), [3]Weihed & Schöberg (1991), [4]Skiöld (1988), [5]Billström & Weihed (1996) and [6]Skiöld *et al.* (1993). (**b**) Simplified schematic stratigraphy in the Skellefte district (modified after Allen *et al.* 1996; Billström & Weihed 1996).

Skellefte district has historically been divided in to three major stratigraphic groups (cf. Kathol & Weihed 2005); the Skellefte, Vargfors and Arvidsjaur Groups (Fig. 1), with different interpreted ages of formation (Table 1).

The Skellefte Group is the lowest stratigraphical unit and is dominated by juvenile volcaniclastic rocks, lavas and porphyritic intrusions with intercalated sedimentary rocks such as mudstone, siltstone, sandstone and breccia–conglomerate (Allen *et al.* 1996). U–Pb zircon dating of volcanic rocks within the Skellefte Group yields an age of 1884 ± 6 Ma (Billström & Weihed 1996), and most of the VMS deposits in the Skellefte district are hosted by the upper part of the Skellefte Group (Fig. 1). The Vargfors Group is overlying the Skellefte Group and comprises mainly fine and coarse-grained clastic sedimentary rocks with locally abundant intercalated volcanic rocks (Allen *et al.* 1996). Zircons from the volcanic rocks within the Vargfors Group yields a U–Pb age of 1875 ± 4 Ma (Billström & Weihed 1996). The Arvidsjaur Group is characterized by subaerial, felsic to intermediate volcanic rocks such as volcaniclastic rocks, ash-fall tuff and ignimbrites yielding a U–Pb zircon age of 1876 ± 3 Ma (Skiöld *et al.* 1993). The Arvidsjaur Group has been interpreted as a subaerial equivalent to the Skellefte Group or a lateral equivalent to the Vargfors Group (Allen *et al.* 1996).

The volcano sedimentary succession was intruded by multiple phases of the JGC at *c.* 1.89–1.86 Ga (Wilson *et al.* 1987; González-Roldán 2010) and by the Gallejaur gabbro-monzonite suite to the west of the JGC (Fig. 1) at *c.* 1.87 Ga (Skiöld 1988). Magma mixing and mingling have formed hybrid rocks between the monzonite and gabbro, and mafic microgranular enclaves are common. Mafic dykes cutting the JGC are tentatively correlated with the Gallejaur magmatism (Kathol & Weihed 2005).

The JGC represent I-type, calc-alkaline, early orogenic granitoids evolved from at least three different initial magmas, derived from a mantle source depleted in light rare earth element (LREE) (Wilson *et al.* 1987). The oldest phase, the GI, occurs in the margins of the batholith and is heterogeneous, but dominated by a coarse grained granodiorite–tonalite composed of quartz, oligoclase, biotite, hornblende, microcline and accessory epidote, apatite, titanite, zircon and opaque minerals. The GI has been dated by the U–Pb method on zircon at 1888 + 40/−14 Ma (Wilson *et al.* 1987) and at 1886 ± 3, 1880 ± 4 and 1885 ± 5 Ma (González-Roldán 2010). The younger units, GII–GIV (Fig. 1) are more felsic in character with granodioritic–granitic compositions (Wilson *et al.* 1987; González Roldán *et al.* 2006) and have been dated by the U–Pb zircon method at 1874 + 45/−26 Ma (GII), 1873 + 18/−14 Ma (GIII) (Wilson *et al.* 1987) and 1874 ± 6 Ma (GII), 1871 ± 4 (GII) and 1863 ± 5 (GIII) (González-Roldán 2010).

Clasts of the GI type in the Abborrtjärn conglomerate indicate that the GI was uplifted and eroded prior to the deposition of the sedimentary rocks in the Vargfors Group (Fig. 1b). The GI was subsequently intruded by the GII–GIV phases, which caused metamorphism, hydrothermal alteration and deformation, not observed within the younger suites (González Roldán *et al.* 2006). Many authors have suggested that the Skellefte Group volcanic rocks and JGC are comagmatic because of similar chemical compositions and similar ages (Lundberg 1980, Claesson 1985, Wilson *et al.* 1987).

The central Skellefte district has experienced at least two major phases of folding; an older phase (D_2) that formed as a response to SW–NE shortening, started at *c.* 1.87 Ga and reached peak metamorphic conditions at *c.* 1.82 Ma, forming tight to isoclinal upright folds with variably plunging fold axis (Claesson & Lundqvist 1995, Romer & Nisca 1995, Bergman Weihed 2001). NW striking shear zones are generally correlated with the D_2 event since they do not affect rocks younger than 1.80 Ga (Weihed *et al.* 1992, Bergman Weihed 2001). A younger deformation phase (D_3) created open folds with steep north to NE striking axial surfaces and fold axis that are coaxial with earlier folds (Weihed *et al.* 1992, Bergman Weihed 2001). The NW striking shear zones are often overprinted by D_3 crenulations, while NNE-trending ductile shear zones related to the D_3 event affect rocks younger than 1.80 Ga (Bergman Weihed 2001). The Vidsel-Röjnoret Shear System situated along the eastern contact of the JGC belongs to the D_3 structures and has a preliminary titanite age of 1.79 Ga (cf. Bergman Weihed 2001).

The older margin of the JGC hosts several different style of mineral deposits, for example, the Tallberg porphyry Cu deposit situated *c.* 3 km NW of the Älgträsk deposit (Weihed *et al.* 1987; Weihed 1992*b*). The Tallberg deposit is hosted by a medium grained tonalite belonging to the GI, which was intruded by several quartz-feldspar porphyritic dykes associated with disseminated and stockwork style mineralization within propylitic and sericitic alteration haloes in the surrounding GI tonalite (Weihed 1992*b*).

Methods

Drill cores from the Älgträsk and, to some extent, also the Tallberg area, have been logged and sampled to characterize the different styles of alteration and rock types observed. The samples were

Table 1. *Age determinations from the Skellefte district. Selected age determinations from the Skellefte, Arvidsjaur and Vargfors Groups together with age data from the Jörn Granitoid Complex (JGC) and other intrusive suites in the Skellefte district*

Name	Rock type	Age	Precision	Methods	Reference
Extrusive suites					
Skellefte Group	Dacite	1869	±15	Pb–Pb zr	Bergman Weihed *et al.* (1996)
Skellefte Group	Rhyolite (quartz porphyry)	1882	±8	U–Pb zr	Welin (1987)
Skellefte Group	Mass flow	1889	±4	U–Pb zr	Billström & Weihed (1996)
Skellefte Group	Rhyodacite/peprite	1884	±6	U–Pb zr	Billström & Weihed (1996)
Skellefte Group	Mass flow	1847	±3	U–Pb zr	Billström & Weihed (1996)
Skellefte Group	Rhyolite (fsp porphyry)	1885	±3	U–Pb zr	Montelius (2005)
Skellefte Group	Rhyolite (pumice unit)	1902	±21	U–Pb zr	Montelius (2005)
Skellefte Group	Rhyolite (qtz-fsp porphyry)	1901	±3	U–Pb zr	Montelius (2005)
Skellefte Group	Rhyolite (qtz-fsp porphyry with pumice)	1885	±6	U-Pb zr	Montelius (2005)
Vargfors Group, Gallejaur Volcanics	Felsic welded ignimbrite	1875	±4	U–Pb zr	Billström & Weihed (1996)
Arvidsjaur Group	Rhyolite	1876	±3	U–Pb zr	Skiöld *et al.* (1993)
Intrusive suits					
GI (JGC)	Granodiorite	1888	+20/−14	U–Pb zr	Wilson *et al.* (1987)
GI (JGC)	Granodiorite	1886	±3	U–Pb zr	Gonzáles Roldán (2010)
GI (JGC)	Granodiorite	1885	±5	U–Pb zr	Gonzáles Roldán (2010)
GI (JGC)	Granodiorite	1880	±4	U–Pb zr	Gonzáles Roldán (2010)
Tallberg porphyry	Diorite/tonalite	1886	−15/−9	U–Pb zr	Weihed & Schöberg (1991)
Antak	Granite	1879	+15/−12	U–Pb zr	Kathol & Persson (1997)
Gallejaur	Gabbro	1876	±4	U–Pb zr	Skiöld *et al.* (1993)
Gallejaur	Monzonite	1873	±10	U–Pb zr	Skiöld (1988)
GII (JGC)	Granodiorite	1874	+48/−26	U–Pb zr	Wilson *et al.* (1987)
GII (JGC)	Granodiorite–Granite	1874	±6	U–Pb zr	Gonzáles Roldán (2010)
GII (JGC)	Granodiorite–Granite	1871	±4	U–Pb zr	Gonzáles Roldán (2010)
GIII (JGC)	Granite	1873	+18/−14	U–Pb zr	Wilson *et al.* (1987)
GIII (JGC)	Granite	1863	±5	U–Pb zr	Gonzáles Roldán (2010)
GIII (JGC)	Granite	1862	±14	Pb–Pb ttn	Lundström *et al.* (1997)
Revsund	Granite	1778	±16	U–Pb zr	Skiöld (1988)

analysed by Acme Analytical Laboratories (Vancouver) Ltd., Canada by Inductively Coupled Plasma Emission Spectrometry (ICP-ES) and Inductively Coupled Plasma Mass Spectrometry (ICP-MS) for major and trace elements respectively. Analytical data of representative samples are presented in Table 2. Major elements and trace elements have been used to determine rock type, affinity and to visualize and calculate chemical changes within different alteration systems. Rare earth element (REE) data from the different igneous units have been plotted in chondrite normalized diagrams (normalized to chondrite values of Nakamura 1974) to visualize trends and signatures. Data were plotted using Geochemical Data toolkit (Janoušek *et al.* 2006).

The Jörn Granitoid Complex and related intrusive suites in the Älgträsk and Tallberg areas

The oldest, *c.* 1.89 Ga, GI phase of the JGC dominates the Älgträsk and Tallberg areas (Figs 2 & 3). The GI is known to be heterogeneous in composition in other areas (cf. Wilson *et al.* 1987), ranging from gabbro to granodiorite with a tholeiitic to calc-alkaline character (Fig. 4). In drillcores from the Älgträsk area, three different intrusive phases have been identified; gabbro, tonalite and granodiorite–granite, all medium to coarse-grained (Figs 4a & 5), though the area is dominated by a coarse grained, often quartz-porphyritic granodiorite, while the gabbro occur in a larger body north–NE of the Älgträsk deposit (Fig. 2). The Tallberg area is dominated by a medium grained tonalite, similar to the tonalite occurring in minor areas in Älgträsk. Both granodiorite and tonalite in the respective areas are cut by numerous felsic quartz feldspar porphyry (QFP) and mafic dykes (Figs 2 & 3). The QFP dykes in the Älgträsk and the Tallberg area share characteristics and are commonly *c.* 10 m wide, but dykes as wide as *c.* 30 m have been encountered. Aplitic dykes are of sparse occurrence and commonly less than one metre wide, with a steep dip and north–NE strike. The mafic dykes are of multiple generations with varying strike and dip, though crosscutting all other lithologies.

The Tallberg porphyry Cu deposit, hosted by the tonalite NW of Älgträsk (Figs 1 & 2), is associated with the intrusion of steeply dipping, east–NE striking dacitic QFP dykes (Weihed *et al.* 1987). The deposit comprises disseminated and quartz vein stockwork hosted pyrite, chalcopyrite and molybdenite with minor electrum $\pm$ telluride minerals associated with mainly propylitic alteration (Weihed 1992*b*). Phyllic alteration zones, commonly deformed and in places barren with respect to sulphide minerals, frequently occurs along porphyry dykes in the southern Tallberg area. Immediately E–NE of the Tallberg area, the Älgliden mafic-ultramafic intrusion occurs with a dyke-like geometry (Figs 1 & 2). It is *c.* 50 m wide, with a NE strike and a steep dip, and can be traced for *c.* 2 km. The lower part of the dyke contains an up to 0.5 m wide semi-massive sulphide lens, containing a Cu and Ni mineralization. The Älgliden intrusion cuts a molybdenite and chalcopyrite bearing quartz-stockwork associated with the Tallberg mineralization, but is in turn cut by unmineralized mafic dykes.

Igneous petrology of the Älgträsk area

The GI phase of the JGC in the Älgträsk area is dominated by a coarse grained granodiorite–granite (Figs 4a & 5a) containing quartz, plagioclase, 5–15 vol% K-feldspar, biotite, amphibole with accessory apatite, calcite, magnetite, hematite, epidote, sericite and calcite. It is commonly quartz porphyritic with concentrically zoned and twinned plagioclase, often rimmed by K-feldspar. The least altered samples are slightly altered with sericite dusting of feldspars and biotite alteration of amphibole and chlorite alteration of ferromagnesian minerals. The quartz-porpyritic granodiorite (QPG) in Älgträsk displays a relatively gentle slope for the LREE, a distinct negative Eu anomaly and a relatively flat heavy rare earth elements (HREE) pattern (Fig. 6a), and has the most fractionated signature of the major plutonic rocks in the Älgträsk area.

The QPG frequently contain mafic microgranular enclaves (MME, Fig. 5b). The enclaves vary from centimetre to metres in size and often have bulging, fine-grained plagioclase rich contacts to the QPG. The MMEs comprise hornblende, biotite, plagioclase, K-feldspar and quartz with accessory apatite, zircon, magnetite, pyrite, and rutile. The REE signatures of the MME (Fig. 6b) mimics the signatures for the least altered QPG, but show a slightly more fractionated REE pattern.

In Älgträsk, a rock with tonalitic composition (dioritic–granodioritic, Fig. 4a) occurs as metre sized enclaves within the quartz porphyritic granodiorite and as a rock of hybrid character close to the contact to the gabbro with mingling textures with the QPG. The tonalite is typically granular and medium grained, comprising plagioclase, quartz, biotite, hornblende with minor (0–5 vol%), K-feldspar (hence tonalite and not diorite, Fig. 5c) with accessory apatite, rutile, magnetite and zircon. The tonalite lacks the macroscopic quartz porphyritic texture of the typical Älgträsk granodiorite and commonly contains less K-feldspar and quartz. Alteration where plagioclase is replaced by

Table 2. *Whole rock geochemical data. Whole-rock data of selected representative samples for the different rock types in the study area and the different styles of alteration within the quartz porphyritic granodiorite.*

Major oxide/ element	Detection limit	146–89 QPG, Älgträsk	27–77 Tonalite, Älgträsk	22–104 Tonalite, Tallberg	48–45 Gabbro, Älgträsk	169–33 QFP, Tallberg	58–38 QFP, Älgträsk	55–130 Aplite	47–80 Älgliden intrusion	99–62 High-Ti mafic dyke	118–45 Low-Ti mafic dyke	83–126 Propylitic alteration	83–125 Phyllic alteration	0701 Silicic alteration	265–84 Sodic alteration	65–140 Quartz-destructive alteration
SiO_2	0.01	71.79	66.71	64.63	51.19	68.55	69.21	76.99	41.90	48.61	48.7	70.68	69.49	56.04	71.08	54.69
Al_2O_3	0.01	13.61	16.23	14.01	16.64	15.05	15.40	12.18	7.68	15.18	14.2	13.46	10.65	2.59	14.11	19.52
$Fe_2O_3^T$	0.04	3.66	3.78	7.61	8.87	2.56	3.04	0.91	19.93	11.60	9.13	3.83	10.16	25.99	1.48	8.03
MgO	0.01	0.79	1.07	2.92	5.50	1.00	1.02	0.18	18.94	5.85	8.7	0.80	1.33	0.14	0.93	1.50
CaO	0.01	3.26	4.72	4.63	8.13	3.25	3.73	2.25	3.93	7.96	8.51	2.35	0.58	0.06	6.14	2.29
Na_2O	0.01	3.35	4.17	2.23	1.39	4.82	4.50	4.78	0.95	2.35	3.01	2.82	0.30	0.04	4.52	7.37
K_2O	0.01	2.01	1.03	1.09	1.32	0.55	1.05	0.93	0.53	1.02	1.27	2.82	2.83	1.41	0.16	1.67
TiO_2	0.01	0.23	0.26	0.41	0.56	0.21	0.23	0.06	0.71	1.72	0.71	0.21	0.21	0.08	0.32	0.28
P_2O_5	0.01	0.027	0.12	0.09	0.11	0.07	0.08	0.004	0.14	0.55	0.265	0.06	0.06	0.01	0.06	0.04
MnO	0.01	0.06	0.05	0.07	0.24	0.03	0.06	0.02	0.22	0.18	0.19	0.10	0.14	0.02	0.01	0.09
Cr_2O_3	0.02	bd	bd	0.005	0.005	bd	0.003	bd	0.306	0.022	0.081	0.002	0.003	0.004	bd	bd
Ni	20	25	bd	bd	21	bd	bd	bd	685.0	60	172	bd	bd	bd	bd	bd
Sc	1	12	5	27	26	4	5	2	17	28	28	11.0	11.0	7.0	12.0	17.0
LOI	0.1	1.1	1.7	2.2	5.8	3.8	1.5	1.6	4.2	4.7	5	2.8	4.1	13.6	1.0	4.4
Sum	0.01	99.89	99.81	99.93	99.70	99.86	99.86	99.88	99.48	99.76	99.76	99.92	99.89	99.95	99.85	99.86
Ba	1	508	389	261	324	334	492	443	189	225	325	575.0	506.0	202.0	83.0	381.0
Be	1	bd	bd	1	bd	bd	2	bd	1	2	bd	1	bd	bd	bd	2
Co	0.2	6.0	3.8	10.6	26.8	2.4	5.0	2.3	126.1	36.8	37.9	4.2	8.3	58.3	1.6	8.3
Cs	0.1	1.2	0.4	1.1	0.6	0.5	1.0	0.3	1.6	3.0	1.8	1.0	1.5	0.2	0.2	0.8
Ga	0.5	12.6	17.7	16.2	17.0	16.8	17.5	11.4	10.9	19.0	15.7	12.9	14.3	4.2	13.5	18.5
Hf	0.1	4.1	2.4	1.5	1.3	1.9	2.2	3.3	1.2	3.2	1.8	3.8	3.7	0.3	4.7	5.4
Nb	0.1	5.9	3.4	3.7	2.7	2.6	3.2	4.2	3.2	9.8	3.5	6.2	6.2	0.8	6.5	7.6
Rb	1	41.3	14.8	25.2	23.6	10.2	17.9	15.5	12.3	43.4	31.6	50.7	59.2	13.8	2.7	29.0
Sn	0.5	1	bd	bd	bd	bd	bd	bd	1	bd	bd	2	6	2	bd	6
Sr	0.5	145.3	508.0	205.6	414.0	508.6	591.0	131.3	192.0	348.2	452.0	100.8	22.3	6.9	265.9	233.7
Ta	0.1	0.5	0.2	0.2	0.2	0.1	0.3	0.3	0.2	0.5	0.2	0.7	0.6	0.1	0.5	0.7
Th	0.2	8.1	1.2	2.7	1.5	1.2	1.3	3.3	0.8	1.4	2.4	5.9	4.2	bd	6.7	7.9
U	0.1	4.6	1.0	2.1	0.9	1.0	0.9	2.9	0.9	0.7	1.8	5.4	5.0	0.2	2.2	6.3

V	8	43	32	155	195	28	32	bd	165	212	179	39	39	38	32	60
W	0.5	0.7	bd	0.7	0.8	1.3	0.6	0.9	1.3	bd	0.7	1.8	7.8	6.3	bd	1.8
Zr	0.1	136.2	83.8	45.0	39.1	56.5	70.9	71.8	38.6	126.2	61.2	110.7	110.1	11.6	157.9	160.7
Y	0.1	18.6	9.2	12.9	11.2	4.8	5.6	16.8	9.6	25.2	11.9	23.3	20.3	2.4	28.2	29.9
La	0.1	29.3	9.2	9.3	8.4	7.8	9.3	13.9	7.4	21.4	17.7	12.7	8.7	1.0	31.4	29.0
Ce	0.1	53.7	19.2	17.7	17.9	16.4	19.0	30.8	15.4	49.6	39.7	27.1	17.9	2.0	60.5	56.9
Pr	0.02	5.9	2.47	2.18	2.40	2.09	2.26	3.96	2.07	6.93	5.62	3.48	2.28	0.24	6.94	6.54
Nd	0.3	20.6	9.8	9.1	10.2	7.7	8.7	14.8	9.3	29.6	23.6	13.4	9.0	0.9	24.1	25.2
Sm	0.05	3.43	1.85	1.79	2.19	1.40	1.48	2.45	1.92	6.04	4.68	3.01	1.90	0.23	4.76	4.39
Eu	0.02	0.40	0.54	0.45	0.64	0.40	0.40	0.11	0.58	2.02	1.13	0.54	0.26	0.03	0.91	1.16
Gd	0.05	3.03	1.67	1.88	2.01	1.10	1.09	2.22	1.87	5.78	3.71	2.99	2.06	0.22	4.37	4.22
Tb	0.01	0.41	0.27	0.34	0.35	0.16	0.17	0.35	0.31	0.90	0.41	0.55	0.41	0.05	0.74	0.76
Dy	0.05	3.23	1.47	2.02	1.93	0.87	0.96	2.49	1.60	4.86	2.44	3.44	2.88	0.39	4.43	4.55
Ho	0.02	0.66	0.31	0.43	0.41	0.17	0.18	0.54	0.33	0.93	0.4	0.73	0.66	0.08	0.92	0.99
Er	0.03	2.01	0.86	1.35	1.17	0.41	0.56	1.62	1.00	2.58	1.02	2.33	2.18	0.24	2.88	3.07
Tm	0.01	0.25	0.15	0.21	0.19	0.07	0.09	0.26	0.16	0.38	0.14	0.41	0.39	0.04	0.46	0.52
Yb	0.05	2.22	0.98	1.40	1.14	0.46	0.57	1.83	0.96	2.32	0.95	2.64	2.45	0.28	3.00	3.30
Lu	0.01	0.33	0.14	0.21	0.18	0.08	0.09	0.29	0.15	0.34	0.15	0.43	0.41	0.04	0.48	0.54
C	0.01	0.05	0.13	0.08	0.64	0.07	0.18	0.30	0.04	0.47	0.61	0.28	0.05	bd	0.16	0.23
S	0.02	0.08	0.04	0.10	0.16	0.04	bd	0.07	0.97	0.05	0.03	0.09	3.11	19.33	0.06	3.42
Mo	0.1	2.5	0.7	1.4	0.2	1.0	0.4	3.9	4.6	0.8	0.6	0.7	1.2	22.9	0.4	2.4
Cu	0.1	6.2	24.6	64.3	52.6	61.0	6.5	24.2	923.1	51.8	69.2	18.4	93.8	157.1	3.4	3.4
Pb	0.1	3.8	0.8	2.0	2.6	2.3	1.9	1.5	4.3	3.2	5.4	1.9	5.6	35.6	2.2	3.6
Zn	1	36	35	69	266	53	55	2	59	122	113	43.0	84.0	89.0	20.0	50.0
Ni	0.1	1.3	1.5	7.6	21.5	6.8	6.1	0.9	618.8	55.1	112.4	1.2	1.5	4.2	0.5	1.5
As	0.5	3.7	2.8	3.3	7.7	5.7	1.8	3.8	bd	52.1	18.2	4.3	22.3	300.6	1.6	4.4
Cd	0.1	bd	bd	bd	bd	bd	bd	bd	0.2	bd	bd	bd	bd	0.2	bd	bd
Sb	0.1	0.2	0.2	0.4	0.3	0.4	0.3	0.4	bd	0.3	0.5	0.4	0.4	2.0	0.3	0.5
Bi	0.1	bd	bd	bd	bd	bd	bd	bd	0.2	bd	bd	bd	1.5	7.2	bd	0.2
Ag	0.1	bd	bd	bd	0.1	bd	bd	bd	0.6	0.1	0.2	bd	1.5	3.1	bd	0.2
Au	0.5	bd	bd	4.5	3.6	2.5	5.3	5.0	64.7	0.9	1.6	14.8	1456	2157	bd	86.9
Hg	0.01	bd	bd	0.02	0.16	bd	bd	0.01	0.01	bd	bd	0.01	0.02	0.15	0.02	0.03
Tl	0.1	0.2	bd	0.1	bd	bd	bd	bd	0.8	0.5	0.4	bd	0.2	bd	bd	bd
Se	0.5	bd	bd	bd	bd	bd	bd	bd	1.2	bd	bd	bd	0.5	3.5	bd	bd
Te	1	bd	bd	bd	bd	bd	bd	bd	bd	bd	bd	bd	2	12	bd	bd

Note: Major oxides are reported in wt%, trace elements in ppm and Au in ppb. Bd, below detection; QPG, quartz porphyritic granodiorite; QFP, quartz feldspar porphyry.

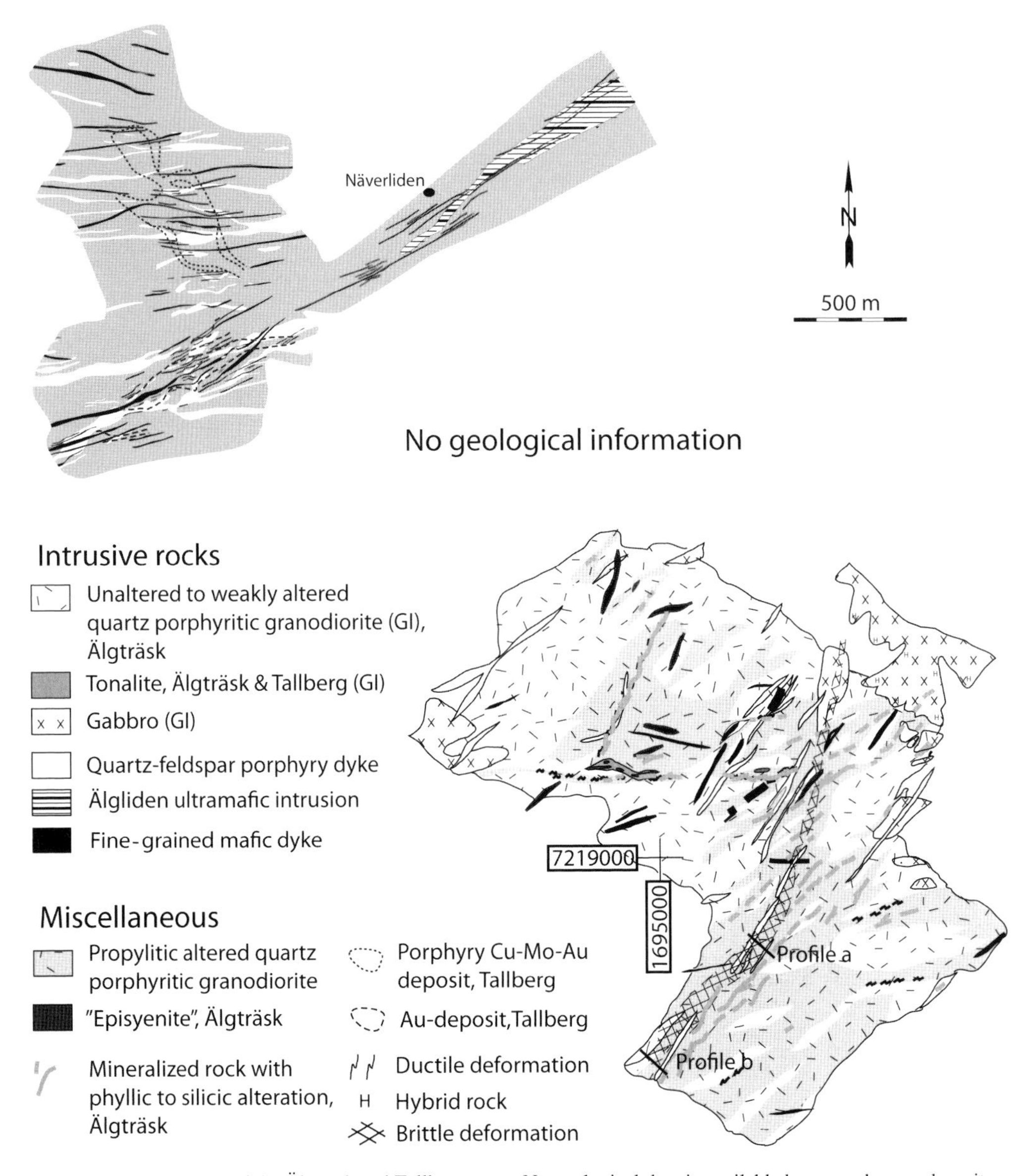

Fig. 2. Geological map of the Älgträsk and Tallberg areas. No geological data is available between the two deposits. Geological map of the Tallberg deposit modified after Weihed *et al.* (1992). Profiles are presented in Figure 3.

sericite ± epidote and biotite and hornblende by chlorite is common. Macroscopically, the tonalite and hybrid rock in Älgträsk share many similarities with the tonalite in Tallberg (Fig. 5c, d). The REE-pattern (Fig. 6d) of the tonalite in the Älgträsk area displays a rather steep negative slope in the LREE without a negative Eu anomaly and a relatively flat HREE pattern, similar to the tonalite in Tallberg (Fig. 6e, f), although the latter is more enriched in HREE. The Näverliden tonalite (Fig. 6g) immediately north of the Tallberg deposit, is generally less fractionated than other tonalites, whereas the Tallberg tonalite data overlap with the other groups.

Many gabbros in the Älgträsk area are altered, which makes the identification of primary mineral assemblages more problematic. The gabbro in the northeastern part of the Älgträsk area (Fig. 2) comprises mainly amphibole and epidote (often zoisite) ±quartz ± biotite/chlorite pseudomorphing olivine, pyroxene and/or plagioclase (Fig. 5e). The REE signature of the gabbro (Fig. 6c) shows a large spread, probably also due to alteration.

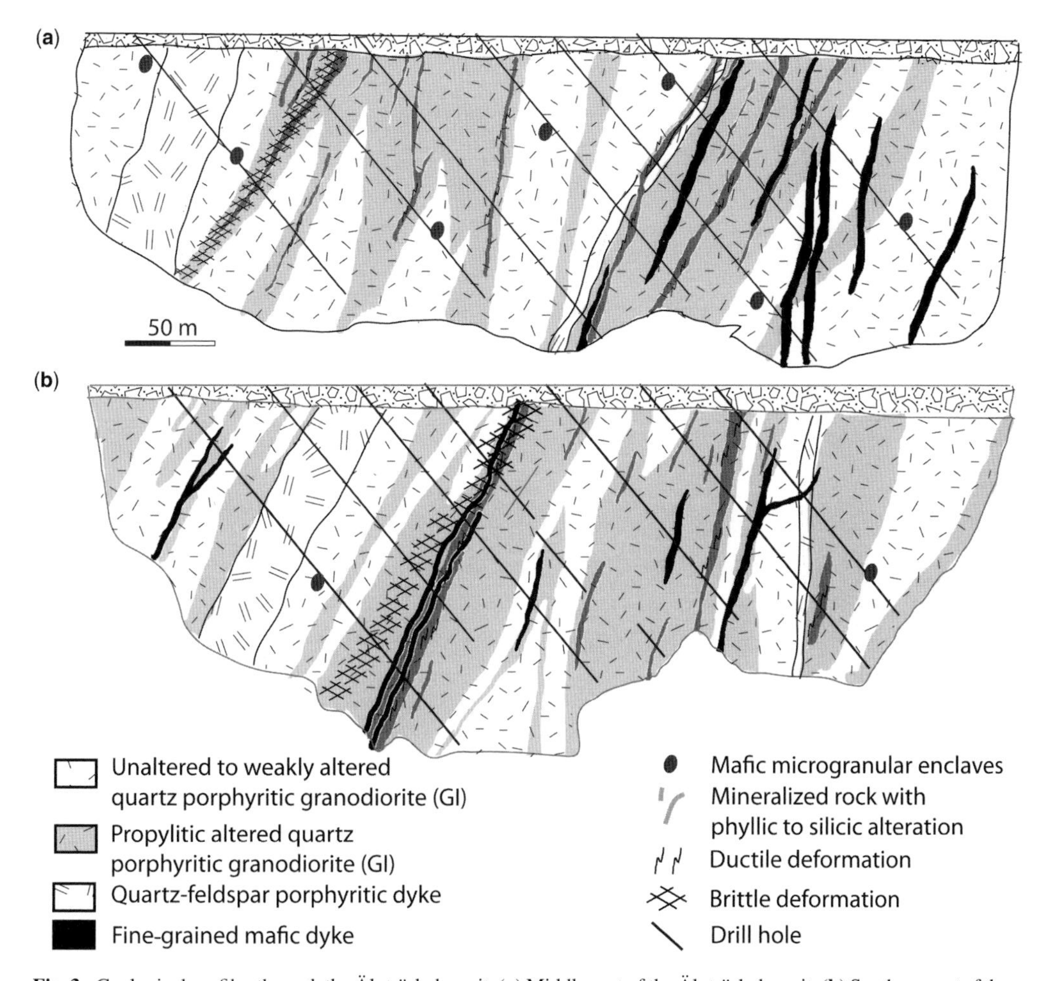

Fig. 3. Geological profiles through the Älgträsk deposit. (**a**) Middle part of the Älgträsk deposit. (**b**) Southern part of the Älgträsk deposit. Location shown in Figure 2.

However, it generally displays a gentle negative LREE slope, no distinct Eu anomaly and a relatively flat HREE trend.

The Tallberg porphyry Cu deposit comprising disseminated and quartz vein hosted chalcopyrite and molybdenite, with minor electrum and telluride minerals (Weihed 1992*b*), is associated with QFP dykes that intruded a tonalite of the GI phase (Weihed *et al.* 1987). Dykes with similar characteristics as in Tallberg also occur in the Älgträsk area with a general NE–SW strike direction and a steep dip (Figs 2 & 3). The dykes in Älgträsk are up to 30 m wide and have a distinct chilled contact to the wall rock, a fine grained matrix, mainly composed of quartz and minor feldspars, and phenocrysts (>20%) aligned sub-parallel to the contact. The phenocrysts are generally up to 5 mm large, constituting about 65% plagioclase and 35% quartz, with accessory biotite and hornblende (Fig. 5f). The plagioclase is commonly euhedral, zoned and complexly twinned while the quartz is subhedral. There is also a group of porphyries where plagioclase dominates the phenocrysts composition, with only 0–5% quartz phenocrysts. A general feature of the QFP dykes is secondary sericite ± epidote replacing plagioclase and epidote + chlorite replacing sub-euhedral amphibole, similar to what is observed in the least altered QPG. The REE pattern of the QFP displays a rather steep LREE slope, no distinct Eu anomaly and a relatively flat to slightly positive HREE slope. This REE pattern is compared to previously published data (Weihed 1992*b*) in Fig. 6e, where the QFP of both the Tallberg and Älgträsk deposits show similar REE signatures. One sample from a Tallberg QFP dyke, sampled during this study,

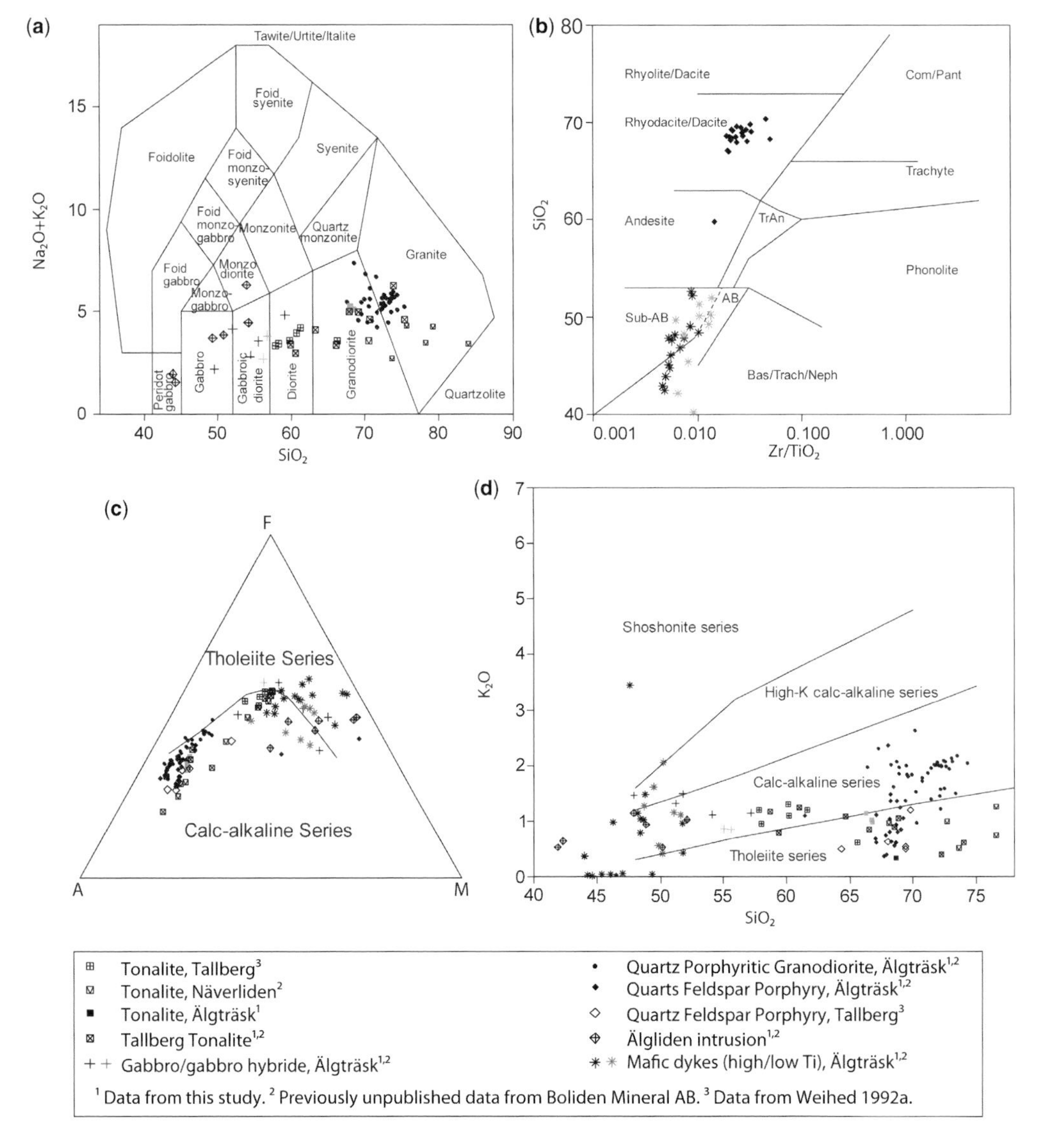

Fig. 4. (**a**) $Na_2O + K_2O$ v. SiO_2 classification diagram for the dykes in the Älgträsk area. (**b**) Winchester & Floyd (1977) classification diagram with quartz feldspar porphyry and mafic dykes. (**c**) AFM diagram for all rock types in the Älgträsk and Tallberg areas. (**d**) K_2O v. SiO_2 classification diagram, for all rock types in the Älgträsk and Tallberg areas. [1]Data from this study. [2]Unpublished data from Boliden Mineral AB. [3]Data from Weihed (1992*b*).

shows similar petrographic characteristics as the dykes in Älgträsk, but is slightly more depleted in LREE compared to other samples from Tallberg and Älgträsk (Fig. 6h, i). The QFP dykes in Tallberg have been dated at 1886 + 15/ − 9 Ma by U–Pb in zircon (Weihed & Schöberg 1991).

The Älgliden mafic-ultramafic intrusion is situated between the Tallberg and Älgträsk areas (Fig. 2). The intrusion shows a dyke-like geometry, strikes NE, is *c.* 2 km long and about 50 m wide with a steep dip. The rock is a medium grained olivine gabbro, ultramafic to mafic in character (Fig. 4a) and comprises mainly plagioclase, orthopyroxene, olivine, hornblende, and biotite with accessory baddeleyite and apatite. Olivine crystals are fractured and commonly replaced by serpentine along fractures. The intrusion contains abundant disseminated pyrrhotite, chalcopyrite and magnetite with minor pentlandite, pyrite and sphalerite. Pyrrhotite generally encloses grains of magnetite and pseudomorphs magnetite along grain-boundaries. A similar sulphide mineral assemblage is found in a 30–50 cm wide massive sulphide lens, commonly situated within the middle-lower part of the intrusion. A few metres from the intrusions contacts, disseminated pyrrhotite gives way to disseminated pyrite. The REE-signature for the Älgliden dyke (Fig. 6j) is variable, but strongly fractionated with no obvious Eu anomaly.

Mafic dykes with variable strike directions are common in the area (Fig. 2). The dykes are fine grained and both non-phyric and phenocryst-rich types occur. In most dykes, the primary minerals are overprinted by fine grained alteration assemblages of calcite + epidote + chlorite $\pm$ hornblende $\pm$ quartz. Since the dykes are very fine grained, and also altered, they are classified in a Winchester & Floyd (1977) diagram based on immobile elements and SiO_2 (Fig. 4b). The dykes plot as sub-alkaline basalts to andesitic basalts. Two groups of dykes can be distinguished based on textural and chemical characteristics: (i) clinopyroxene-phyric with 0.6–1.1 wt% TiO_2 (Fig. 5h) and (ii) plagioclase phyric dykes with disseminated titanomagnetite (skeletal) and 1.5–1.7 wt% TiO_2 (Fig. 5i). The latter displays variable chlorite and/or sericite alteration of plagioclase phenocrysts. In a chondrite normalized REE diagram (Fig. 6l) it shows a fractionated and generally a gently dipping trend without any distinct anomalies. Many of the dykes carry up to 3 mm rounded calcite grains rimmed by quartz (Fig. 5i) interpreted as amygdules.

The aplitic dykes are fine grained and generally comprise albite, K-feldspar, quartz and biotite (Fig. 5j) with a characteristic myrmekitic texture. Aplites have only been observed in the quartz porphyritic granodiorite. REE signatures of the aplitic dykes are similar to the QPG, but with a more pronounced negative Eu anomaly (Fig. 6k).

The rocks in the area generally lack any pronounced tectonic fabric. Ductile deformation zones are in most places coincident with strongly altered $\pm$ mineralized rocks, but also locally occur in mafic dykes. Later brittle deformation is in some places associated with a displacement of rock units.

The Älgträsk Au deposit

The mineralization in the Älgträsk area is situated in the southern part of the JGC and is mainly hosted within the QPG belonging to the oldest intrusive phase (GI) of the JGC. Mineralization is mainly structurally controlled, north–east striking, and composed of weak to strong dissemination and veins of sulphides within zones of propylitic, phyllic and silicic alteration. The mineralized zones comprise mainly pyrite with accessory chalcopyrite, sphalerite, Au (electrum), and Te-minerals such as hessite and calaverite. The alteration zones are commonly deformed in a ductile manner, with pronounced lineation of alteration assemblages and primary igneous minerals. The mineralized zones crosscut both the gabbro and quartz feldspar porphyry dykes in the Älgträsk area and are associated with a similar style of alteration as in the QPG. The mineralization is cut by the mafic dykes, which in places, are also deformed. A sodic–calcic alteration and a quartz destructive alteration have also been encountered, sharing at least a spatial relationship with the mineralized zones. The alteration types described below have been characterized based on chemical and mineralogical composition. Photographs and key features of the different styles of alterations are presented in Figure 7 and REE diagrams of the different styles of alteration in Figure 8.

Hydrothermal alteration types in the Älgträsk area

Based on core logging and thin section studies, six different types of hydrothermal alteration have been identified in the Älgträsk area: (1) feldspar destructive; (2) propylitic; (3) phyllic; (4) silicic; (5) sodic–calcic; and (6) quartz-destructive alteration. Of these alteration types, the feldspar-destructive alteration is most widespread; the propylitic, phyllic and silicic alteration is associated with the gold mineralization in Älgträsk, while the sodic–calcic and quartz destructive alteration are only known to share a spatial relationship with the mineralization. However, feldspar destructive to propylitic alteration has been observed in other parts of the JGC, and is interpreted as an expression of regional metamorphism (Wilson *et al.* 1987; González-Roldán 2010). The styles of alteration associated with mineralization may be simultaneously developed in different parts of the systems, representing a coeval evolution during mineralization, not necessarily describing a temporal evolution. The styles of alteration are hence described in terms of progressive intensity towards gold mineralization. However, the Na–Ca and the quartz destructive alteration will be described at the end due to their uncertain temporal relationship with mineralization.

Feldspar destructive alteration. The textural relationships in what is defined as the least altered granodiorite (Fig. 6a) suggest that selective replacement of plagioclase by sericite $\pm$ epidote has

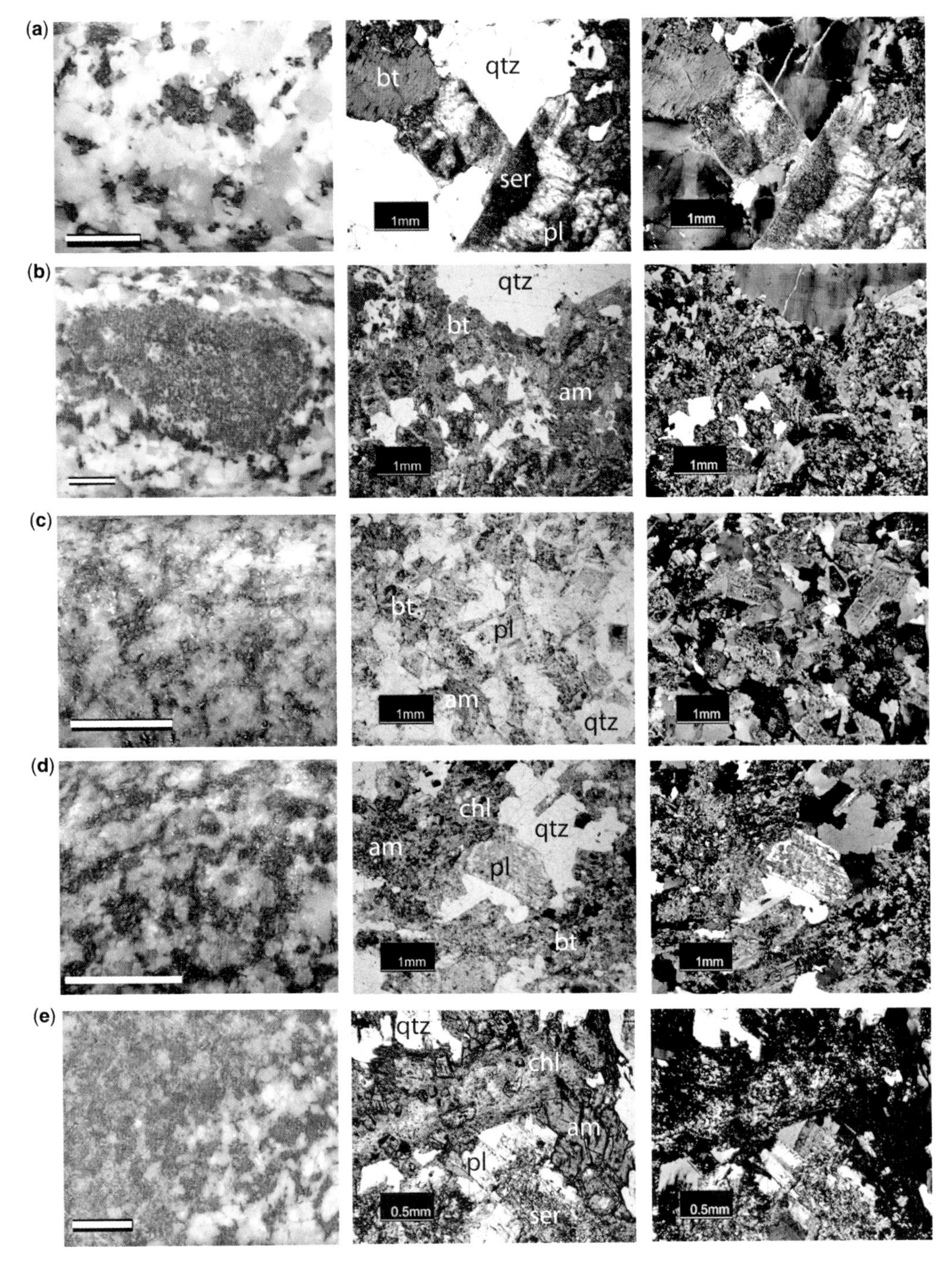

Fig. 5. Igneous rock types in the study area. Each rock is shown with drill core photographs followed by microphotograph in polarized and cross-polarized light. (**a**) Quartz porphyritic granodiorite (**b**) Microgranular mafic enclave in quartz porphyritic granodiorite, Älgträsk. (**c**) Tonalite, Älgträsk. (**d**) Tonalite, Tallberg. (**e**) Gabbro, Älgträsk. (**f**) Quartz-feldspar porphyry, Älgträsk. (**g**) Älgliden intrusion. (**h**) Low-Ti mafic dyke, Älgträsk. (**i**) High-Ti mafic dyke with amygdules, Älgträsk. (**j**) Aplite, Älgträsk. Scale bar for drill core photographs is 1 cm. Abbreviations: am, amphibole; bt, biotite; cc, calcite; chl, chlorite; kfs, K feldspar; mag, magnetite; ol, olivine; pl, plagioclase; px, pyroxene; qtz, quartz; ser, sericite.

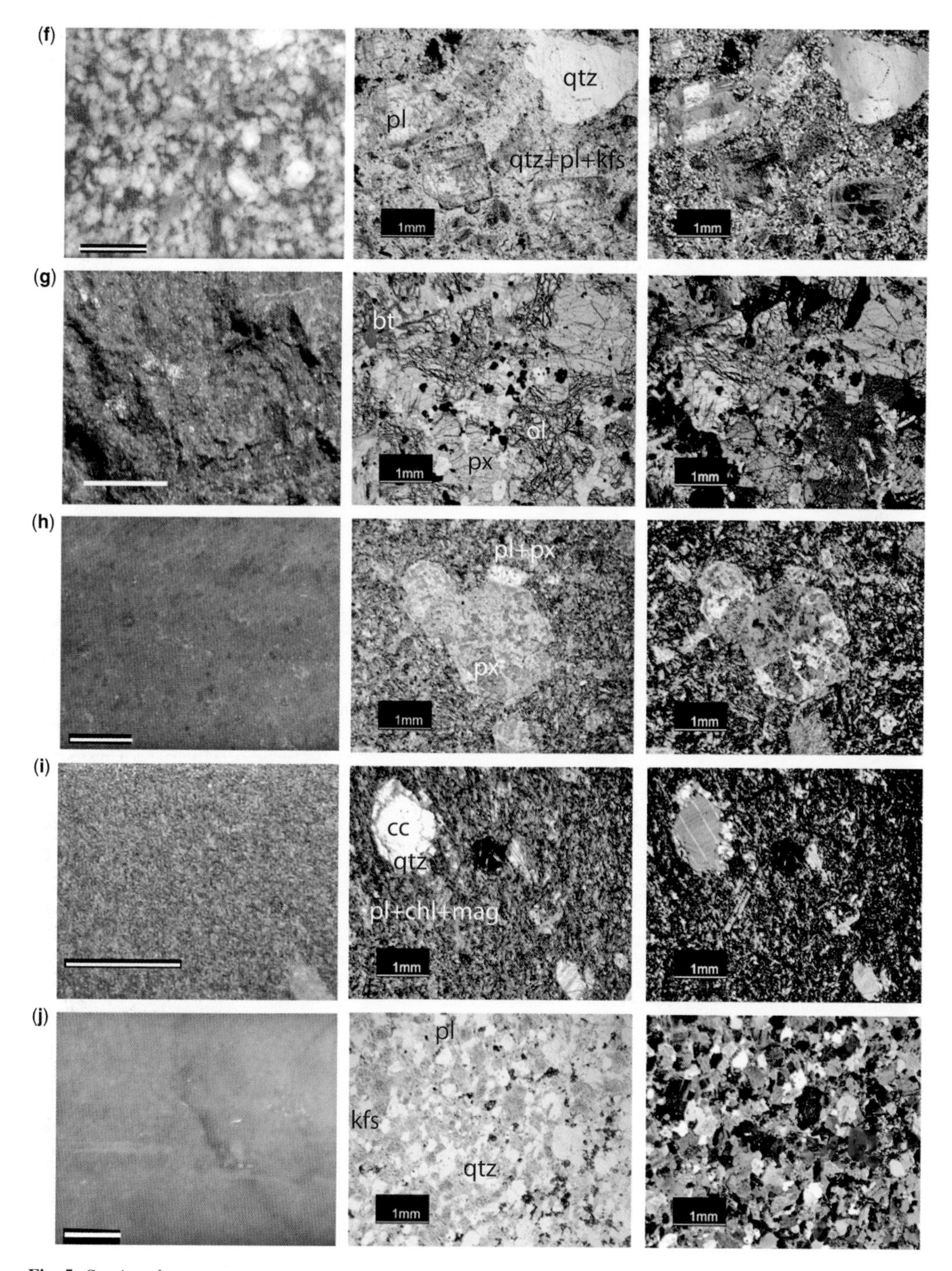

Fig. 5 *Continued.*

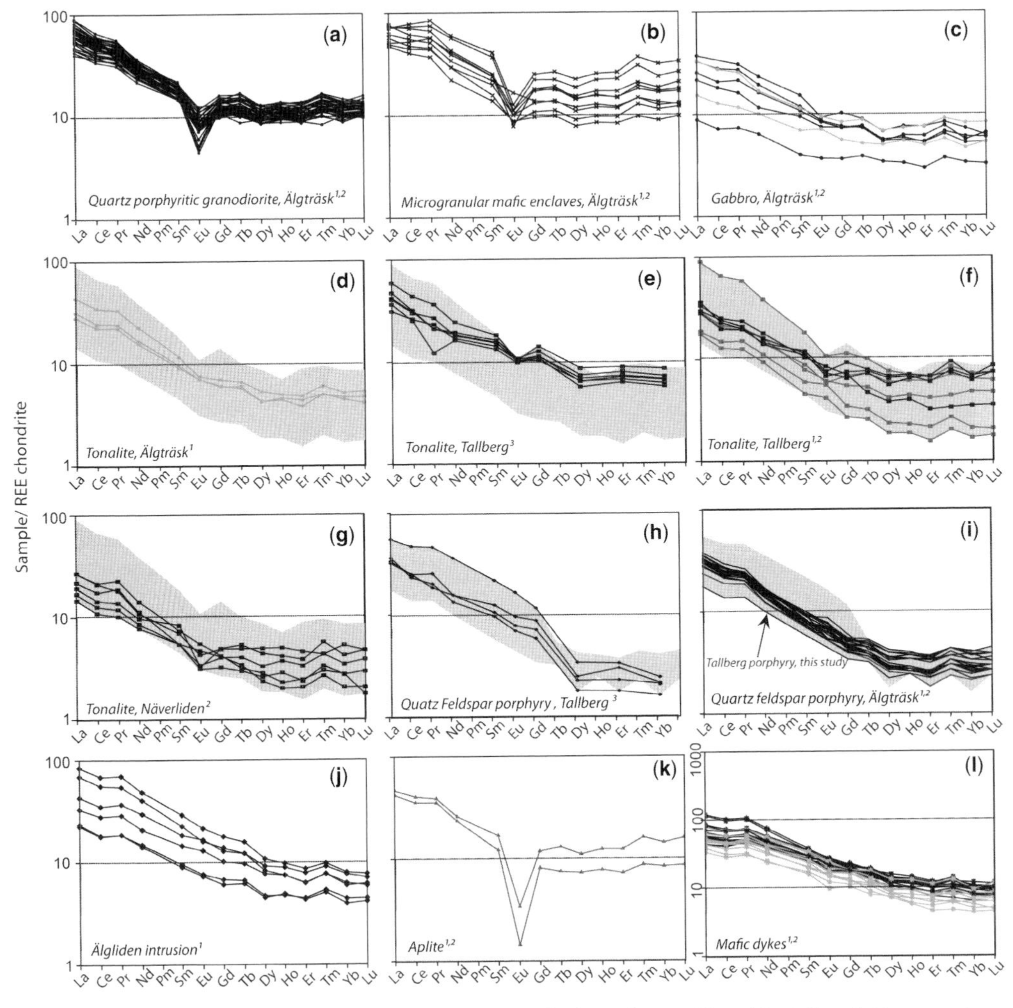

Fig. 6. Chondrite normalized REE signatures of igneous rocks in the study area. Note: (**d**–**g**) Grey field represents combined tonalite data. (**h**–**i**) Grey field represents all quartz-feldspar porphyry dyke data. (**l**) Grey lines represent low-Ti, black lines high-Ti dykes. Note the scale on *y* axis. [1]Data from this study. [2]Unpublished data from Boliden Mineral AB. [3]Data from Weihed (1992*b*).

occurred, together with selective replacement of ferromagnesian minerals by chlorite. Corroded epidote occurs as inclusions in magmatic biotite and biotite ± chlorite ± epidote pseudomorphing hornblende. Generally, quartz phenocrysts are polycrystalline and chlorite replacement of biotite is common. This style of alteration does not only affect the least altered rocks of the area, it also overprints the quartz-destructive and sodic–calcic alteration. It is interpreted to be spatially distal to mineralization, representing an incipient propylitic alteration. However, since greenschist facies metamorphism has been recorded in the area (Wilson *et al.* 1987; González-Roldán 2010) it can not be excluded that the observed alteration is due to a lower greenschist facies metamorphism.

Propylitic alteration. Propylitic alteration, generally characterized by sericite, chlorite, epidote and albite (e.g. Beane 1982 and references therein), is widespread in the area and zones >30 m wide are not uncommon. It is associated with chlorite and carbonate + quartz veinlets with sericitic alteration envelopes, carrying mainly pyrite. Rocks that display propylitic alteration (Fig. 7a) to a various degree show an extensive sericitic replacement of plagioclase while K-feldspar normally is not altered, preserving the original rock texture in

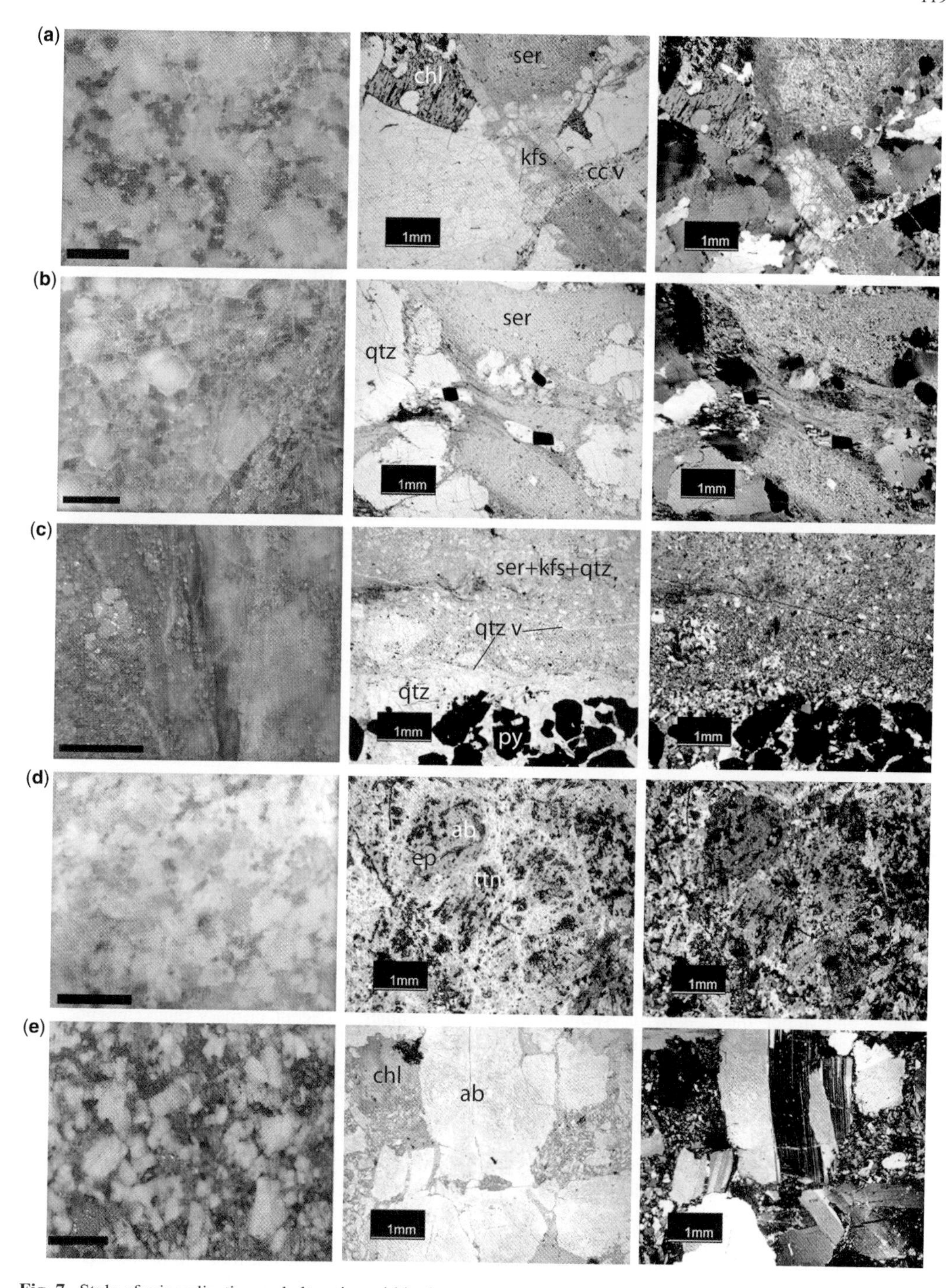

Fig. 7. Style of mineralization and alteration within the quartz porphyritic granodiorite. (**a**) Propylitic alteration, alteration mineral assemblage calcite, chlorite, sericite, pyrite with minor epidote chalcopyrite, sphalerite, magnetite, gold and telluride minerals. (**b**) phyllic alteration consisting mainly sericite, quartz, calcite, pyrite with minor epidote, chalcopyrite, sphalerite, magnetite, gold and telluride minerals (**c**) Silicic alteration consisting mainly quartz, sericite, pyrite, muscovite K-feldspar with minor calcite, chalcopyrite, sphalerite, gold and telluride minerals. (**d**) Sodic–calcic alteration dominated by sodic plagioclase, titanite, actinolite with minor epidote and garnets. (**e**) Quartz destructive alteration, 'episyenite', dominated by sodic plagioclase and chlorite with minor titanite, epidote and pyrite. Abbreviations: ab, albite; cc, calcite; chl, chlorite; ep, epidote; kfs, K feldspar; py, pyrite; qtz, quartz; ser, sericite; ttn, titanite; v, vein.

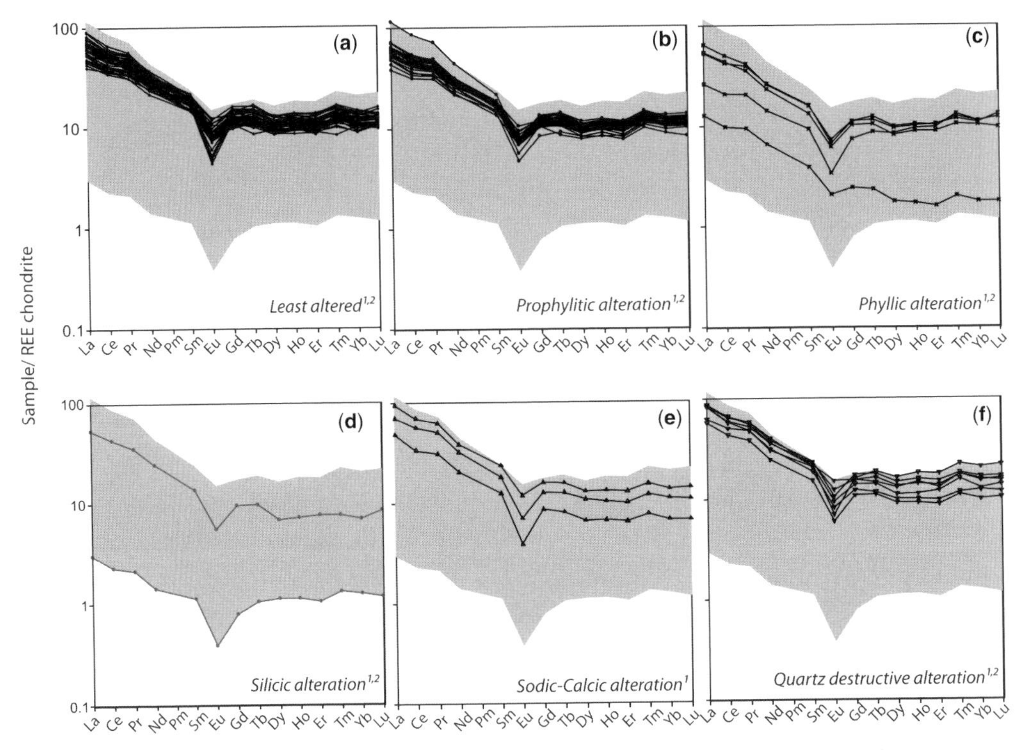

Fig. 8. Chondrite normalized REE diagrams of the alteration zones in the quartz porphyric granodiorite. [1]Data from this study. [2]Unpublished data from Boliden Mineral AB.

non-deformed samples. Mafic minerals are replaced by chlorite ± Fe–Ti oxides. Relicts of quartz phenocrysts often show grain boundary migration textures and are in some cases elongated and pass in to domains of fine grained recrystallized quartz while corroded epidote observed as inclusions in biotite in the least altered samples are replaced by chlorite + sericite ± calcite. Unmineralized calcite veinlets and minor quartz + pyrite veinlets with chloritic alteration are common. Disseminated pyrite is abundant, normally associated with minor epidote, chalcopyrite and sphalerite, often close or in sites of ferromagnesian minerals. Carbonate + quartz veinlets with sulphides and minor epidote are often found crosscutting this alteration style. Many propylitic zones are plastically deformed, where quartz forms strain fringes on euhedral pyrite, and epidote along pyrite edges is fractured. Minor Fe-oxides are commonly fractured and deformed. Calcite ± carbonate filled micro-fractures and micro breccias crosscut silicate minerals, and thin shear zones are visible in sericite-rich parts. Veins of pyrite ± chalcopyrite ± sphalerite ± tellurides ± Au with a gangue of quartz ± calcite enveloped by sericitic alteration are common within the propylitic alteration zones. The REE signatures from the propylitically altered QPGs are slightly less fractionated than those for the least altered QPG (Fig. 8b).

Phyllic alteration. Quartz, sericite and pyrite are generally considered to characterize zones of proximal phyllic alteration (e.g. Beane 1982 and references therein). In Älgträsk, sericite replaces all other minerals except primary quartz phenocrysts (Fig. 7b). Disseminated pyrite, chalcopyrite ± sphalerite is common and sulphide grains are commonly rimmed with a millimetre wide zone of epidote. The original rock texture is commonly well preserved outside deformation zones. Veinlets of pyrite ± chalcopyrite ± sphalerite ± arsenopyrite ± tellurides ± Au with a gangue of calcite ± quartz ± K-feldspar without alteration envelopes are observed in places. The REE signatures of the least altered rock are preserved within this style of alteration, though seemingly more depleted (Fig. 8c).

Silicic alteration. Silicic alteration is generally dominated by minerals such as quartz, adularia, mica and pyrite (Thompson & Thompson 1996 and references therein). This style of alteration is the most intense alteration associated with the gold mineralization. It occurs within the propylitic and phyllic alteration and the igneous texture is in most places obliterated (Fig. 7c). The alteration assemblage is composed of quartz and pyrite ± calcite ± sericite ± muscovite ± K-feldspar, with ore minerals sphalerite, chalcopyrite, arsenopyrite (±pyrrhotite), gold and Te-minerals (e.g. hessite, calavarite). Some silicic alteration zones are almost barren of sulphide minerals. Concentrations of euhedral arsenopyrite overgrown by single crystal arsenopyrite have been observed. Rare thin quartz veinlets crosscut the silicic alteration. Pyrite is normally the dominating sulphide mineral, commonly surrounded by mica. The REE signatures of samples from silicified QPG are similar to the signatures of the least altered samples, but with a less fractionated trend. In the intensely silicified samples, there is also a stronger depletion in LREE (Fig. 8d).

Sodic–calcic alteration. Sodic–calcic alteration is known from several porphyry Cu deposits, and is characterized by alkali exchange replacement of K-feldspar by sodic plagioclase and replacement of for example, biotite by actinolite (cf. Carten 1986). The alteration in the Älgträsk area referred to as sodic–calcic alteration is characterized macroscopically by a pale colour, with stains of red and green (Fig. 7d) and mostly occur in the eastern periphery of the area. It is commonly medium to coarse-grained and composed of sodic plagioclase, diopside, epidote, titanite and quartz with accessory actinolite. Garnets have been observed within quartz and amphibole veinlets, but also within the rocks in the alteration zone. Sites of ferromagnesian minerals are commonly occupied by titanite + quartz or actinolite. In intensely altered samples, the original rock texture is obliterated, with a mineral assemblage composed of sodic plagioclase, quartz, epidote and titanite. This alteration is commonly overprinted by later sericite + epidote ± clinozoisite alteration, corresponding to the feldspar destructive alteration described above. Veinlets observed within this type of alteration zones consist of amphibole ± quartz ± garnets enveloped by amphibole, or epidote enveloped by albite and epidote. In places, up to 30 cm in drillcore, the rock is composed entirely of albite. The REE signature of the Na–Ca alteration is very similar to the unaltered QPG, and hence REE's seem to have remained immobile (Fig. 8e).

Quartz destructive alteration. The quartz destructive alteration in the Älgträsk area is focused on the quartz porphyritic granodiorite, commonly preceded by intense chlorite veining enveloped by albite alteration. These alteration zones are dominated by sodic plagioclase with interstitial chlorite, titanite, epidote, quartz, micas and calcite, with accessory monazite (Fig. 7e). Vugs are commonly observed in the textural position of quartz. Some of the alteration zones are sulphide bearing and pyrite is then the dominating sulphide phase, filling vugs after quartz. Locally, quartz, micas and carbonates as well as sericite and epidote alteration of interstitial minerals and plagioclase is noted together with sulphides. This is, however, not everywhere associated with gold mineralization. The REE pattern in the altered rock preserves the characteristics of the quartz-porphyritic granodiorite, with a less pronounced negative Eu anomaly (Fig. 8f).

The QPG and the QFP display the same styles of alteration and mineralization. Where the gabbro is mineralized, the distal alteration assemblage comprises chlorite, sericite, epidote and quartz, while the proximal alteration assemblage comprises sericite and minor epidote, both obliterating the original rock texture.

Setting of gold. High gold content in the Älgträsk deposit generally shows a strong correlation with the intensity of alteration, with higher grades in the phyllic and silicic alteration compared to the propylitic alteration. Gold mainly occurs as grains of electrum, with size variation from <1 to 150 μm, commonly around 5–10 μm (Fig. 9). In phyllic and silicic alteration zones, individual grains of gold can be up to a few mm in size. Microprobe analyses on electrum commonly indicate 85–92% Au, while a few grains show contents as low as 45% Au. Traces of mercury have also been recorded in a few of the electrum grains. The gold is strongly correlated with pyrite, chalcopyrite ± sphalerite ± telluride bearing quartz + calcite veinlets within the propylitic and phyllic alteration zones and in similar sulphide mineral assemblages within the silicified zones. It is mainly hosted within fractures, on grain boundaries or as inclusions in pyrite or rare arsenopyrite (Fig. 9), also within the silicic zones. Telluride minerals such as tellurobismuthite, calaverite and hessite are often observed together with grains of gold and in similar textural positions as gold (Fig. 9).

Alteration chemistry within the QPG

The chemical changes within the granodiorite associated with the progressive stages of alteration close to mineralizations have been studied using isocon (Fig. 10), mass change calculations (Fig. 11) and molar ratio plots (Fig. 12). The isocon diagrams and calculation of mass changes have been made using representative samples from the least altered QPG and the different alteration types (Table 2). Immobile elements are generally used to monitor chemical changes occurring

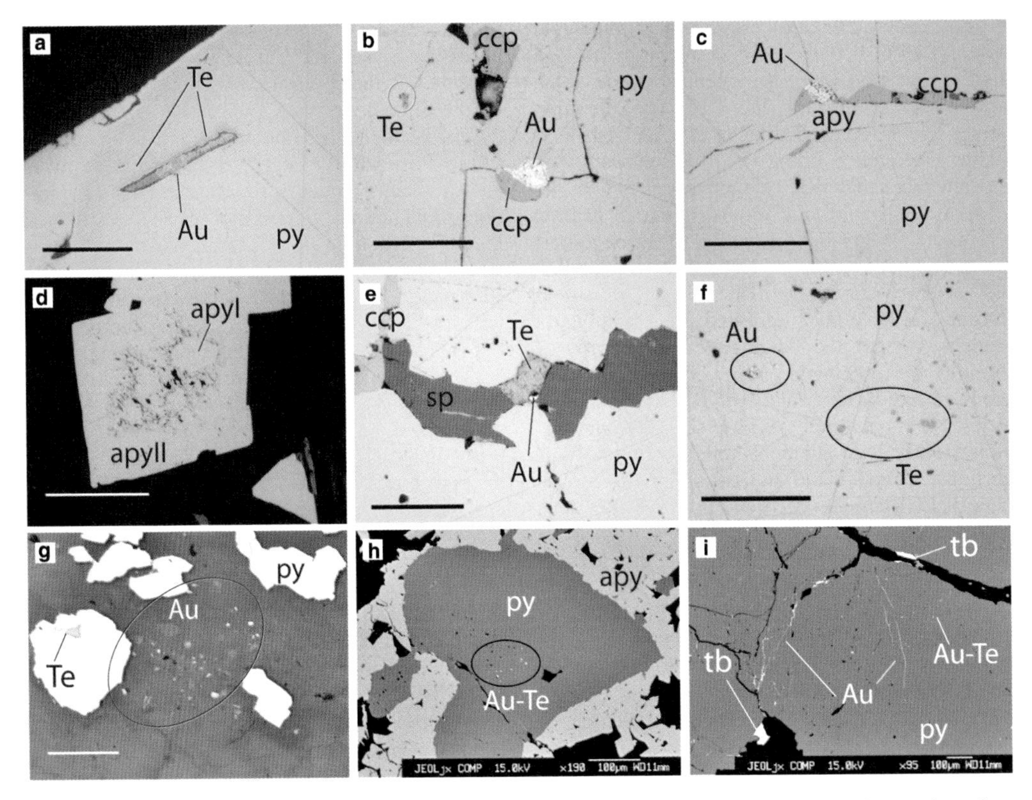

Fig. 9. Setting of gold. (**a**) Gold and telluride inclusions in pyrite. (**b**) Gold and chalcopyrite filling a fracture in pyrite. (**c**) Gold, chalcopyrite and arsenopyrite filling a fracture in pyrite. (**d**) Gold and two generations of arsenopyrite. (**e**) Gold and telluride mineral with sphalerite and chalcopyrite in a fracture in pyrite. (**f**) Gold and tellurides as inclusions in pyrite. (**g**) Gold with gangue minerals as quarts and K-mica. (**h**) BSE picture of arsenopyrite enclosing earlier pyrite with inclusions of Au and Au–Te mineral assemblages. (**i**) BSE picture of pyrite with gold in fractures, as inclusions and along relict grain boundaries in pyrite together with Au–Te–Bi minerals (scale bar 0.2 mm). Mineral abbreviations: apy, arsenopyrite; ccp, chalcopyrite; py, pyrite; sp, sphalerite; tb, tellurobismuthite; Te, telluride mineral. Scale bar in microphotographs other than (**g**) is 50 μm.

during alteration connected to hydrothermal mineralization, for example, in VMS deposits. Mobility of elements is determined by plotting the concentration of element X in the least altered *vs.* its altered equivalent (e.g. Grant 1986). If the element has been immobile during the alteration, the data display a linear trend through the origin. This line, the isocon, defines the equivalent masses before and after the alteration (Grant 1986, 2005). Using elements such as Zr, TiO_2, Al_2O_3, La and Y which are usually relatively immobile during alteration (cf. Grant 1986; MacLean & Kranidiotis 1987; Baumgartner & Olsen 1995), the different ratios can be used to determine if addition or loss of elements and volume changes have affected the rock during alteration, presuming a constant ratio of immobile elements. An element that plots above the isocon has experienced addition and an element plotting below the isocon has experienced loss (Grant 1986). A decreasing slope of the isocon compared with unaltered samples will correspond to an addition of mass, while an increasing slope will correspond to a mass decrease. Isocon diagrams for the different styles of alteration are shown in Figure 10. Calculations of the absolute mass change (ΔC) have been done. The method of MacLean (1990) given by equation 1, has been used in the calculations, using Zr as the immobile element.

$$\Delta C = C_{1(\mathrm{immobile})}/C_{a(\mathrm{immobile})} \times C_{1(\mathrm{mobile})} - C_{a(\mathrm{mobile})} \quad (1)$$

ΔC = absolute mass change (g/100 g for major oxides, or g/ton for trace elements and ppb for

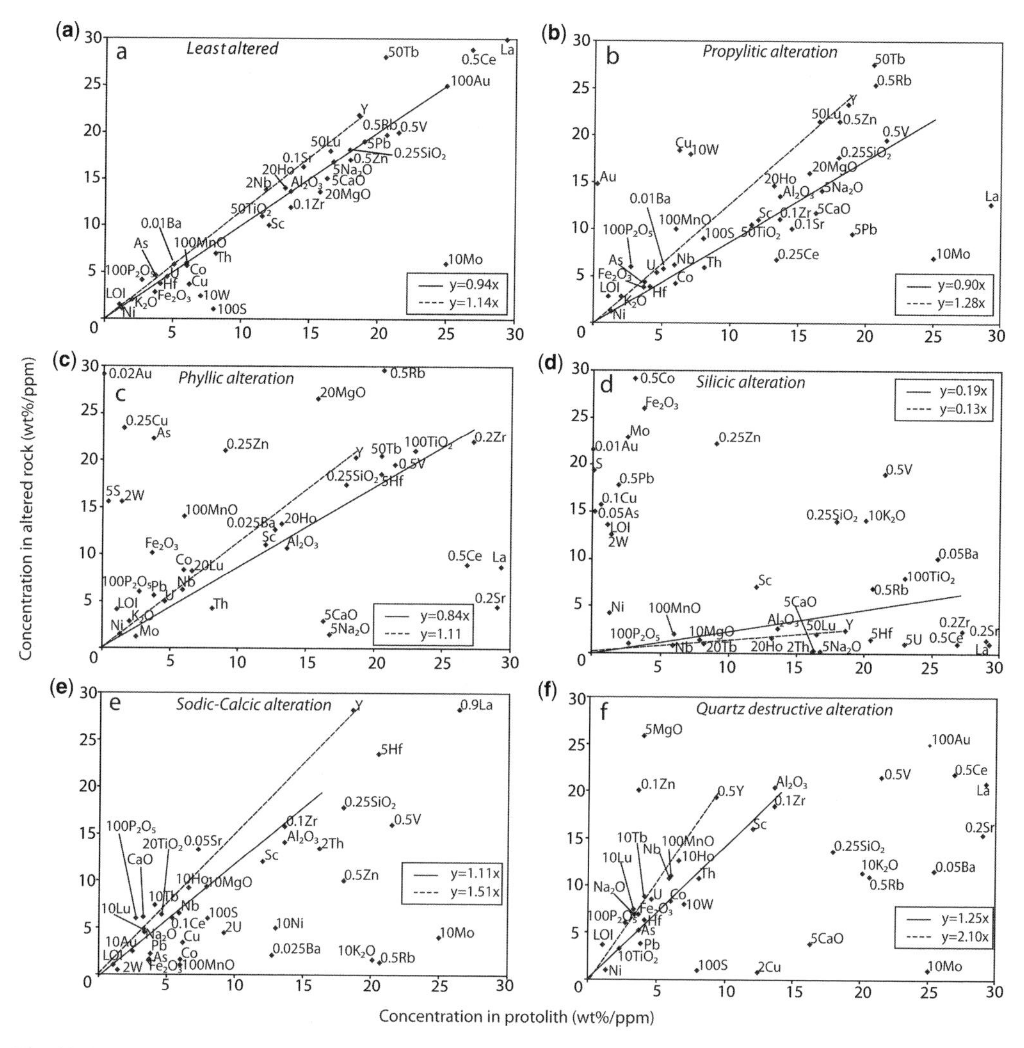

Fig. 10. Isocon diagrams from representative samples of the different styles of alteration in the quartz porphyritic granodiorite. Analytical data from samples in Table 2. Some samples are multiplied by a given constant to fit a common scale on the diagram. Major oxides are given in wt%, trace elements in ppm and Au in ppb. Solid black lines represent an isocon based on constant ratios of Zr, TiO_2 and Al_2O_3 while dashed line represent an isocon based on immobile Lu, Y. Elements below the isocon are depleted in the altered rock whereas elements above were enriched during alteration. An increase in the slope of the isocon represents a mass loss, while a decrease in the slope corresponds to a mass gain.

Au); C_l = the concentration of an element in least altered rock; C_a = the concentration of an element in its altered equivalent.

The results from the calculations of absolute mass change on selected elements are presented in Table 3 and plotted in Figure 11. Molar element ratios were calculated from the whole rock geochemistry and plotted in diagrams using the technique of Warren *et al.* (2007), which was developed to define vectors toward ore in epithermal deposits. However, these calculations do not visualize the changes in Si in the altered rock. The results will be discussed below.

Discussion

Alteration related to mineralization

The chondrite normalized REE signatures of the alteration zones commonly are similar to the signatures of the least altered samples suggesting total rare earth elements (tREE) immobility. However,

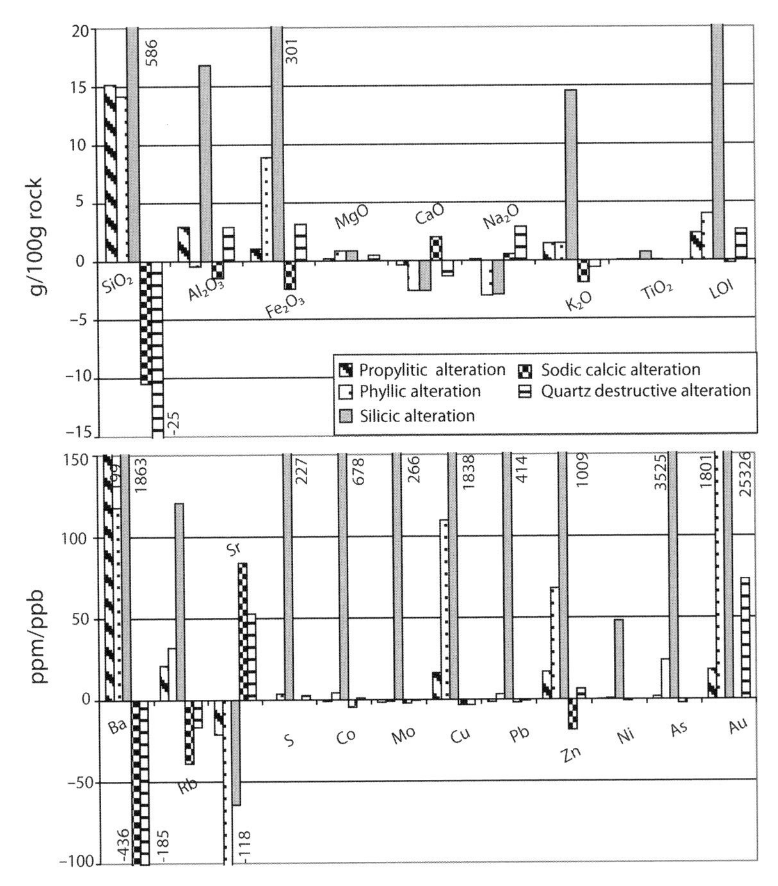

Fig. 11. Gains and losses for selected elements from the different styles of alteration in the quartz porphyritic granodiorite. Major oxides are given in wt%, trace elements in ppm and Au in ppb per 100 g rock. Calculations where made using Zr as the immobile element.

in the most intensely silicified samples, there is a slight depletion in the LREE which indicates LREE mobility. In the propylitic, phyllic and silicic alterations types, a decrease in tREE abundance is visible in the QPG while the signature of the least altered rock is preserved. This is likely an effect of dilution of the REE and not an effect of REE mobility.

Zr, Y, Lu, Al_2O_3 and TiO_2 are generally considered as immobile during hydrothermal alteration in for example, VMS systems (MacLean & Kranidiotis 1987). The correlation between these elements in isocon diagrams is, however, not as good as expected if immobility is assumed for example, in silicified zones. Zr, Al_2O_3 and TiO_2 have in spite of this been used to get a semi-quantitative impression of larger gains or losses associated with alteration, and are plotted together with an isocon based on Lu–Y for comparison (Fig. 10). As seen in Figure 10a, there is heterogeneity among the least altered samples, but since the textural patterns are very similar, it is suggested that these variations are due to natural variation within the rock. Molybdenum and copper shows the strongest variation among the least altered samples, with variations between 0.6–9.0 ppm Mo and 1.4–269 ppm Cu.

Overprinting where minerals formed in an earlier stage of alteration survives a more intense style of alteration should also be considered. Epidote or pyrite formed during the propylitic alteration might still be present even if an overprinting of phyllic–silicic alteration occur, creating a mixed mineral assemblage. Relict minerals will likely affect the mass balance calculations by contributing with elements that otherwise might have been removed, but the calculations are still useful for illustrating major changes within the rock.

The feldspar destructive alteration is most likely representing an incipient propylitic alteration, observed mainly in the least altered rocks. Textures

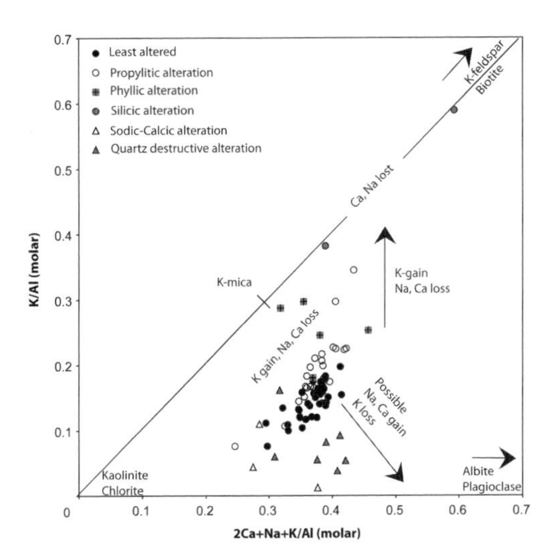

Fig. 12. Molar ratios of K/Al v. 2Ca + Na + K/Al. Samples from the least altered and the different styles of alterations within the quartz porphyritic granodiorite. Data from this study and previously unpublished data from Boliden Mineral AB.

such as K-feldspar rims on plagioclase and biotite replacing hornblende are commonly seen in porphyry deposits which might indicate an earlier stage of weak potassic alteration. However, the K-feldspar/albite coronas on plagioclase could also represent a late igneous texture, whereas biotite replacing hornblende as well as fresh biotite enclosing corroded epidote may represent a later overprint together with sericite replacing plagioclase. This event could represent the formation of a spatially distal and temporally younger, mineralization belonging to the same hydrothermal system. The overprinting could also represent a local expression of regional metamorphism as seen in other parts of the JGC (Wilson *et al.* 1987; González-Roldán 2010). However, the typical decussated biotite on magmatic biotite interpreted to have formed during regional greenschist metamorphism (González-Roldán 2010) has not been observed in the quartz-porphyritic granodiorite during this study. The alteration assemblage of sericite + epidote + clinozoizite in feldspars within the sodic–calcic alteration is suggested to represent the feldspar destructive alteration overprinting the sodic–calcic alteration.

In the propylitic alteration zones the geochemical data suggest an increase in SiO_2, together with an increase in Au, Cu, Zn, S (Figs 10b & 11) which is consistent with micro-veinlets of mainly quartz ± calcite with pyrite ± chalcopyrite ± sphalerite. Losses of Sr are likely due to primary feldspar destruction. The plagioclase is generally completely replaced by sericite ± epidote and ferromagnesian minerals are partly or fully replaced by chlorite, epidote and Fe–Ti oxides (Fig. 7a). K-feldspar seems to be one of the least altered minerals in the rock, though perthitic exsolutions and sericitic dusting might indicate later alteration. Ca from the destruction of plagioclase and ferromagnesian minerals may contribute to the formation of calcite filled microfractures. According to the isocon plot (Fig. 10b), no larger changes in rock volume during the propylitic alteration has taken place. The geochemical data plotted in the isocon diagram (Fig. 10) and the mass change histograms (Fig. 11) suggest an addition of SiO_2, $Fe_2O_3{}^T$, MgO, K_2O, Ba, S, Cu, Zn, As, Au and decrease in mainly CaO, Na_2O, Sr within the phyllic alteration. This is consistent with a mineral alteration assemblage dominated by quartz, sericite ± calcite ± pyrite ± muscovite ± chlorite ± K-feldspar and absence of plagioclase (though often pseudomorphed by sericite). Data plotted in a cationic molar ratio diagram (Fig. 12) show a significant shift from the propylitic alteration, also indicating K-enrichment and depletion of Na and Ca.

Isocon diagrams as well as histograms (Figs 10d & 11) indicate a strong enrichment of for example, SiO_2, K_2O, $Fe_2O_3^T$, S, Cu, Zn, Au, As and Mo with a depletion of mainly Na_2O, CaO and Sr within the silicic alteration. This is also consistent with mineralogical evidence of silicification, abundant pyrite, chalcopyrite, sphalerite ± muscovite ± K-feldspar and absence of plagioclase and ferromagnesian minerals. Molar ratios for the two samples from silicified QPG plot on the trend for Na and Ca loss (Fig. 12), while K has been added. The diagram suggests that the K-bearing minerals in the rock would constitute K-mica and K-feldspar, which is also indicated from petrographic studies, though the exact composition of the mica is unknown. The isocon fit for the Zr, Al_2O_3 and TiO_2 is not as good within the silicic alteration zone as in other styles of alterations, which may indicate mobility of one or more of the elements during silicification, but both isocon slopes still indicate a large addition of mass (Fig. 10d). Gold shows the highest mass change in this style of alteration and also good correlation with addition of base metals. Dilution of the REE's by a change in mass might also explain why the REE signature of this alteration gives the impression to be less fractionated (Fig. 8d). REE patterns of the silicic alteration show the characteristic shape of the QPG, although with a less steep pattern of the LREE which might indicate a small mobility of the LREE, while the HREE remained immobile.

The isocon diagrams and mass change histograms (Figs 10e & 11) indicate an increase of CaO, Na_2O and Sr and a depletion of SiO_2,

Table 3. *Mass balance. Results from mass change calculations from the altered zones in the quartz porphyritic granodiorite.*

Major oxide/ element	83–126 Propylitic alteration	83–125 Phyllic alteration	0701 Silicic alteration	265–84 Sodic alteration	65–140 Quartz-destructive alteration
SiO_2	15	14	586	−10	−25
Al_2O_3	3	0	17	−1	3
$Fe_2O_3{}^T$	1	9	301	−2	3
MgO	0	1	1	0	0
CaO	0	−3	−3	2	−1
Na_2O	0	−3	−3	1	3
K_2O	1	1	15	−2	−1
TiO_2	0	0	1	0	0
LOI	2	4	159	0	3
Ba	199	118	1864	−436	−185
Rb	21	32	121	−39	−17
Sr	−21	−118	−64	84	53
V	5	5	403	−15	8
W	2	9	73	−1	1
S	0	4	227	0	3
Co	−1	4	679	−5	1
Mo	−2	−1	266	−2	0
Cu	16	110	1838	−3	−3
Pb	−1	3	414	−2	−1
Zn	17	68	1009	−19	6
Ni	0	1	48	−1	0
As	2	24	3526	−2	0
Au	18	1801	25326	0	73

Note: Major oxides given in wt%, trace elements in ppm and Au in ppb per 100 g rock

$Fe_2O_3{}^T$, Cu, S, Ba, K_2O, Mo and Zn within the sodic–calcic alteration. This is consistent with mineralogical evidence of an increase of plagioclase ± calcite and the absence of K-feldspar, pyrite, chalcopyrite and sphalerite. The REE patterns of the sodic–calcic alteration still have the characteristic shape of the QPG which suggests immobility of REE (Fig. 8e). Molar element ratios suggest that Na and Ca is added during alteration (Fig. 12), also consistent with the observed mineralogy. In porphyry Cu deposits, sodic–calcic alteration is interpreted as a deep seated zone of recharge, as seen in for example, the Yerington district (Dilles & Proffett 1995). This remains to be proven for the Älgträsk deposit.

The quartz destructive alteration is similar to 'episyenitization', a process known to take place in evolved granites from Si-undersaturated fluids, likely of magmatic origin (cf. Petersson 2002). Within the quartz destructive alteration, both isocon diagrams (Fig. 10f) and mass change histograms (Fig. 11) indicate an increase in Na_2O, $Fe_2O_3{}^T$, MgO, S, Au, Zn and a decrease of SiO_2, K_2O, CaO, Cu and Ba. The chemical data are supported by the presence of mainly albite and chlorite ± pyrite ± sphalerite. As suggested by the isocon diagram, and textural evidence (vugs), there is likely a decrease in mass during the formations of this alteration style. The normalized REE pattern for this alteration type shows a characteristic QPG pattern, although some samples appear to be more fractionated than the least altered samples. This is likely to be due to mass loss during the formation of this style of alteration or due to the concentration of feldspars relative to the content in the altered rock. The molar element ratios (Fig. 12) suggest a loss in K and an addition of Na and Ca, consistent with mineralogical changes. Episyenites documented in the Bohus granite, southwestern Sweden, are characterized by an initial stage of quartz leaching, volume decrease and enrichment of Na (Petersson & Eliasson 1997), similar to that seen in Älgträsk (Figs 10 & 11). The quartz destructive alteration in Älgträsk is suggested to have formed by processes similar to 'episyenitization', and is hence the second described occurrence of episyenite from Sweden.

Feldspar destructive alteration as well as propylitic alteration have been observed in other parts of the GI of the JGC, interpreted as due to greenschist facies metamorphism creating a similar mineral paragenesis (González-Roldán 2010). Gonzáles-Roldán (2010) have observed metamorphic biotite overprinting hydrothermal chlorite in southern

JGC, but does not mention any increase in either sulphide minerals or gold in the host rock related to alteration, as is evident in the propylitic alteration in the Älgträsk area. The similarity between the described regional greenschist facies metamorphic mineral assemblages and distal alteration assemblages related to mineralization makes the interpretation of alteration difficult, and it cannot be excluded that at least some of the feldspar destructive alteration represents regional metamorphism. If greenschist facies metamorphism has affected the deposit, this would obstruct the interpretation of any primary alteration mineral assemblages related with mineralization. The K-micas observed, might hence constitute an alteration product of primary alteration minerals, for example, potassium rich clay minerals observed in for example, epithermal systems.

Sericitic dusting and epidote alteration of albite in the Na–Ca alteration, together with calcite veinlets in places cutting and displacing quartz veins with pyrite as well as microscopic pyrite bearing shear zones in sericite-rich material indicate a post-mineralization alteration and deformation. This phase of feldspar destructive alteration might represent a distal expression of alteration related to a parallel mineralization, or represent a metamorphic overprint.

Petrogenetic aspects of the host intrusives

The intrusive rocks in the study area are all calc-alkaline (Fig. 4d). The REE signatures normalized to chondrite show that the gabbro and tonalite in the Älgträsk area and the tonalite in the Tallberg area (including Näverliden) all have a moderate to steep LREE slope, a weak to none existing Eu anomaly and a flat HREE signature with a small positive Tm anomaly (Fig. 6d–g).

The Tallberg tonalite and the tonalite/hybridic rock observed in Älgträsk have a similar REE pattern and plot within the gabbroic signature, whereas the Näverliden tonalites situated further to the north seem to be slightly less fractionated compared to the tonalites and the gabbro. The REE signatures for tonalite in Tallberg are different with one type similar to the less fractionated Näverliden tonalite and one type similar to the more fractionated tonalite and gabbro. However, the more depleted signature of the Näverliden tonalite and the more depleted signatures among tonalites in Tallberg may be the result of either magma mixing or of volume change in the rock, and hence REE dilution due to quartz brecciation, which is indicated by the quartz rich character of the tonalite in Näverliden (Fig. 4a). The tonalitic tREE signatures are similar to the gabbro although the tonalite generally seems to have a less fractionated HREE signature. This might indicate a genetic relationship where the two intrusive phases share a magmatic source, or magmas of separate sources mingled during emplacement. The REE signatures for both these rocks are similar to the Gallejaur intrusive rocks and the GII phase of the JGC (cf. Kathol & Weihed 2005), although slightly less fractionated.

A different REE signature is observed in data from the QPG in Älgträsk, which is more fractionated and has a pronounced negative Eu anomaly (Fig. 6a), likely formed by plagioclase fractionation. The different signature of the QPG suggests that it is not genetically related to the tonalite or gabbro, and that the QPG has a separate magmatic source. Alternatively, the QPG represents the upper part of the magma chamber that crystallized after the main phases giving rise to a Eu-depleted and more fractionated character. Mingling textures observed in drill-core indicate that a partly solidified magma (e.g. the QPG) interacted with a hotter more mafic (gabbro/tonalite) magma at depth. Similarities in REE signatures between the MME (in the QPG) and the QPG suggest that the MME might represent restite material from a gradual separation of solid source residue (White & Chappell 1977), as seen in other granitoid intrusions, for example, the Vinga intrusion SW Sweden (Årebäck *et al.* 2008). If the MME represent mingling textures between the QPG and the gabbro, this would likely result in a less fractionated REE pattern compared to the QPG, more similar to the signature of the gabbro. The normalized REE signatures for the QFP dykes in Älgträsk are similar to signatures from the QFP in Tallberg (data from Weihed 1992*a*, and one sample from this study), therefore these rocks are suggested to be genetically related. Similar petrological features as well as similar REE signatures may indicate that the porphyry dykes originate from the same magmatic source. The dissimilarities between the REE signature of the QPG in Älgträsk and the QFP dykes and the fact that the QPG is more fractionated than the porphyritic dykes, excludes the QPG as a potential magmatic source for the porphyries. In general, the quartz-feldspar porphyritic dykes are among the least fractionated in the area, and compared with the tonalite, the REE signatures are very similar. It is therefore probable that the porphyritic dykes represent a late offshoot of the magmatic activity creating the tonalite.

The ultramafic Älgliden dyke situated NE of the Tallberg deposit share REE characteristics with the tonalitic and gabbroic rocks in the area. The more fractionated REE signature of the Älgliden intrusion is also similar to the REE signature of the Gallejaur intrusives (cf. Kathol & Weihed 2005). Since the ultramafic intrusion display REE signatures similar to the tonalite and gabbro, but is generally more fractionated than most of the igneous rock in the area, and lack deformation and alteration, it is suggested that the intrusion is younger and formed after the mineralizing events in the Älgträsk and Tallberg area. The intrusion of a younger ultramafic dyke into a generally intermediate-felsic domain indicates a tectonically active period with crustal extension. The ultramafic dyke is in turn cut by at least one mafic dyke, and a quartz-poor porphyry dyke (the andesitic sample among porphyry data in Fig. 4b), indicating that more than one generation of porphyry dykes are present in the area, and that the Älgliden intrusion is probably older than the mafic dykes.

The mafic dykes can be classified as sub-alkaline to andesitic basalts (Winchester & Floyd 1977) and are divided in to at least two different types, based on the Ti content and textural relationship, in which the Ti-poor basalt contains 0.6–1.1 wt% TiO_2 while the Ti-rich basalt contains 1.5–1.7 wt% TiO_2 (Table 2). The titanium is likely hosted by disseminated Ti-magnetite and ilmenite in the high-Ti rocks. Chondrite normalized REE signatures for the basaltic dykes show similar trends as basalts from both the Petiknäs and Maurliden deposits (Montelius *et al.* 2007; Schlatter 2007) situated 20–30 km SE and W of the Älgträsk area. Due to the mafic nature of the rock, the rounded-subrounded, about 1–5 mm large, calcite $\pm$ quartz and quartz grains are suggested to be amygdales. The dykes containing the amygdules are therefore interpreted to be rather surface near intrusions. The Ti-poor basalt is comparable to altered basaltic andesites in the Petiknäs south VMS deposit, described by Schlatter (2007) and high Ti–Zr basalts in the Maurliden VMS deposit (Montelius *et al.* 2007) although the Zr content of the dykes in Älgträsk is slightly lower. Mafic volcanic rocks of the Varuträsk formation belonging to the Vargfors Group described by Bergström (2001) show similar characteristics with the high Ti-basalts. REE signatures also suggest that both groups are similar to basalts from the Boliden area, and basalt-andesites from the Gallejaur formation (cf. Bergström 2001). The high-Ti basalts might hence be correlated with the Arvidsjaur Group or Vargfors Group volcanic rocks. The dyking event(s) are thus tentatively tied to the younger period of volcanism in the Skellefte district, which has implications for the time–space relationships between VMS and porphyry type deposits in the area.

The strike of the Älgliden intrusion follows the main NE–SW direction, also followed by quartz feldspar porphyry dykes and mineralizing systems in Älgträsk (Fig. 2). This indicates a strong tectonic control on both mineralization and the dyking events in the area. The alteration zones in the Älgträsk deposits commonly exhibit ductile deformation, tentatively correlated with the D_2 or D_3 event. Alteration, mineralization and ductile deformation are overprinted by a later post D_3, brittle deformation, striking NNE–SSW, in the Älgträsk are (Figs 2 & 3). Dykes trending in a NE direction indicates a NW–SE extension, compatible with the NE–SW shortening during D_2 (Bergman Weihed 2001). In Älgträsk the mineralized zones show a tendency to bend into the late NNE structures in a sinistral manner, indicating that the late brittle structures were preceded by ductile deformation related to the D_3 event. This in turn indicates a prolonged event tentatively spanning over 70 million years.

Smaller euhedral–anhedral arsenopyrite grains enveloped by a single arsenopyrite crystal (Fig. 9d), indicate two generations of sulphide precipitation. In this sample, gold has been observed mainly as fracture fillings in pyrite and arsenopyrite and between sulphide grains, but also as inclusions in arsenopyrite, tentatively belonging to the second generation. Similar textures are also observed in pyrite. Larger pyrite grains with enclosed smaller arsenopyrite crystals often have inclusions of Au–Te minerals (Fig. 9h), supporting the existence of two sulphide generations, but could also be due to sudden changes in physical–chemical parameters or fluid composition. Strain fringes growing on pyrite in samples from for example, low strain zones in phyllic and silicic alteration indicate pyrite precipitation before deformation (Fig. 7b). Back scattered electron (BSE) images, however, show that gold generally occur as drop-like inclusions in undeformed pyrite (Fig. 9h) and more often in relict grain boundaries and along fractures in pyrite affected by brittle deformation (Fig. 9i). This might imply an incipient gold remobilization. Due to the generally small grain size of gold in Älgträsk and since the two sulphide generations can not always be observed in the same sample, it has not yet been possible to determine which of the two generations of pyrite and/or arsenopyrite the gold is primarily associated with. However, unmineralized mafic dykes that cut, and in places also displace, the mineralization have also experienced a later phase of ductile deformation. There are at least two ways to interpret this: (1) the mineralizations at Älgträsk was controlled by large scale tectonic structures, which were later reactivated as ductile deformation zones focused on altered rocks and possibly also causing remobilization overprinting earlier formed mineralization. This event was also followed by later tectonic events, causing deformation of younger mafic dykes, or (2) of the proposed two ductile deformation events, the first was related to fluids carrying precious and base metals, which deposited sulphides and precious metals within shear zones, while a subsequent event caused ductile deformation of mafic dykes and possibly also remobilization/precipitation of sulphides.

Two genetic models may be considered for the formation of the Älgträsk Au-deposit: (1) an orogenic gold model, or (2) a shallow level hydrothermal deposit, possibly connected with the porphyry system of Tallberg or another deeper lying porphyry system.

Epithermal (≤ 300 °C) shallow level systems (<2 km) have been recognized in several porphyry settings (Arribas *et al.* 1995; Hedenquist *et al.* 1996; Einaudi *et al.* 2003; Sillitoe & Hedenquist 2003). The sulphide mineral assemblage in the mineralized zones at Älgträsk are similar to the low–intermediate sulphidation mineral assemblage of these hydrothermal systems (Einaudi *et al.* 2003). Intermediate sulphidation systems generally have alteration mineralogy consisting of sericite, quartz, rhodochrosite, barite and anhydrite while low sulphidation systems typically have proximal illite, chalcedony, adularia and calcite, dependent on temperature (Hedenquist *et al.* 1996; Einaudi *et al.* 2003). Both styles are also surrounded by extensive propylitic alteration. Metal enrichment in this type of system is known to vary with tectonic setting, but generally the fluids are enriched in Ag, As, Au, Bi, Cu, Pb, Sn, Te and Zn (Arribas *et al.* 1995; Hedenquist *et al.* 1996; Einaudi *et al.* 2003; Sillitoe & Hedenquist 2003).

Orogenic gold deposits are structurally controlled and commonly hosted in brittle to ductile structures near large scale compressional deformation zones (Groves *et al.* 1998). The alteration mineralogy and enrichment in metals in these deposits are known to vary with host rocks and formation pressures and temperatures. In greenschist facies, a general addition of SiO_2, K, Rb $\pm$ Ba $\pm$ Na $\pm$ B occur together with an increase in Au $\pm$ Ag $\pm$ As $\pm$ Sb $\pm$ Te $\pm$ W with low contents of Pb, Zn, Cu (cf. Groves *et al.* 1998). There is generally a large scale vertical zonation present in orogenic gold systems where different metal assemblages occur on different levels, from hypozonal Au–Ag, mesozonal Au–As–Te, epizonal Au–Sb to more shallow level Hg–Sb and Hg (Groves *et al.* 1998). The Björkdal intrusive hosted Au deposit situated only 30 km west of the Älgträsk deposit has been described as an orogenic Au deposit (Weihed *et al.* 2003). The Björkdal deposit is hosted within a quartz-monzodiorite. The gold is associated with multiple centimetre–metre sized quartz veins with weak alteration envelopes, carrying Te-minerals, pyrrhotite and chalcopyrite besides gold (Nysten 1990; Broman *et al.* 1994; Weihed *et al.* 2003; Billström *et al.* 2008). There is no agreement whether this deposit formed during ductile-brittle deformation and trust duplex structures present in the deposit (Weihed *et al.* 2003), or by magmatic fluids associated with the intrusion (Billström *et al.* 2008). The structural control on both the Björkdal deposit and the Älgträsk deposit is consistent with an orogenic gold type genesis. However, structural control is also important in epithermal and porphyry systems since fluids, irrespective of origin, may be canalized within shear zones favouring mineralization.

The Älgträsk deposit shares many characteristics with both the shallow level hydrothermal and orogenic models, for example, the alteration mineralogy, metal enrichment and structural control. The metal enrichment in Älgträsk (Au, Ag, Te, Zn, and Cu) and zonation (silicic, phyllic, propylitic) with a proximal addition of mainly Si and K and loss of Na and Ca are characteristic of epithermal deposits. However, if deformation affected earlier formed porphyry Cu mineralization, remobilization of metals from this system might cause enrichment in base metals, normally not abundant in orogenic gold deposits. In any case, the mineralizing event(s) must have taken place before the emplacement of the mafic dykes, and before the tectonic event causing subsequent deformation of these dykes. If, and how much, remobilization has taken place during the deformation(s), and the timing of the deformation(s) in relation to the mineralization, remains controversial.

Conclusions

The Älgträsk deposit is hosted by Palaeoproterozoic *c.* 1.89 Ga intrusive rocks. The mineralized zones in the Älgträsk deposit and its related styles of alteration were subject to a net mass change. The hydrothermal fluids caused enrichment of SiO_2, $Fe_2O_3{}^T$, K_2O, Cu, Zn, Au, Te and As and a loss of Na_2O and CaO. The chemical changes are also reflected in new growth of quartz, sericite, K-feldspar and sulphides minerals such as pyrite, sphalerite, chalcopyrite, gold (electrum) and tellurides. A quartz-destructive alteration ('episyenitization') sharing a spatial relationship with the mineralized zones in Älgträsk are characterized by a gain of mainly Na_2O, but also by loss of SiO_2, K_2O and CaO. Sodic–calcic style of alteration to the east of the mineralization is characterized by a net gain in CaO, Na_2O and a loss of SiO_2, K_2O, $Fe_2O_3{}^T$ and base metals. This zone might represent a 'recharge' zone for fluids related to the hydrothermal system forming the Älgträsk deposit.

The REE signatures from the alteration zones preserve the signature of the least altered rocks, suggesting that REE's have been essentially immobile during the different alteration processes affecting the granodiorite. Isocon diagrams add evidence that mass changes within the rock might account for dilution of REE's creating the less fractionated signature in the most intense alteration zones.

Textures in QFP dykes suggest that similar styles of alteration and mineralization have affected these dykes and the QPG in Älgträsk. The REE signatures and petrography of similar porphyry dykes associated with porphyry style Cu mineralization in Tallberg (described by Weihed 1992*b*) suggest that the porphyries belong to the same magmatic system. Similarities in the REE signatures of the tonalite and porphyry dykes suggest a genetic relationship. The granodiorite in Älgträsk displays a more fractionated signature compared to the tonalite and QFP, a pronounced Eu anomaly, and is suggested to be genetically unrelated to the porphyry dykes. The REE signatures of the Tallberg tonalite and QFP dykes are however more similar signatures of the GII and Gallejaur intrusive suites. Structural evidence suggests that the Au deposit in Älgträsk is hosted by D_2 to D_3 structures and thus is younger than the porphyry Cu deposit in Tallberg.

Titanium content and rock textures suggest that there are at least two generations of basaltic dykes present in the Älgträsk area, one with a high Ti content, and one with a medium (low) Ti content, the latter in many places pyroxene phyric. Subrounded carbonate-quartz likely represents amygdules, which indicates a surface near emplacement for some of the dykes. The mafic dykes are interpreted to be younger than the mineralization and therefore deformation within the mafic dykes suggests that at least one phase of ductile deformation is post-mineralization.

Boliden Mineral AB is greatly acknowledged for providing geological information on the Älgträsk and Tallberg deposits. The study has been funded by New Boliden, The Geological Survey of Sweden and Luleå University of Technology.

References

ALLEN, R. L., WEIHED, P. & SVENSON, S. A. 1996. Setting of Zn–Cu–Au–Ag massive sulfide deposits in the evolution and facies architecture of a 1.9 Ga marine volcanic arc, Skellefte district, Sweden. *Economic Geology*, **91**, 1022–1053.

ÅREBÄCK, H., BARRETT, T. J., ABRAHAMSSON, S. & FAGERSTRÖM, P. 2005. The Palaeoproterozoic Kristineberg VMS deposit, Skellefte district, northern Sweden, part I: geology. *Mineralium Deposita*, **40**, 351–367.

ÅREBÄCK, H., ANDERSSON, U. B. & PETERSSON, J. 2008. Petrological evidence for crustal melting, unmixing, and undercooling in an alkali-calcic, high level intrusion: the late Sveconorwegian Vinga intrusion, SW Sweden. *Mineralogy and Petrology*, **93**, 1–46.

ARRIBAS, A., HEDENQUIST, J. W., ITAYA, T., OKADA, T., CONCEPCION, R. A. & GARCIA, J. S. 1995. Contemporaneous formation of adjacent porphyry and epithermal Cu–Au deposits over 300 ka in northern Luzon, Philippines. *Geology*, **23**, 337–340.

BARRETT, T. J., MACLEAN, W. H. & ÅREBÄCK, H. 2005. The Palaeoproterozoic Kristineberg VMS deposit, Skellefte district, northern Sweden. Part II: chemostratigraphy and alteration. *Mineralium Deposita*, **40**, 368–395.

BAUMGARTNER, L. P. & OLSEN, S. N. 1995. A least-squares approach to mass-transport calculations using the isocon method. *Economic Geology and the Bulletin of the Society of Economic Geologists*, **90**, 1261–1270.

BEANE, R. E. 1982. Hydrothermal alteration in silicate rocks. *In*: TITELY, S. R. (ed.) *Advances in Geology of the Porphyry Copper Deposits*. The University of Arizona Press, Tucson, Arizona, 117–137.

BERGMAN WEIHED, J. 2001. Palaeoproterozoic deformation zones in the Skellefte and Arvidsjaur areas, northern Sweden. *In*: WEIHED, P. (ed.) *SGU Economic Geology Research, 1999–2000*. **1 (C833)**. Swedish Geological Survey, Uppsala, 46–68.

BERGMAN WEIHED, J., BERGSTRÖM, U., BILLSTROM, K. & WEIHED, P. 1996. Geology, tectonic setting, and origin of the Paleoproterozoic Boliden Au–Cu–As deposit, Skellefte District, northern Sweden. *Economic Geology*, **91**, 1073–1097.

BERGSTRÖM, U. 2001. Geochemistry and tectonic setting of volcanic units in the northern Västerbotten county, northern Sweden. *In*: WEIHED, P. (ed.) *SGU Economic Geology Research, 1999–2000*. **1 (C833)**, Swedish Geological Survey, Uppsala, 69–92.

BILLSTRÖM, K. & WEIHED, P. 1996. Age and provenance of host rocks and ores in the Paleoproterozoic Skellefte district, northern Sweden. *Economic Geology*, **91**, 1054–1052.

BILLSTRÖM, K., BROMAN, C., JONSSON, E., RECIO, C., BOYCE, A. J. & TORSSANDER, P. 2008. Geochronological, stable isotopes and fluid inclusion constraints for a premetamorphic development of the intrusive-hosted Björkdal Au deposit, northern Sweden. *International Journal of Earth Sciences*, **98**, 1027–1052.

BROMAN, C., BILLSTRÖM, K., GUSTAVSSON, K. & FALLICK, A. E. 1994. Fluid inclusions, stable isotopes and gold deposition at Björkdal, northern Sweden. *Mineralium Deposita*, **29**, 139–149.

CARTEN, R. B. 1986. Sodium–calcium metasomatism; chemical, temporal, and spatial relationships at the Yerington, Nevada, porphyry copper deposit. *Economic Geology*, **81**, 1495–1519.

CLAESSON, L.-Å. 1985. The geochemistry of early Proterozoic metavolcanic rocks hosting massive sulphide deposits in the Skellefte district, northern Sweden. *Journal of the Geological Society*, **142**, 899–909.

CLAESSON, S. & LUNDQVIST, T. 1995. Origins and ages of proterozoic granitoids in the bothnian basin, central Sweden – isotopic and geochemical constraints. *Lithos*, **36**, 115–140.

DILLES, J. H. & PROFFETT, J. M. 1995. Metallogenesis of the Yerington Batholith, Nevada. *In*: PIERCE, F. W. & BOLM, J. G. (eds) Porphyry copper deposits of the American Cordillera. *Arizona Geological Society Digest*, **20**, 306–315.

Duckworth, R. C. & Rickard, D. 1993. Sulfide Mylonites from the Renstrom VMS Deposit, Northern Sweden. *Mineralogical Magazine*, **57**, 83–91.

Einaudi, M. T., Hedenquist, J. H. & Inan, E. E. 2003. Sulfidation state of fluids in active and extinct hydrothermal systems: transitions from porphyry to epithermal environments. *Society of Economic Geologists Special Publication*, **10**, 285–313.

González-Roldán, M. 2010. *Mineralogy, petrology and geochemistry of syn-volcanic intrusions in the Skellefte mining district, Northern Sweden*. PhD thesis, University of Huelva.

González-Roldán, M. J., Allen, R. L., Donaire, T. & Pascual, E. 2006. Secuencia de Emplazamiento, Alteración Hidrotermal y Metamorfismo en el Complejo Intrusivo de Jörn, Distrito Minero de Skellefte, Norte de Suecia. *Geogaceta*, **40**, 115–118.

Grant, J. A. 1986. The isocon diagram – a simple solution to Gresens' equation for metasomatic alteration. *Economic Geology*, **81**, 1976–1982.

Grant, J. A. 2005. Isocon analysis: a brief review of the method and applications. *Physics and Chemistry of the Earth*, **30**, 997–1004.

Groves, D. I., Goldfarb, R. J., Gebre-Mariam, M., Hagemann, S. G. & Robert, F. 1998. Orogenic gold deposits: a proposed classification in the context of their crustal distribution and relationship to other gold deposit types. *Ore Geology Reviews*, **13**, 7–27.

Hannington, M. D., Kjarsgaard, I. M., Galley, A. G. & Taylor, B. 2003. Mineral–chemical studies of metamorphosed hydrothermal alteration in the Kristineberg volcanogenic massive sulfide district, Sweden. *Mineralium Deposita*, **38**, 423–442.

Hedenquist, J. H., Izawa, E., Arribas, A. & White, N. C. 1996. *Epithermal Gold Deposits: Styles, Characteristics, and Exploration*. Society of Resource Geology, Komiyama Printing Co., Tokyo.

Janoušek, V., Farrow, C. M. & Erban, V. 2006. Interpretation of whole-rock geochemical data in igneous geochemistry: introducing Geochemical Data Toolkit (GCDkit). *Journal of Petrology*, **47**, 1255–1259.

Kathol, B. & Persson, P.-O. 1997. U–Pb zircon dating of the Antak granite, northeastern Västerbotten county, northern Sweden. *In*: Lundqvist, T. (ed.) *Radiometric Dating Results*, **3**. Geological Survey of Sweden, Uppsala, 6–13.

Kathol, B. & Weihed, P. 2005. *Description of regional geological and geophysical maps of the Skellefte District and surrounding areas*, Serie Ba **57**. Geological Survey of Sweden, Uppsala.

Lundberg, B. 1980. Aspects of the geology of the Skellefte fiield, northern Sweden. *Geologiska Föreningens i Stockholms Förhandlingar*, **102**, 156–166.

Lundström, I., Vaasjoki, M., Bergstrom, U., Antal, I. & Strandman, F. 1997. Radiometric age determinations of plutonic rocks in the Boliden area, the Hobergsliden granite and Stavaträsk diorite. *In*: Lundqvist, T. (ed.) *Radiometric Dating Results 3*. Geological Survey of Sweden, Uppsala, 20–30.

MacLean, W. H. 1990. Mass change calculations in altered rock series. *Mineralium Deposita*, **25**, 44–49.

MacLean, W. H. & Kranidiotis, P. 1987. Immobile elements as monitors of mass transfer in hydrothermal alteration: Phelps Dodge massive sulfide deposit, Matagami, Quebec. *Economic Geology*, **82**, 951–962.

Montelius, C. 2005. *The genetic relationship between rhyolitic volcanism and Zn–Cu–Au deposits in the Maurliden volcanic centre, Skellefte district, Sweden: Volcanic facies, Lithogeochemistry and Geochronology*. PhD thesis, Luleå University of Technology.

Montelius, C., Allen, R. L., Svenson, S. A. & Weihed, P. 2007. Facies architecture of the Palaeoproterozoic VMS-bearing Maurliden volcanic centre, Skellefte district, Sweden. *GFF*, **129**, 177–196.

Nakamura, N. 1974. Determination of REE, Ba, Fe, Mg, Na and K in carbonaceous and ordinary chondrites. *Geochimica et Cosmochimica Acta*, **38**, 757–775.

NEWBOLIDEN. 2010. Mineral resources on 31st December 2009, http://www.boliden.com.

Nysten, P. 1990. Tsumoite from the Björkdal gold deposit, Västerbotten county, northern Sweden. *Geologiska Föreningens i Stockholms Förhandlingar*, **112**, 56–60.

Petersson, J. 2002. *The genesis and subsequent evolution of episyenites in the Bohus granite, Sweden*. PhD thesis, University of Gothenburg.

Petersson, J. & Eliasson, T. 1997. Mineral evolution and element mobility during episyenitization (dequartzification) and albitization in the postkinematic Bohus granite, southwest Sweden. *Lithos*, **42**, 123–146.

Rickard, D. T. & Zweifel, H. 1975. Genesis of Precambrian sulfide ores, Skellefte District, Sweden. *Economic Geology*, **70**, 255–274.

Romer, R. L. & Nisca, D. H. 1995. Svecofennian crustal deformation of the Baltic Shield and U–Pb age of late-kinematic tonalitic intrusions in the Burtrask Shear Zone, Northern Sweden. *Precambrian Research*, **75**, 17–29.

Schlatter, D. M. 2007. *Volcanic Stratigraphy and Hydrothermal Alteration of the Petiknäs South Zn–Pb–Cu–Au–Ag Volcanic-hosted Massive Sulfide Deposit, Sweden*. PhD thesis, Luleå University of Technology.

Sillitoe, R. H. & Hedenquist, J. H. 2003. Linkages beween volcanotectonic settings, ore-fluid compositions, and epithermal precious metal deposits. *Society of Economic Geologists Special Publication*, **10**, 315–343.

Skiöld, T. 1988. Implications of new U–Pb zircon chronology to Early Proterozoic crustal accretion in northern Sweden. *Precambrian Research*, **38**, 147–164.

Skiöld, T. & Rutland, R. W. R. 2006. Successive similar to 1.94 Ga plutonism and similar to 1.92 Ga deformation and metamorphism south of the Skellefte district, northern Sweden: substantiation of the marginal basin accretion hypothesis of Svecofennian evolution. *Precambrian Research*, **148**, 181–204.

Skiöld, T., Öhlander, B., Markkula, H., Widenfalk, L. & Claesson, L. A. 1993. Chronology of Proterozoic orogenic processes at the Archean continental-margin in northern Sweden. *Precambrian Research*, **64**, 225–238.

Thompson, A. J. B. & Thompson, J. F. H. (eds) Dunne, K. P. E. (MDD Series ed.) 1996. Atlas of Alteration. *A Field and Petrographic Guide to Hydrothermal Alteration Minerals*. Geological Association of Canada, Mineral Deposit Division, Vancouver, **119**.

WARREN, I., SIMMONS, S. F. & MAUK, J. L. 2007. Whole-rock geochemical techniques for evaluation hydrothermal alteration, mass changes, and compositional gradients associated with epithermal Au–Ag Mineralization. *Economic Geology*, **102**, 923–948.

WEIHED, P. 1992*a*. *Geology and genesis of the Early Proterozoic Tallberg porphyry-type deposit, Skellefte district, northern Sweden*. PhD thesis, University of Gothenburg.

WEIHED, P. 1992*b*. Lithogeochemistry, metal and alteration zoning in the proterozoic Tallberg Porphyry-Type deposit, northern Sweden. *Journal of Geochemical Exploration*, **42**, 301–325.

WEIHED, P. & SCHÖBERG, H. 1991. Age of Porphyry-type deposits in the Skellefte District, northern Sweden. *Geologiska Föreningens i Stockholms Förhandlingar*, **113**, 289–294.

WEIHED, P., ISAKSSON, I. & SVENSSON, S.-Å. 1987. The Tallberg porphyry copper deposit in northern Sweden: a preliminary report. *Geologiska Föreningens i Stockholms Förhandlingar*, **109**, 47–53.

WEIHED, P., BERGMAN, J. & BERGSTRÖM, U. 1992. Metallogeny and tectonic evolution of the Early Proterozoic Skellefte district, northern Sweden. *Precambrian Research*, **58**, 143–167.

WEIHED, P., BERGMAN WEIHED, J., SORJONEN-WARD, P. & MATSSON, B. 2002*a*. Post-deformation, sulphide-quartz vein hosted gold ore in the footwall alteration zone of the Palaeoproterozoic Langdal VHMS deposit, Skellefte District, northern Sweden. *GFF*, **124**, 201–210.

WEIHED, P., BILLSTRÖM, K., PERSSON, P. O. & BERGMAN WEIHED, J. 2002*b*. Relationship between 1.90–1.85 Ga accretionary processes and 1.82–1.80 Ga oblique subduction at the Karelian craton margin, Fennoscandian Shield. *GFF*, **124**, 163–180.

WEIHED, P., BERGMAN WEIHED, J. & SORJONEN-WARD, P. 2003. Structural evolution of the Björkdal gold deposit, Skellefte district, northern Sweden: implications for Early Proterozoic mesothermal gold in the late stage of the Svecokarelian orogen. *Economic Geology*, **98**, 1291–1309.

WELIN, E. 1987. The depositional evolution of the Svecofennian supracrustal sequence in Finland and Sweden. *Precambrian Research*, **35**, 95–113.

WHITE, A. J. R. & CHAPPELL, B. W. 1977. Ultrametamorphism and granitoid genesis. *Tectonophysics*, **43**, 7–22.

WILSON, M. R., SEHLSTEDT, S. *ET AL*. 1987. Jörn: an early Proterozoic intrusive complex in a volcanic-arc environment, north Sweden. *Precambrian Research*, **36**, 201–225.

WINCHESTER, J. A. & FLOYD, P. A. 1977. Geochemical discrimination of different magma series and their differentiation products using immobile elements. *Chemical Geology*. **20**, 325–343.

Geological setting, alteration, and fluid inclusion characteristics of Zaglic and Safikhanloo epithermal gold prospects, NW Iran

SUSAN EBRAHIMI[1]*, SAEED ALIREZAEI[2] & YUANMING PAN[3]

[1]*School of Mining, Petroleum and Geophysics Engineering, Shahrood University, Shahrood, Iran*

[2]*Faculty of Earth Sciences, University of Shahid Beheshti, Tehran, Iran*

[3]*Department of Geological Sciences, University of Saskatchewan, Saskatoon, S7N 5E2, Canada*

**Corresponding author (e-mail: ebrahimisusan@shahroodut.ac.ir)*

Abstract: The Zaglic and Safikhanloo epithermal gold prospects are located in the Arasbaran zone, to the west of the Cenozoic Alborz-Azarbaijan magmatic belt in NW Iran. Mineralization is mainly restricted to quartz and quartz -carbonate veins and veinlets. Pyrite is the main sulphide, associated with subordinate chalcopyrite and bornite. Gold occurs as microscopic and submicroscopic grains in quartz and pyrite.

The country rocks are Tertiary intermediate–mafic volcanic and volcaniclastic rocks of andesite to trachy-andesite composition intruded by a composite granitic to syenitic pluton. They are medium- to high-K, calc-alkaline and alkaline rocks and display fractionated REE (rare earth element) patterns, with light rare earth elements (LREE) significantly enriched relative to the heavy rare elements (HREE). On primitive mantle normalized plots, they display depletions in Nb, Ti and P, and enrichments in Pb, which are common characteristics of arc-related magmas worldwide. Hydrothermal alteration minerals developed in the wall rocks include quartz, calcite, pyrite, kaolinite, montmorillonite, illite, chlorite, and epidote. Minor alunite occurs in Safikhanloo. Gold is locally enriched in the altered rocks immediate to the veins.

The ore-stage quartz from both prospects is dominated by liquid-rich fluid inclusions; vapour-rich inclusions are rare. The homogenization temperature varies between 170–230 and 170–330 °C and salinity varies between 1.4 to 9.5 and <1 to 6.7 wt% NaCl equivalent, for Safikhanloo and Zaglic, respectively. The occurrence of hydrothermal breccias, bladed calcite, adularia, and rare coexisting vapour- and liquid-dominant inclusions suggest that boiling occurred in the course of the evolution of the ore fluids. The large variations in Th and the salinity values can be explained by boiling and/or mixing.

Lack of sulphate minerals in the veins suggests that sulphides and gold precipitated from a reduced, H_2S-dominant fluid. Calculated $\delta^{34}S$ values for the ore fluid vary between -4.6 and -9.3‰. Sulphur could have been derived directly from magmatic sources, or leached from the volcanic and plutonic country rocks. Ore formation in Zaglic and Safikhanloo occurred in response to mixing, boiling, and interactions with wall rocks. Considering the intermediate-argillic alteration, the low contents of base metal sulphides, and the overall low salinities, the Zaglic and Safikhanloo can be classified as low-sulphidation epithermal systems.

Regional geological and geochemical exploration programs during 1990–2002, conducted by the Geological Survey of Iran, led to the discovery of many gold occurrences, mostly associated with three Cenozoic magmatic belts in the north (known as Alborz-Azarbaijan magmatic belt, AAMB), west-central (known as Urumieh-Dokhtar magmatic belt, UDMB) and east Iran (Fig. 1).

A highly promising area lies to the west of AAMB, where several porphyry style and skarn type base metal deposits were already known (Fig. 2). The area, known as Arasbaran zone, or Arasbaran metallogenic zone, hosts the world class Sungun porphyry Cu–Mo deposit (Kalagari *et al.* 2001), and Anjerd, Sungun, and Mazraeh Cu skarn deposits (Karimzadeh Somarin *et al.* 2002; Karimzadeh Somarin 2004). The new discoveries include Masjed-Daghi Cu–Au, Sonajil Cu, and Haftcheshmeh Cu–Mo porphyry deposits (Karimzadeh Somarin *et al.* 2002; Mohamadi & Borna 2006; Zarnab Company 2007), and several epithermal style gold occurrences, including Zaglic, Safikhanloo, Sharafabad, Masjed-Daghi, Sarikhanloo, Mivehrood, and Khoynehrood (Fig. 2).

The evolution of the Arasbaran zone (AZ) has been a controversial issue. While some authors consider the AZ as an integral part of the UDMB, based on similarities in the geochemistry of the

From: Sial, A. N., Bettencourt, J. S., De Campos, C. P. & Ferreira, V. P. (eds) *Granite-Related Ore Deposits.* Geological Society, London, Special Publications, **350**, 133–147.
DOI: 10.1144/SP350.8 0305-8719/11/$15.00

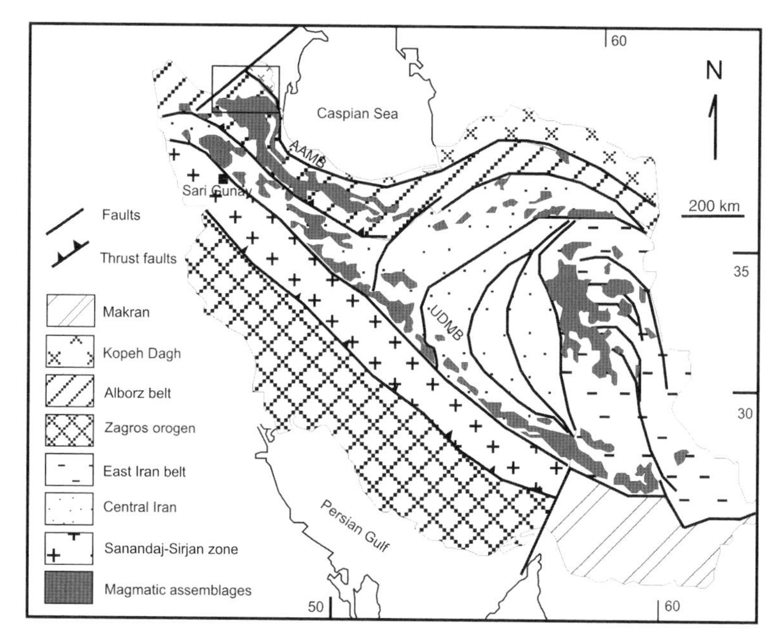

Fig. 1. A simplified map showing the main geological divisions, and the distribution of the Cenozoic magmatic assemblages, in Iran (after Stöcklin 1968; Alavi 1996). The square shows the location of the Zaglic and Safikhanloo prospects and filled square shows the location of Sari Guany deposit.

volcanic–plutonic rocks (Nogol-Sadat 1993), others argue for a rift setting (Riou 1979) or rift- and collision-related setting (e.g. Moayyed *et al.* 2008). Hassanzadeh *et al.* (2002) presented evidence for an intra-arc rifting in UDMB during Oligocene–Miocene, leading to the separation and northward movement of what is now known as Alborz-Azarbaijan magmatic belt (see Fig. 1).

The UDMB is dominated by calc-alkaline volcanic and plutonic rocks and is considered to be an Andean type magmatic belt generated by NW-dipping subduction of Neo-Tethyan oceanic crust

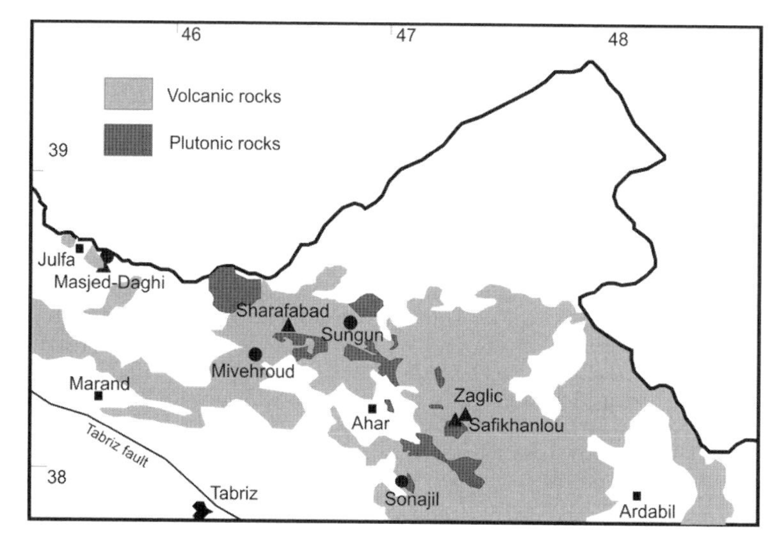

Fig. 2. Simplified map of NW Iran showing the distribution of Cenozoic magmatic rocks. Filled triangles: epithermal deposits; filled circles: porphyry deposits.

beneath the Central Iranian micro-continent, and the collision of the African and Eurasian plates during the Alpine orogeny in the Tertiary (Berberian & King 1982; Stampfli *et al.* 2001). The UDMB hosts many porphyry style deposits and occurrences, including the world-class Sarcheshmeh [>1200 MT of ore at 0.8% Cu and 0.025% Mo; (Shahabpour 1982)].

The recent discovery of epithermal and porphyry style mineralization in AZ suggests that the area is potentially productive and merits further investigations. The epithermal systems are covered by detailed geological mapping at 1:5000 and 1:2000 scales, several hundred metres of trenches and drill holes, and assays for Au, Ag, Cu, Pb and Zn (Heydarzadeh 2005; Mohamadi 2006). Earlier works indicated that the epithermal systems share some similarities, including the host rocks, structural controls, and the quartz-pyrite dominant vein materials. However, significant variations exist with respect to the associated minerals in the veins, the alteration assemblages, and the fluid T_H and salinity values (Ebrahimi 2008; Alirezaei *et al.* 2008). The present study focuses on two of the epithermal systems, Zaglic, Safikhanloo, in an attempt to better understand the geochemistry and tectonic setting of the host rocks, the nature and possible source of the ore fluids, and the mechanisms of ore formation.

Geological background

The Zaglic and Safikhanloo areas lie in the Ahar quadrangle in Arasbaran zone that is characterized by extensive outcrops of Cretaceous flysch type sediments and Cenozoic volcanic and plutonic rocks. Riou (1979) distinguished four magmatic events in the Ahar quadrangle during Upper Jurassic–Tertiary, starting with dominantly silica-undersaturated, alkaline and shoshonitic intermediate–mafic volcanic rocks, followed by normal calc-alkaline and alkaline felsic-intermediate volcanic–plutonic assemblages. Riou (1979) argued for a rift-related setting for the Ahar quadrangle and the surrounding areas during Paleocene–Oligocene times.

The Zaglic and Safikhanloo occurrences, only 4 km apart, lie in an area covered by Eocene–Miocene volcanic, pyroclastic and intrusive rocks (Figs 3 & 4). The oldest rocks include dark grey to green, porphyritic and microlitic andesite, basaltic andesite, porphyritic trachy-andesite to latite-andesite of Upper Eocene age (Ean). The rocks consist mainly of plagioclase (andesine–oligoclase), hornblende, clinopyroxene and subordinate biotite and quartz. The dominantly andesitic lava flows are covered by light to dark green, fine-grained tuffaceous materials (Evt) of intermediate compositions, as the scattered crystals of plagioclase, hornblende, pyroxene and biotite suggest. The unit gradually changes into andesitic tuff breccias and andesitic breccias.

The volcanic–pyroclastic sequence in Safikhanloo was intruded by a hypabyssal intrusion (Osy) composed dominantly of syenite associated with subordinate granite and monzosyenite. The rocks consist mainly of plagioclase, alkali feldspars, quartz, hornblende, and biotite. Apatite and Fe–Ti oxides are common accessory minerals. An Oligocene age is proposed for the intrusion by analogy with similar intrusions in the Ahar quadrangle. Numerous acidic to intermediate dykes intruded into the Upper Eocene volcanic and pyroclastic

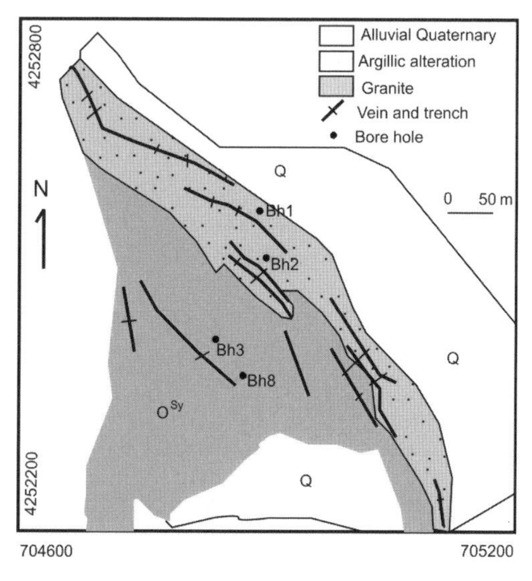

Fig. 3. Geological map of the Safikhanloo prospect, simplified after Mohamadi (2006).

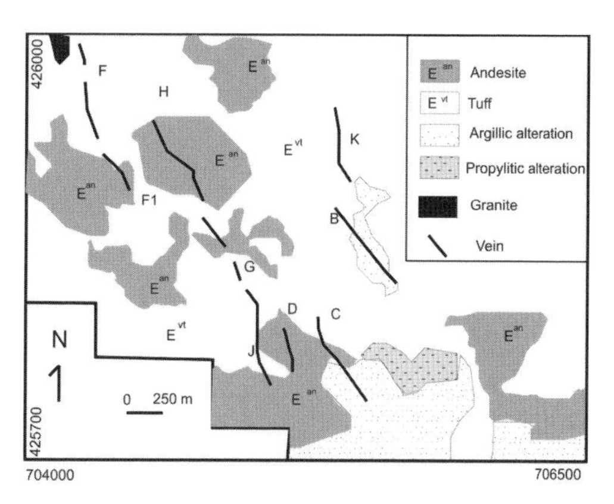

Fig. 4. Geological map of the Zaglic prospect, simplified after Heydarzadeh (2005).

Table 1. *Analysis of representative samples from Safikhanloo and Zaglic areas*

Sample no. Lithology (total alkali-silica)	SH08 Andesite	SH47 Trachy-andesite	SH55 Trachy-andesite	DH5-S43 Basaltic-trachy-andesite	DH5-S48 Basaltic-trachy-andesite	DH6-S16 Rhyolite
Weight %						
SiO_2	59.2	54.9	58.1	48.1	50.4	69.2
Al_2O_3	15.7	16	16.8	16.6	17.4	14.2
CaO	6.06	5.34	6.11	6.15	6.03	1.86
MgO	2.51	3.3	2.47	7	2.39	1.46
Na_2O	2.75	4.69	3.75	3	4.85	3.36
K_2O	2.72	2.94	2.35	3.48	2.3	4.54
Fe_2O_3	5.2	5.52	6.41	8.16	8.18	3.33
MnO	0.1	0.1	0.14	0.15	0.11	0.06
TiO_2	0.66	0.87	0.75	1.17	1.18	0.58
P_2O_5	0.25	0.4	0.31	0.63	0.64	0.28
LOI	4.7	4.45	2.45	4.1	5.1	2.05
Sum	100.1	98.7	99.8	98.8	98.8	98.1
ppm						
Li	49.3	29.4	29.7	53.4	31.8	35.1
Sc	16	20	20	23	19	12
V	111	113	139	181	196	54
Rb	50	86	50	92	90	145
Sr	691	744	675	1123	1121	484
Y	17	19	19	20	20	16
Zr	150	286	154	141	155	198
Nb	22.2	47.6	19.5	23.1	23.9	42.7
Mo	0.79	0.99	1.12	0.75	0.58	2.89
Cs	3.34	3.98	0.67	10.57	8.08	3.64
Ba	1231	623	753	998	554	652
La	34	54	35	43	48	55
Ce	61	104	65	85	95	100
Pr	6.7	11.5	7.1	10.2	10.9	10.9
Nd	22.3	37.5	24.7	37.1	40.2	35.2
Sm	3.9	6.1	4.3	6.7	7.4	5.6
Eu	1.24	1.70	1.34	2.11	2.20	1.23
Gd	3.8	5.5	4.3	6.0	6.7	4.7
Tb	0.48	0.62	0.57	0.73	0.75	0.56
Dy	3.0	3.7	3.5	4.0	4.0	3.2
Ho	0.58	0.69	0.70	0.73	0.76	0.56
Er	1.85	2.08	2.07	2.07	2.16	1.63
Tm	0.28	0.28	0.29	0.27	0.26	0.21
Yb	1.84	1.98	2.11	1.75	1.67	1.31
Lu	0.28	0.24	0.28	0.23	0.21	0.17
Hf	3.83	7.41	3.88	3.50	3.61	5.43
Ta	1.58	3.05	1.25	1.24	1.31	2.08
Tl	0.37	0.37	0.20	0.23	0.41	1.17
Pb	15.4	15.8	10.7	9.8	10.4	13.6
Th	13.66	24.45	8.47	5.28	7.08	14.88
U	3.51	6.49	2.31	1.54	1.96	4.72
P	1093	1827	1463	2808	2935	1290
Ti	3815	5226	4632	6919	7254	3517
Cr	19	80	17	194	15	20
Co	37	28	31	32	32	40
Ni	11	41	11	93	16	16
Cu	18	55	34	21	66	87
Zn	81	84	90	93	95	52
Ge	0.76	0.38	0.86	0.84	0.65	0.22
As				26	18	77
Ag	0.41	0.39	0.40	0.40	0.48	2.33
Sn	2.01	4.06	1.79	1.84	2.07	4.12
Sb	0.69	1.06	0.81	2.30	2.64	1.94
W	210	122	172	55	93	412
Ga	13	15	15	13	16	12

rocks and the Oligocene intrusive rocks. The dykes vary in length and width between 50–500 and 1–20 m, respectively.

The geochemistry of the country rocks

Representative samples from the least altered country rocks were selected for whole rock analysis. Major oxides were determined by the X-ray fluorescence spectrometry (XRF) at SGS Assay Labs., Don Mills, Ontario. Trace elements were analyzed by inductively coupled plasma mass spectrometry (ICP-MS), using a Perkin Elmer Elan 5000, at the Department of Geological Sciences, University of Saskatchewan. The sample locations are indicated in Figures 3 and 4, and a full list of the analysis is shown in Table 1.

On the total alkali-silica diagram (LeBas *et al.* 1986) the samples plot within andesite, trachy-andesite, and basaltic trachy-andesite fields (Fig. 5). The sample lying in the rhyolite domain represents the intrusive body in Safikhanloo. On the calc-alkaline–tholeiitic discrimination diagram of Irvine & Baragar (1971), all samples plot within the calc-alkaline domain (Fig. 6), consistent with data for a larger set of samples from Sharafabad (Ebrahimi & Alirezaei 2008). On R1-R2 discrimination diagram (De La Roche *et al.* 1980), the samples fall in alkaline and sub-alkaline domains and display a bimodal character (Fig. 7).

Volcanic rocks from both Safikhanloo and Zaglic areas display similar REE patterns, with LREE significantly enriched relative to the HREE (Fig. 8). The La_N/Yb_N ratios vary between 11–21. No distinct Eu anomaly is displayed by the volcanic rocks. The only sample with negative Eu anomaly belongs to the granodiorite intrusion in Safikhanloo that appears to have been crystallized from a fractionated magma.

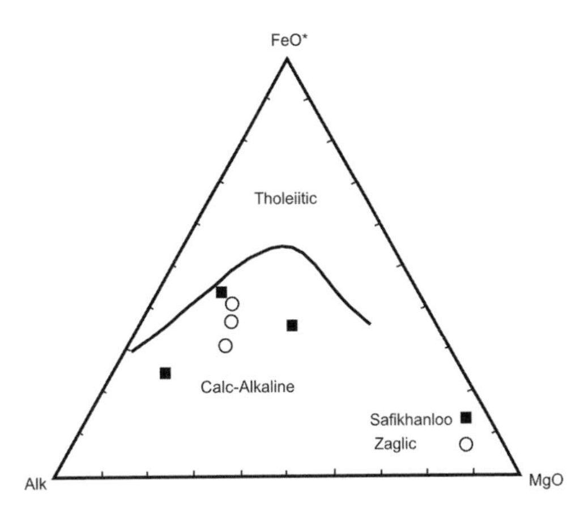

Fig. 6. Plots of samples on the calc-alkaline–tholeiitic discrimination diagram (Irvine & Baragar 1971) showing a calc-alkaline affinity for the country rocks.

The relatively high La_N/Yb_N ratios can be attributed to a low degree of partial melting in the source area (Riou *et al.* 1981), and/or a high degree of fractionation of the parent magma. The absence of negative Eu anomalies in the country rocks provides evidence against fractionation as the main cause of the high La_N/Yb_N ratios, as fractionation of plagioclase is expected to result in distinct negative Eu anomalies. The low contents of HREE might be attributed to the presence of garnet in the source area.

A multi-element spider diagram is shown in Figure 9. General features shown by all samples are enrichments in large-ion lithophile element

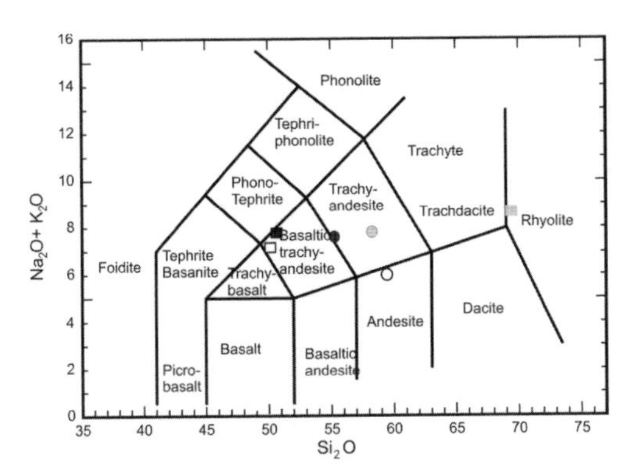

Fig. 5. Plots of representative rocks from the Zaglic (circle) and Safikhanloo (square) on total alkali v. SiO_2 diagram of Le Bas *et al.* (1986).

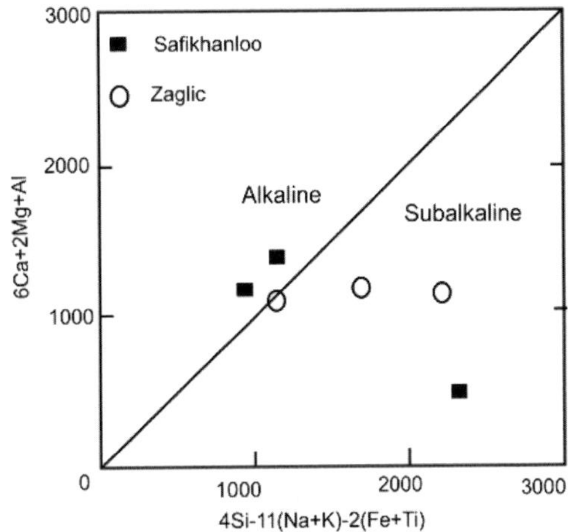

Fig. 7. Plots of samples on the R1–R2 alkaline–subalkaline discrimination diagram of De La Roche *et al.* (1980).

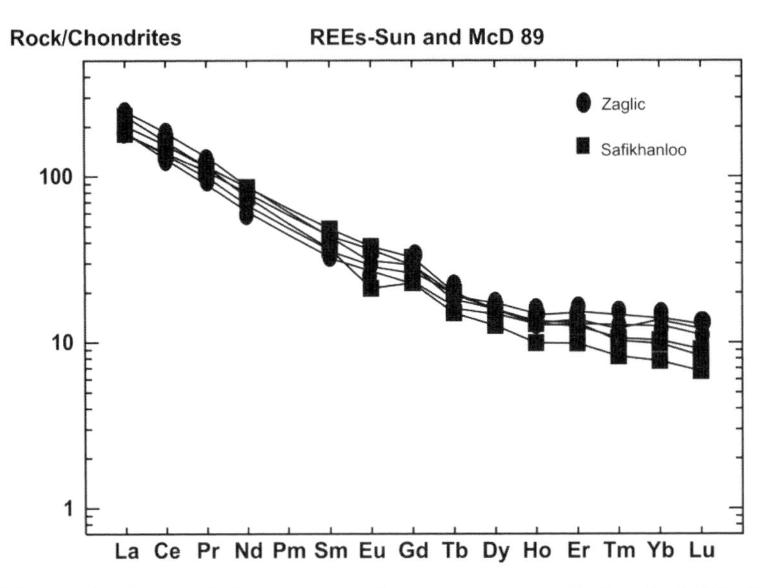

Fig. 8. Chondrite-normalized rare earth elements patterns for representative samples from the Safikhanloo and Zaglic areas.

(LILE) and LREE relative to HREE and HFSE (high field strength elements), relative depletions in Nb, Ti and P, and enrichments in Pb which are common characteristics of many arc- and rift-related magmas worldwide (e.g. Gill 1981; Pearce 1983; Richards 2009).

A comparison is made with the Cenozoic, dominantly arc-related, Urumieh-Dokhtar belt and with the middle Miocene country rocks from the newly discovered Sari Gunay epithermal gold deposit in west Iran (Fig. 9) for which a collisional or transitional tectonic setting has been proposed (Richards *et al.* 2006). The country rocks at Zaglic and Safikhanloo are comparable to those from Sari Gunay. The association of alkaline and calc-alkaline rocks in the study area is consistent with a transitional arc-rift setting, or a back-arc extension. A similar setting has been reported for many similar

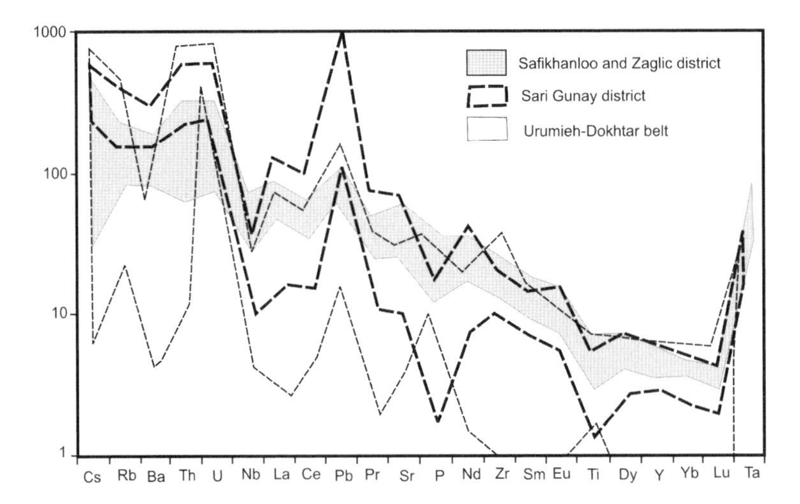

Fig. 9. Primitive mantle-normalized spider diagram for representative samples from the Safikhanloo and Zaglic areas. Plots for Urumieh-Dokhtar belt, and country rocks from Sari Gunay epithermal deposit (Richards *et al.* 2006) are shown for comparison normalization data from Sun & McDonough (1989).

Fig. 10. An auriferous vein composed mostly of quartz and Fe-oxides–hydroxides, and silicified wall rocks, in Zaglic prospect. The Fe oxides–hydroxides are the products of oxidation and decomposition of original pyrite.

(low-sulphidation) epithermal precious metal deposits worldwide (Sillitoe & Hedenquist 2003).

Alteration

The country rocks at Zaglic and Safikhanloo were variably affected by regional argillic, propylitic and silicic alterations prior to the vein formation (Figs 3 & 4). The alterations are common features in volcanic and pyroclastic rocks in the Arasbaran zone; argillic alteration has locally resulted in the development of workable deposits of clay minerals. Alteration associated with the mineralization is confined to thin halos of silicified rocks adjacent to the veins, bordered by argillic and propylitic zones outward. The alteration minerals, as determined from petrographic and X-ray diffraction studies, include opal, cristobalite, microcrystalline quartz, pyrite sericite, kaolinite, illite, montmorillonite, chlorite, epidote and calcite. Minor hypogene alunite occurs associated with clay minerals at Safikhanloo.

Mineralization

Gold mineralization occurs in quartz and quartz-calcite veins and veinlets containing minor sulphides (Fig. 10). Some 11 and 10 veins have been mapped in Zaglic and Safikhanloo, 125–850 m long, 3–8 m wide, and 150–800 m long, 2–10 m wide, respectively. The two occurrences display overall similar ore mineralogy and textures. Pyrite is the main ore mineral, associated with minor chalcopyrite, covellite, bornite, and trace molybdenite in both prospects. Gold occurs mostly as microscopic grains in quartz and pyrite. Silica occurs as grey, white, and clear quartz, as well as amorphous silica. The main textures displayed by the vein materials include massive, crustiform banding, vuggy, and breccias.

Induced Polarization (IP) and Resistivity (RS) exploration techniques on the auriferous veins suggested limited vertical extensions for the sulphide bearing materials in Safikhanloo and Zaglic (Poureh 2004). This was further supported by results from dipole–dipole arrays.

Bulk samples from the veins and the immediate mineralized wall rocks yielded 0.1–17 ppm Au, 1–34 ppm Pb, 1–130 ppm Zn, 1–150 ppm Cu, 1–190 ppm Ag, and $<$0.01–0.03 ppm Hg in Safikhanloo, and 0.1–16.5 ppm Au, 1–764 ppm Ag, and 1–800 ppm Cu, locally up to 3%, in Zaglic. The variations of elements appear to be different in the two prospects; however, when considering the average values, no significant differences can be distinguished. The two areas have been estimated to contain *c*. 2 metric tons of recoverable gold.

Paragenesis and paragenetic sequence

Based on the mineral paragenesis and crosscutting relationships, four stages can be distinguished at both prospects: pre-main mineralization, main mineralization, post-main mineralization, and supergene (Fig. 11).

Pre-main mineralization

A dark gray, microcrystalline quartz, rich in microscopic pyrite, formed early in the evolution of the veins, followed by brecciation and precipitation of abundant fine-grained gray quartz (Fig. 12a). Pyrite occurs mostly as fine- to medium-grained subhedral to anhedral crystals in the grey quartz, and as disseminations in the wall rocks. Minor magnetite and rutile, and trace molybdenite form at this stage.

Main mineralization

The main stage of gold mineralization is represented by white-gray, sulphide-rich quartz breccias. The sulphides include pyrite, associated with minor chalcopyrite, bornite, covellite, and trace cubanite and tetrahedrite (Fig. 12d). Gold occurs as scattered

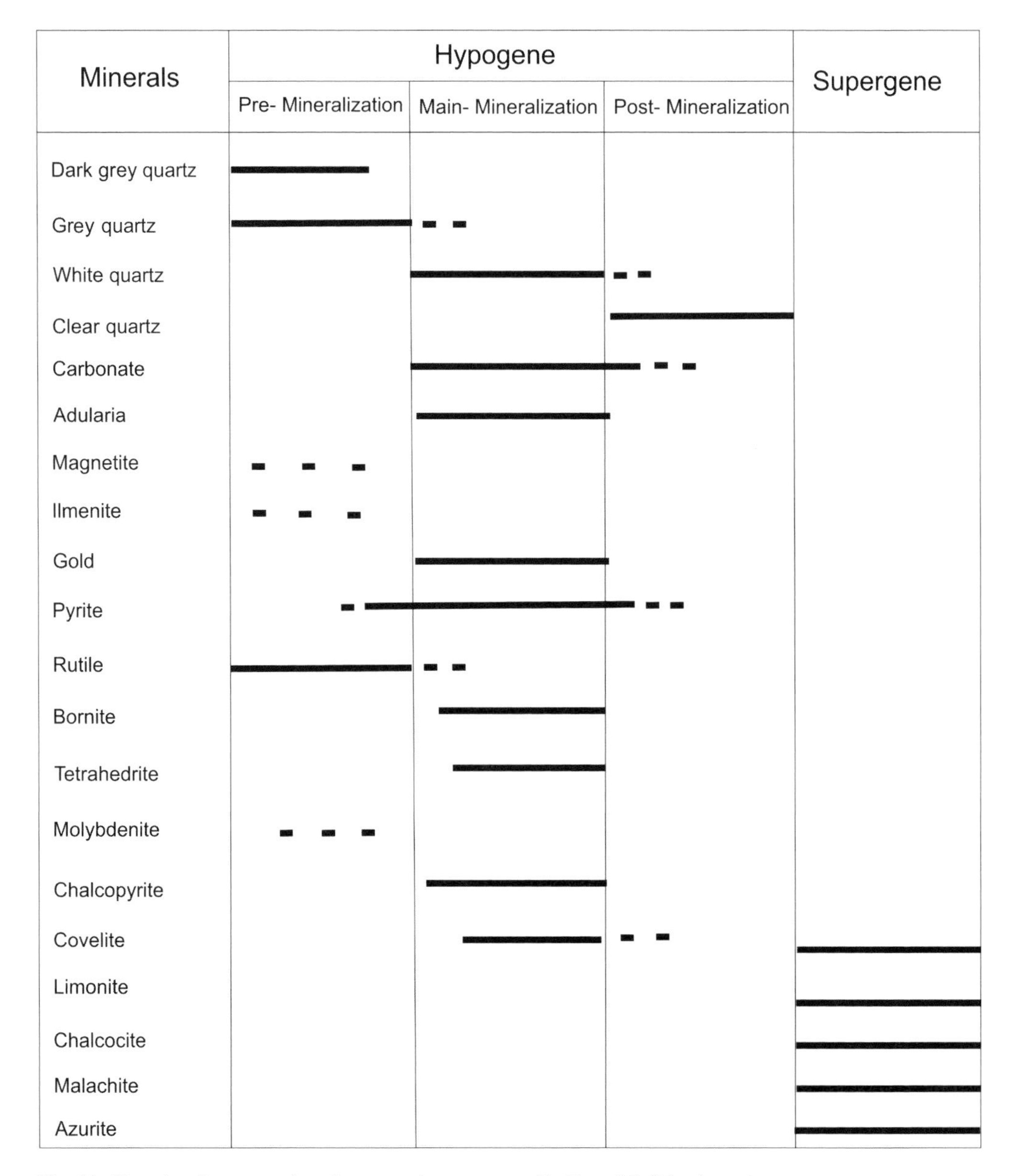

Fig. 11. The mineral paragenesis and paragenetic sequence at Zaglic and Safikhanloo vein systems.

microscopic grains in quartz and pyrite, as well as submicroscopic particles in pyrite (Fig. 12e). Gold grains vary in size from 10–130 μm.

Carbonates are common in Safikhanloo occurring as cementing materials in the hydrothermal breccias. Calcite locally occurs as bladed crystals (Fig. 12f); most bladed calcite is partially to totally replaced by quartz, and to a lesser extent, by adularia. Carbonates are not so common in Zaglic. Minor adularia occurs as euhedral rhombic grains associated with quartz in the breccias.

Post-main mineralization

This stage is represented by euhedral, coarse, clear quartz, associated with medium- to coarse-grained euhedral pyrite. The quartz often fills cavities in the vein materials. Deposition of calcite continued

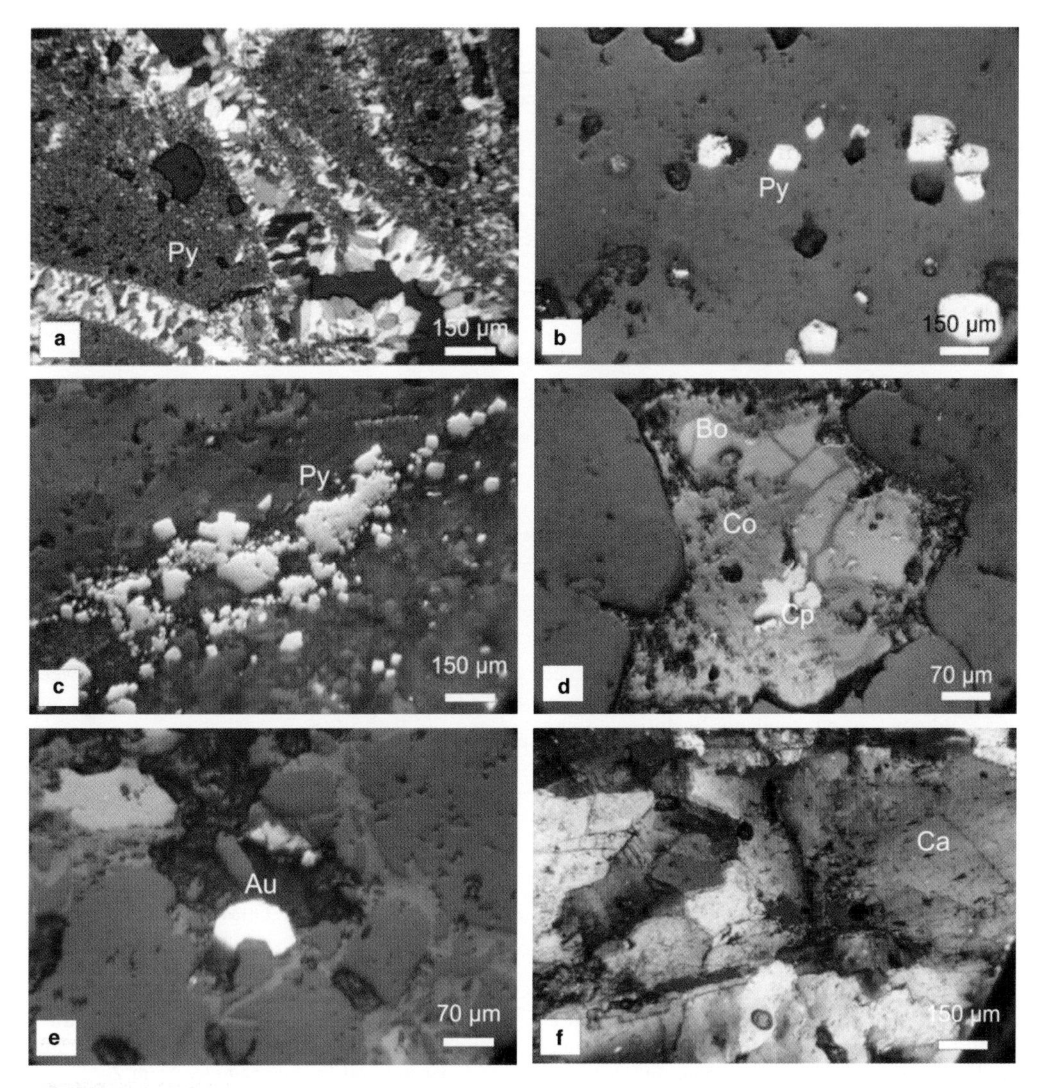

Fig. 12. Characteristic features of vein materials from Safikhanloo and Zaglic prospects. (**a**) Lenticular dark gray quartz (dark grey), rich in microscopic pyrite (black spots). (**b** and **c**). Pyrite from the main stage of mineralization occurring as disseminations (b) and veinlets (c) in silica. (**d**) Bornite (Bo) and chalcopyrite (Cp) replaced by covellite during supergene processes. (**e**) microscopic gold grain (Au) in quartz. (**f**) Coarse-grained (bladed) calcite (Ca) from the main stage of mineralization. a, b from Zaglic and c–f from Safikhanloo.

into this stage at Safikhanloo occurring in microfractures as well as filling open spaces, particularly in the upper parts of the veins. Calcite is a minor constituent in Zaglic. No gold and base metal sulphides are associated with this stage.

Supergene stage

Sulphide minerals at Zaglic and Safikhanloo have been oxidized to shallow depths. The supergene mineral assemblages consist of chalcocite, covellite, azurite, malachite, and iron oxides (mostly limonite and goethite). Covellite occurs both as a primary and as a secondary supergene mineral (Heydarzadeh 2005).

Fluid inclusion studies

Doubly polished thin sections, 50 µm thick, were prepared for fluid inclusion studies from ore-stage quartz and calcite collected from trenches and drill holes. Fluid inclusions in calcites were found to be

Table 2. *Fluid inclusion data from Zaglic and Safikhanloo prospects*

Area, vein and trench number, drill hole and depth of samples	Mineral	Type	N (Th)	Th range	N (Tm)	Tm range	wt% NaCl equiv. range	Comment
Zaglic-D- TP1	Quartz	P	53	190–250	10	−0.5, −4.2	0.87–6.7	L > V
Zaglic-F- TP8	Quartz	P, PS	50	210–330	10	−0.1, −1.6		L > V
Zaglic-D- TP3	Quartz	P	50	190–250	10	−0.1, −1.6	0.17–2.7	L > V
Safikhanlou-D1-S-29	Quartz	P	49	170–230	10	−0.8, −6.2	0.17–2.7	L > V
Safikhanlou-D1-S-43	Quartz	P, PS	49	170–230	10	−0.8, −3.7	1.4–9.5	L > V
Safikhanlou-D1-S-48	Quartz	P	48	170–230	10	−1.8, −4.4	1.4–6	L > V
							3–7	

Th, homogenization temperature; Tm, ice-melting temperature; P, primary fluid inclusion; PS, pseudosecondary fluid inclusion; N, number; L, liquid; V, vapour; L > V= liquid-rich inclusion.

mostly small (<5 μm) and difficult to characterize. However, most quartz samples contained abundant fluid inclusions of appropriate sizes for microthermometric studies. A short description of the samples, and their locations, is presented in Table 2.

Fluid inclusion data were obtained using a Fluid Inc.-adapted USGS gas flow heating and freezing system at the Department of Geological Science University of Saskatchewan, Canada. The stage was calibrated using synthetic inclusions (pure water and fluorite). The measurements are considered to be accurate to ± 2 °C for homogenization temperatures (Th) and ± 0.2 °C for melting temperatures (Tm). Salinities are in NaCl wt% equivalant using Tm values and the salinity-freezing point depression table of Bodnar (1993).

Primary, pseudosecondary and secondary fluid inclusions were distinguished using the criteria of Bodnar *et al.* (1985). The fluid inclusions are irregular, spherical, or rod-shaped, and range in size from 5–100 μm, independent of origin and occurrence. Microthermometric measurements were made mostly on liquid-rich inclusions that homogenized by disappearance of the vapour bubble. Homogenization temperatures were determined on 298 inclusions, 60 of which were also examined for the temperature of final melting.

The primary inclusions occur parallel to growth zones in quartz or occur in clearly isolated positions. Most inclusions are two-phase (liquid + vapor), liquid-rich, containing 70–90 vol% liquid and 10–30 vol% vapour, and homogenize to a liquid phase upon heating. Coexisting liquid-rich and vapour rich inclusions are rare. No evidence of liquid or gaseous CO_2 was found in the investigated inclusions. In the absence of CO_2 in the observed inclusions, the measured salinities represent the maximum values. The fluid inclusion data are summarized in Table 2.

For Safikhanloo, homogenization temperatures (Th) vary between 170–230 °C (Fig. 13a). No significant variations in Th values were found for various depths. Final melting temperatures (Tm) are between −0.8 to −6.2 °C, and calculated values of apparent fluid salinities vary between 1.4 to 9.5 wt% NaCl equivalent (Fig. 14b). For Zaglic, Th and Tm values vary between 190–331 and −0.1 to −4.2 °C, respectively (Fig. 13b). Calculated values of apparent fluid salinities are between 0.17 to 6.7 wt% NaCl equivalent (Fig. 14a).

Sulphur isotope ratios

Representative pyrite-rich samples from the ore stage vein materials were analyzed for sulphur isotope ratios at the G. G. Hatch Stable Isotope Laboratories, University of Ottawa, using a Thermo Finnigan Delta Plus.

Sulphur isotope data were obtained for two samples from Zaglic and three samples from Safikhanloo (Table 3). The $\delta^{34}S$ values for three samples from Safikhanloo are −2.9, −5.2 and −7.6 per mil and for two samples from Zaglic are −3.3 and −5.7 per mil. This range of the $\delta^{34}S$ values is comparable to that from other epithermal gold deposits in Iran (Fig. 15).

Using fractionation factor of Ohmoto & Rye (1979) and average temperatures of 230 and 200 °C for Zaglic and Safikhanloo, respectively, the calculated $\delta^{34}S$ values of H_2S in equilibrium with pyrite are in the range −5 to −7 per mil and −4.6 to −9.3 per mil for the two prospects.

The absence of sulphate minerals in the veins suggest that sulphur was transported in a reduced state, most likely as HS^-, and that the negative $\delta^{34}S$ values can not be attributed to fractionation processes. The $\delta^{34}S$ values do not point to a specific source for sulphur. The negative $\delta^{34}S$ values do not necessarily rule out a direct magmatic source, as a wide spread in $\delta^{34}S$ values, −10 to over +10 per mil has been indicated for many arc- and rift-related

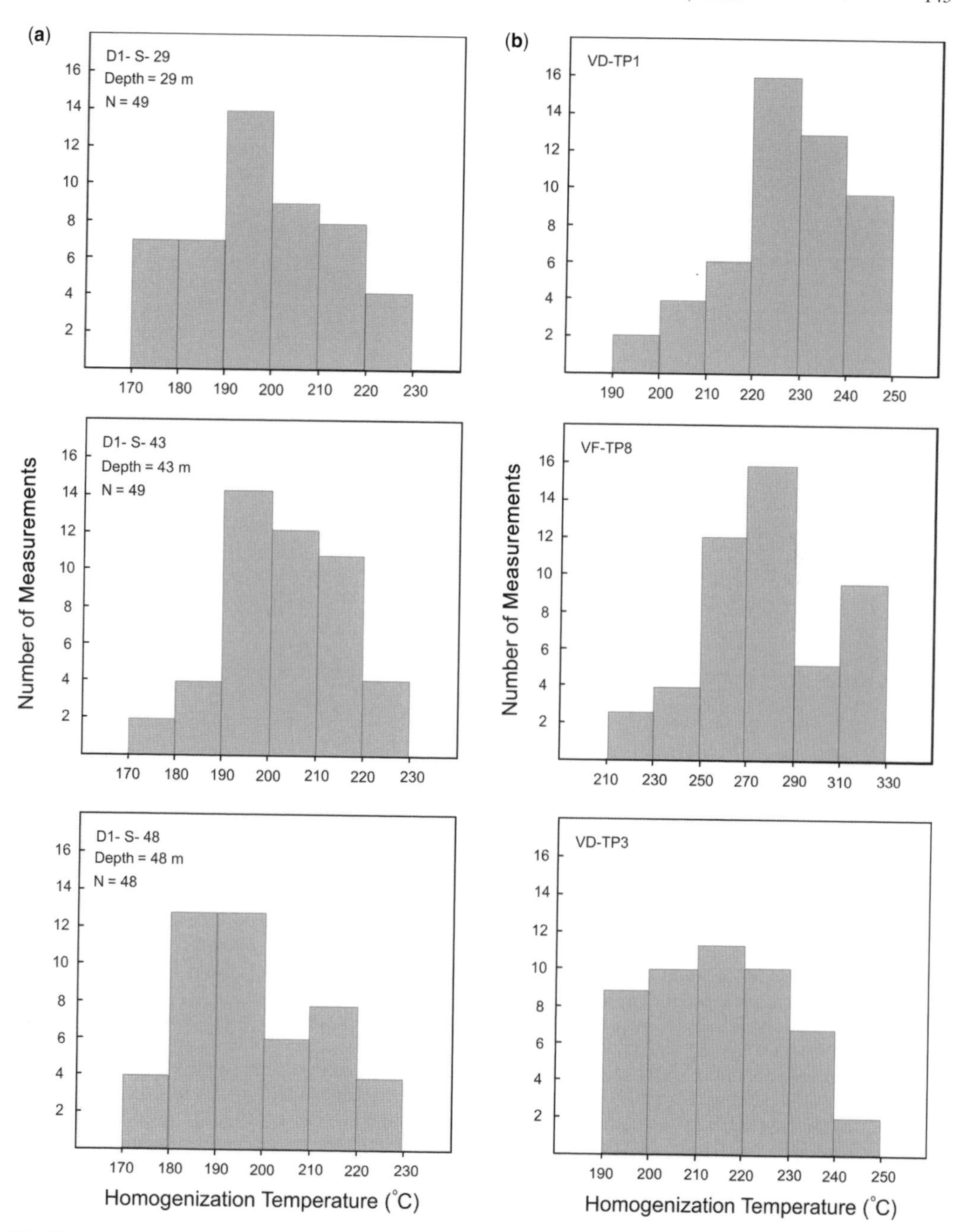

Fig. 13. (**a**) Distribution of Th data for samples from three different depths across borehole D-1 in Safikhanloo prospect. (**b**) Distribution of Th data for samples from Zaglic prospect. The samples were collected from trenches.

magmas (Hoefs 2009). The relatively low salinities argue against a direct magmatic source for fluids. A possible source for ^{34}S-depleted sulphur would be sedimentary, sulphide bearing rocks. However, such rocks are not identified in the area around the prospects. They are also missing in Sharafabad, Gandi and Chahmesi (Shamanian *et al.* 2004; Ebrahimi 2008; Modrek 2009). We suggest that sulphur was likely supplied through leaching of the older volcanic and plutonic rocks. Oxygen and hydrogen

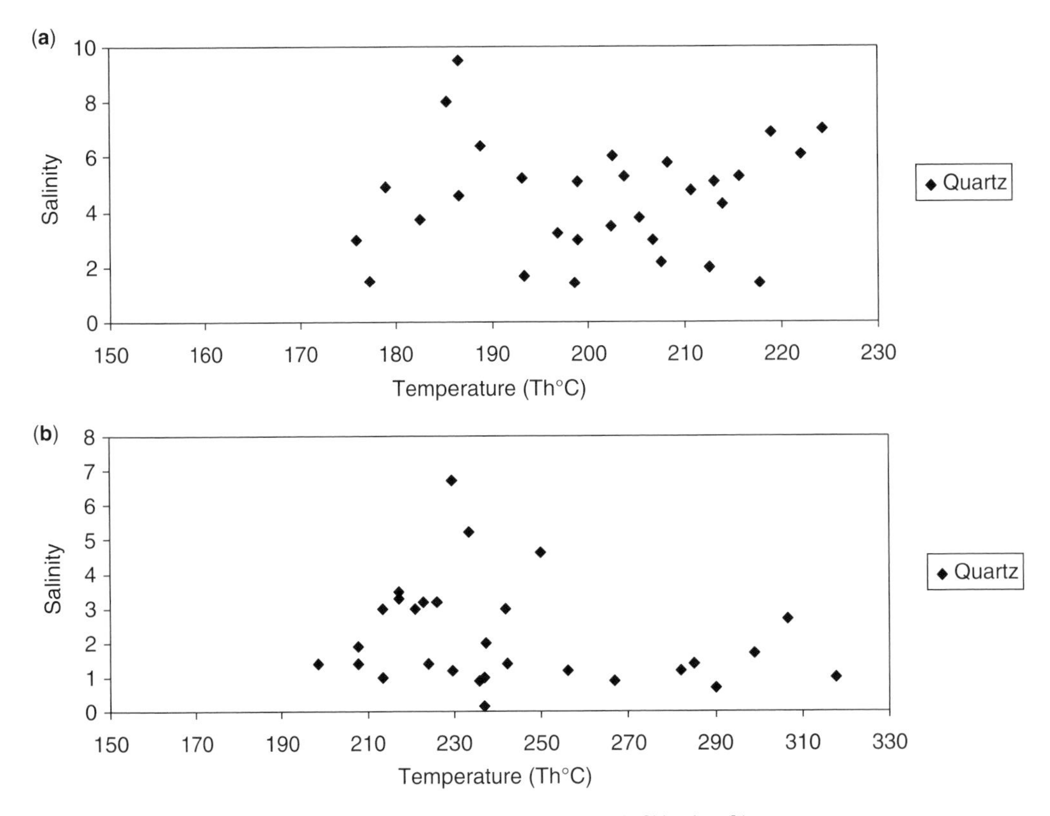

Fig. 14. Distribution of Th v. salinity in samples from Zaglic (**a**) and Safikhanloo (**b**).

isotope data are required to be able to further discuss the source of the ore fluids.

Discussion and conclusion

Gold-bearing veins at Zaglic and Safikhanloo prospects occur in an area covered by the Cenozoic felsic-intermediate volcanic, pyroclastic and intrusive rocks. The least altered country rocks display a trend from calc-alkaline to alkaline, and feature more typical of continental arc/rift magmas (i.e. distinct enrichments in LILE, fractionated REE patterns, depletions in Nb, Ti and P, and enrichments in Pb). The occurrence of Palaeogene alkaline volcanism, particularly in Arasbaran Zone to the north of the Urumieh-Dokhtar magmatic belt, has led some authors to argue for a deep faulting and rifting phase during an overall compression regime in UDMB (Riou *et al.* 1981; Berberian & King 1982).

Wall rock alteration is characterized by the occurrence of calcite, illite, montmorillonite, chlorite, sericite, epidote, and pyrite in both prospects, a mineral assemblage typical of low-sulphidation epithermal deposits (Henley 1985; Hedenquist *et al.* 2000). Minor kaolinite and alunite, common

Table 3. *Sulphur Isotope data from Zaglic and Safikhanloo area*

Number of sample	Area	Mineral	$\delta^{34}S$ CDT	$\delta^{34}S$ H_2S*
D-TP1	Zaglic	Pyrite	−3.3	−5.0
F-TP8	Zaglic	Pyrite	−5.7	−7.1
S-D1-29	Safikhanloo	Pyrite	−2.9	−4.6
S-D1-43	Safikhanloo	Pyrite	−7.6	−9.3
S-D1-48	Safikhanloo	Pyrite	−5.2	−7.0

*Calculated composition of hydrothermal fluid in equilibrium with pyrite using fractionation factors of Ohmoto & Rye (1979).

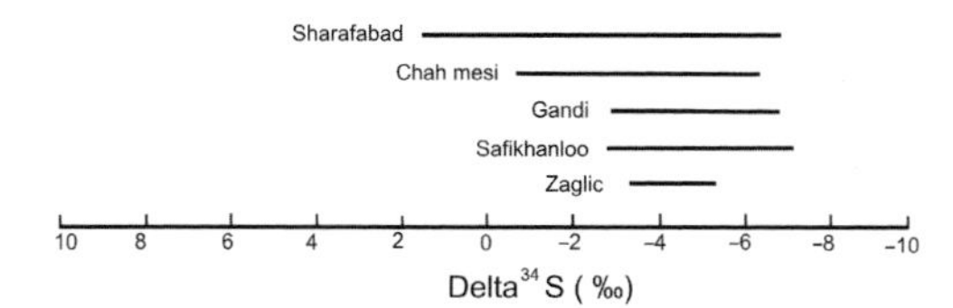

Fig. 15. Variations of $\delta^{34}S$ values for Zaglic and Safikhanloo. Variations for Sharafabad, Gandi and Chah mesi are shown for comparison.

products of steam-heated acid-sulphate alteration, occur at Safikhanloo, implying that water table at this prospect was lying at deeper levels compared to that of the Zaglic (cf. Hedenquist & Browne 1989). Steam heated acid-sulphate water forms only within the vadose zone above the water table (e.g. Hedenquist *et al.* 2000). The occurrence of illite in country rocks in both prospects, and adularia in Safikhanloo, indicate that the pH of the solutions was near neutral and that the fluids were strongly affected by boiling and CO_2 exsolution (cf. Giggenbach 1997). The relatively wide variations in the salinity values (0.17–6.7 and 1.4 to 9.5 wt% NaCl equiv. for Zaglic and Safikhanloo, respectively) could be explained by extensive boiling and vaporization of a low salinity fluid (cf. Simmons & Browne 2000). Alternatively, it could be attributed to mixing with an exotic brine of magmatic origin, basinal fluids, or fluids equilibrated with evaporate strata (e.g. Simmons *et al.* 1988). Involvement of a fluid of magmatic origin is not supported by the alteration and the vein mineral assemblages, or the fluid inclusion data. A contribution from basinal brines or evaporate sediments is ruled out by the absence of sedimentary, salt bearing strata in the Zaglic-Safikhanloo area and the surroundings.

The $\delta^{34}S$ values for pyrites from the main stage of mineralization fall in the range −2.9 to −7.6‰. In the absence of sulphate minerals in the veins, it is assumed that H_2S was the dominant sulphur species in the ore fluids. The sulphur isotope ratios could be explained by derivation from a magmatic source, or leaching of sulphides from the country rocks. Recent studies have indicated that sulphur isotope ratios in arc magmas display a significant shift from the conventional magmatic values, due to the contamination of the source areas, or the ascending magmas, by crustal materials (Hoefs 2009).

The alteration assemblages and the vein materials suggest that the hydrothermal fluids in both Zaglic and Safikhanloo were neutral to slightly acidic, and that H_2S was the dominant sulphur species in the fluids. Under these circumstances, gold is expected to be transported mostly as a bisulphide complex (Seward 1973; Shenberger & Barnes 1989; Hayashi & Ohmoto 1991).

In low-sulphidation environments, the principal control on fluid pH is the concentration of CO_2 in solution (Henley *et al.* 1984). Boiling and loss of CO_2 to the vapour would result in an increase in the pH, and this, in turn, causes a shift from illite to adularia stability. The loss of CO_2 also leads to the deposition of calcite. This explains the common occurrence of adularia and bladed calcite in both prospects.

The occurrence of adularia, quartz- and adularia-pseudomorphs after platy calcite, and coexisting fluid-rich and vapour-rich inclusions in the ore stage quartz indicate that boiling occurred in the course of the fluid evolution in both prospects (cf. Browne 1978; Simmons & Christenson 1994), and that this process might have contributed to the precipitation of gold from ore fluids.

With regards to the dominant intermediate argillic alteration, low contents of base-metal sulphides, homogenization temperatures, and the overall low salinities of the fluids, the Safikhanloo and Zaglic prospects formed in a low-sulphidation epithermal environment.

We are grateful to B. Borna and B. Mohamadi from Geological Survey of Iran for access to the drill cores and facilities for the field work. We thank J. Fan, University of Saskatchewan, for XRF analysis, and colleagues at GG-Hatch stable isotope laboratories, University of Ottawa, for sulphur isotope analysis. Financial support for the work was supplied by a research grant to Y. Pan from National Science and Research Council of Canada, and a grant to S. Ebrahimi from Ministry of Sciences, Researches and Technology of Iran. The manuscript benefited significantly from a review by M. K. Pandit and F. Valderez.

References

Alavi, M. 1996. Tectonostratigraphy synthesis and structural style of the Alborz Mountain system in northern Iran. *Geodynamics Journal*, **21**, 1–33.

Alirezaei, S., Ebrahimi, S. & Pan, Y. 2008. *Fluid Inclusion Characteristics of Epithermal Precious Metal Deposits in the Arasbaran Metallogenic Zone, Northwestern Iran*. Asian Current Research on Fluid Inclusions (ACROF1-2) Kharagpur, India.

Berberian, M. 1982. The Southern Caspian: a compressional depression floored by a trapped, modified oceanic crust. *Canadian Journal of Earth Sciences*, **20**, 163–183.

Berberian, M. & King, G. C. P. 1982. Towards a paleogeography and tectonic evolution of Iran. *Canadian Journal of Earth Sciences*, **18**, 210–265.

Bodnar, R. J. 1993. Revised equation and table for determining the freezing point depression of H_2O–NaCl solutions. *Geochimica et Cosmochimica Acta*, **57**, 683–684.

Bodnar, R. J., Reynolds, T. J. & Kuehn, C. A. 1985. Fluid inclusion systematics in epithermal systems. *Reviews in Economic Geology*, **2**, 73–97.

BROWNE, P. R. L. 1978. Hydrothermal alteration in active geothermal field. *Annual Review of Earth and Planetary Sciences*, **6**, 229–250.

BROWNE, P. R. L. & ELLIS, A. J. 1970. The Ohaaki – Broadlands hydrothermal area, New Zealand: mineralogy and related geochemistry. *American Journal on Science*, **269**, 97–131.

DE LA ROCHE, H., LETERRIER, J., GRANDE CLAUDE, P. & MARCHAL, M. 1980. A classification of volcanic and plutonic rocks using R1–R2 diagrams and major elements analyses – its relationships and current nomenclature. *Chemical Geology*, **29**, 183–210.

DEWEY, J. F., PITMAN, W. C., RYAN, W. B. F. & BONNIN, J. 1973. Plate tectonics and the evolution of the Apian system. *Geological Society of American Bulletin*, **84**, 3137–3180.

EBRAHIMI, S. 2008. *Mineralogy, alteration geochemistry and mechanism of ore formation in gold-bearing veins at Sharafabad, Eastern Azerbaijan, Iran*. PhD thesis, University of Shahid Beheshti, Tehran, Iran.

GIGGENBACH, W. F. 1997. Origin and evolution of the fluids in magmatic–hydrothermal systems. *In*: BARNES, H. L. (ed.) *Geochemistry of Hydrothermal Ore Deposits*. 3rd edn, Wiley, New York, 737–796.

GIGGENBACH, W. F. & STEWART, M. K. 1982. Processes controlling the isotope composition of steam and water discharges from steam vents and steam – heated pools in geothermal areas. *Geothermics*, **11**, 71–80.

GILL, J. B. 1981. *Orogeneic Andesites and Plate Tectonics*. Springer, Berlin.

HAAS, J. L. 1971. The effect of salinity on the maximum thermal gradient of a hydrothermal system at hydrostatic pressure. *Economic Geology*, **66**, 940–946.

HASSANZADEH, J., GHAZI, A. V., AXEN, G. & GUEST, B. 2002. Oligo-Miocene mafic alkaline magmatism in north and northwest of Iran: evidence for the separation of the Alborz from the Urumieh-Dokhtar magmatic arc. *Geological Society of America*, Abstract.

HAYASHI, K. I. & OHMOTO, H. 1991. Solubility of gold in NaCl- and H_2S-bearing aqueous at 250–350 °C. *Geochimica et Cosmochimica Acta*, **55**, 2111–2126.

HEDENQUIST, H. W. & BROWNE, P. R. L. 1989. The evolution of the Waitapu geothermal system, New Zealand, based on the chemical and isotope composition of its fluids, minerals and rocks. *Geochimica et Cosmochimica Acta*, **53**, 2235–2257.

HEDENQUIST, H. W. & HENLEY, R. W. 1985. Effect of CO_2 on freezing point depression measurements of fluid inclusions: evidence from active systems and application to epithermal studies. *Economic Geology*, **80**, 1379–1406.

HEDENQUIST, H. W., ARRIBAS, A. & GONZALES-URIEN, E. 2000. Exploration for epithermal gold deposits. *Reviews in Economic Geology*, **13**, 245–277.

HENLEY, R. W. 1985. The geothermal framework of epithermal deposits. *Reviews in Economic Geology*, **2**, 1–24.

HENLEY, R. W., TRUESDELL, A. H. & BARTON, P. B., JR. 1984. Fluid mineral equilibria in hydrothermal systems. Society of Economic Geologists. *Reviews in Economic Geology*, **1**, 267.

HEYDARZADEH, E., MEHRPARTOU, M., LOTFI, M. & BABAKHANI, A. R. 2005. *Study of economic geology and controlling factors of Au–Cu mineralization in Zaglic area, East of Iran*. MSc Thesis, Institute for Earth Science, Geological Survey of Iran.

HOEFS, J. 2009. *Stable Isotope Geochemistry*. Springer-Verlag, Berlin.

IRVINE, T. N. & BARAGAR, W. R. A. 1971. A guide to the classification of the common volcanic rocks. *Canadian Journal of Earth Sciences*, **8**, 435–458.

KALAGARI, A. A., POLYAD, A. & PATRICK, R. A. D. 2001. Veinlets and micro-veinlets studies in Sungun porphyry deposit, east Azarbaijan, Iran. *Geosciences Spring-Summer*, **10**, 70–79.

KARIMZADEH SOMARIN, A. 2004. Geochemical effects of endoskarn formation in the Mazraeh Cu–Fe skarn deposit in northwestern Iran. *Geochemistry: Exploration, Environment, Analysis*, **4**, 307–315.

KARIMZADEH SOMARIN, A. & HOSSEINZADEH, G. 2002. *Mineralogy of the Anjerd Skarn Deposit, Ahar Region, NW Iran*. International Mineralogical Association, Edinburgh, Scotland.

KARIMZADEH SOMARIN, A., MOAYYED, M. & HOSSEINZADEH, G. 2002. *Ore Mineralization in the Sonajil Porphyry Copper Deposit, Herris Region, NW Iran*. International Mineralogical Association, Edinburgh, Scotland.

LE BAS, M. J., LEMAITRE, R. W., STRECKEISEN, A. & ZANETTIN, B. 1986. A chemical classification of volcanic rocks based on the total alkali-silica diagram. *Journal of Petrology*, **27**, 745–750.

MOAYYED, M., AMERI, A. & VOSOUGHI ABEDINI, M. 2008. Petrogenesis of Plio-Quaternary basalts in Azarbaijan, NW Iran and comparisons them with similar basalts in the east of Turkey. *Iranian Journal of Crystallography and Mineralogy, Summer*, **16**, 327–340.

MODREK, H., 2009. *Characteristic of the mineralogy, alteration and mechanism of ore formation in the Cah Meci polymetallic deposit and its relationship to Midook copper porphyry*. MSc thesis, Shahid Beheshti University.

MOHAMADI, B. 2006. *Semi detailed exploration of the Safikhanloo Area*. Geological Survey of Iran, Iran.

MOHAMADI, M. & BORNA, B. 2006. *Report of Geology and Drilling in the Masjed Daghi Area*. National Iranian Copper Industries Company (NICICO).

NOGOL-SADAT, 1993. *Geology of Iran, A. Aghanabati, 2005*. Geological Survey of Iran, Iran.

OHMOTO, M. & RYE, R. O. 1979. Isotopes of sulfur and carbon. *In*: BARNES, H. L. (ed.) *Geochemistry of Hydrothermal Ore Deposits*, 2nd edn. John Wiley & Sons, 509–567.

PEARCE, J. A. 1983. Role of the sub-continental lithosphere in magma genesis at active continental margins. *In*: HAWKESWORTH, C. J. & NORRY, M. J. (eds) *Continental Basalts Mantel Zenoliths*. Shiva Press, Nantwich, UK, 230–249.

POUREH, D. 2004. *Geophysical Study in the Zaglic Area*. Geological Survey of Iran, Iran.

RICHARDS, J. P. 2009. Postsubduction porphyry Cu–Au and epithermal Au deposits: Products of remelting of subduction-modified lithosphere. *Geology*, **37**, 247–250.

RICHARDS, J. P., WILKINSON, D. & ULLRICH, T. 2006. Geology of the Sari Gunay epithermal gold deposit, northwest Iran. *Economic Geology*, **101**, 1455–1496.

RIOU, R. 1979. *Petrography and Geochemistry of the Volcanic and Plutonic Rocks of the Ahar Quadrangle (Eastern Azarbaijan Iran)*. University of Saarland.

RIOU, R., DUPUY, C. & DOSTAL, J. 1981. Geochemistry of coexisting alkaline and calk-alkaline volcanic rocks from Northern Azerbaijan (N. W. Iran). *Journal of Volcanology and Geothermal Research*, **11**, 253–275.

ROEDDER, E. 1984. Fluid inclusions. Reviews mineralogy. *Mineralogy Society of America*, **12**.

SEWARD, T. M. 1973. Thio complexes of gold and the transport of gold in hydrothermal ore solutions. *Geochimica et Cosmochimica Acta*, **37**, 370–399.

SHAHABPOUR, J. 1982. *Aspects of alteration and mineralization at the Sarcheshmeh Cu–Mo deposits, Kerman Iran*. PhD thesis, University of Leeds.

SHAMANIAN, G. H., HEDENQUIST, J. W., HATTORI, K. H. & HASSANZADEH, J. 2004. The Gandy and Abolhassani epithermal prospects in the Alborz magmatic arc, Semnan province, Northern Iran. *Economic Geology*, **99**, 691–712.

SHENBERGER, D. M. & BARNES, H. L. 1989. Solubility of gold in aqueous sulfide solution from 150 to 350 °C. *Geochemica et Cosmochimica Acta*, **53**, 269–278.

SILLITOE, H. R. & HEDENQUIST, H. W. 2003. Linkage between volcanotectonic settings, ore-fluid compassions, and epithermal precious-metal deposits. *In*: JOHN, D. A., HOFSTRA, A. H. & THEODORE, T. G. (eds) *Regional Studies and Epithermal Deposits. Society of Economic Geology, Special Publication*, **10**, 315–343.

SIMMONS, S. F. & BROWNE, P. R. L. 2000. Hydrothermal minerals and precious metals in the Broadlands-Ohaaki geothermal system: implication for understanding low-sulfidation epithermal environments. *Economic Geology*, **95**, 971–999.

SIMMONS, S. F. & CHRISTENSON, B. C. 1994. Origins of calcite in a boiling geothermal system. *American Journal of Science*, **295**, 361–400.

SIMMONS, S. F., GEMMELL, B. & SAWKINS, F. J. 1988. The Santa Nino silver–lead–zinc vein, Fresnillo district, Zacatecas, Mexico: Part 2. Physical and chemical nature of one-forming solutions. *Economic Geology*, **83**, 1619–1641.

STAMPFLI, G., MOSAR, J., FAVRE, P., PILLEVUIT, A. & VANNAY, J.-C. 2001. Permo-Mesozoic evolution of the western Tethyan realm: the Neotethys/East-Mediterranean connection. *In*: CAVAZZA, W., ROBERTSON, A. H. F. & ZIEGLER, P. A. (eds) *Peri-Tethyan Rift/Wrench Basins and Passive Margins*. IGCP 369, Bulletin du Muséum National d'Histoire Naturelle, Paris, **186**, 51–108.

STÖCKLIN, J. 1968. Structural history and tectonics of Iran: a review. *American Petrology Geological Bulletin*, **52**, 1220–1258.

SUN, S. S. & MCDONOUGH, W. F. 1989. Chemical and isotopic systematics of oceanic basalts: implication for mantel composition and processes. *In*: SAUNDERS, A. D. & NORREY, M. J. (eds) *Magmatism in the Ocean Basins*. Geological Society, London, Special Publications, **42**, 313–345.

ZARNAB COMPANY. 2007. *Geology and Alteration Studies of the Haftcheshmeh Area*. National Iranian Copper Industries Company (NICICO).

Magma mixing and unmixing related mineralization in the Karacaali Magmatic Complex, central Anatolia, Turkey

OKAN DELIBAŞ[1]*, YURDAL GENÇ[1] & CRISTINA P. DE CAMPOS[2]

[1]*Department of Geological Engineering, Hacettepe University, 06800 Beytepe, Ankara, Turkey*

[2]*Department of Earth and Environmental Sciences, LMU, University of Munich, Theresienstr.41/III, D-80333 Munich, Germany*

**Corresponding author (e-mail: delibaso@gmail.com)*

Abstract: The calc-alkaline Karacaali Magmatic Complex (KMC), in the Central Anatolian Crystalline Complex, is an example of an Upper Cretaceous post-collisional I-type, plutonic–volcanic association. Volcanic rocks grade from basalt to rhyolite, whilst coeval plutonic rocks range from gabbro to leucogranite. In this paper we document evidence for the occurrence of both mixing and unmixing during the evolution of this igneous complex.

Mixing of mafic and felsic magmas was observed in petrographic properties at microscopic to macroscopic scales and is further supported by mineral chemistry data.

The occurrence of unmixing is evidenced in the Fe and Cu–Mo mineralization hosted in the KMC. The iron mineralization in basaltic-andesitic rocks consists mostly of magnetite. Magnetite has been grouped into four settings: (1) matrix type; (2) vein-filling type; (3) breccia matrix type; and (4) vesicle-filling type. In contrast, Cu–Mo mineralization is related to vertical north–south trending quartz-, quartz-calcite-, and quartz-tourmaline veins crosscutting monzonitic and granitic rocks.

We propose that the intrusion of an oxidized, Fe- and Cu-rich basic magma into a partially crystallized acid magma resulted in partial mixing and may have triggered the abrupt separation of an iron-oxide-rich melt.

Our results highlight the importance of magma mixing and metal unmixing, possibly associated with stress relaxation during post-collisional evolution.

Supplementary material: Electron microprobe analyses of plagioclase in monzonitic rocks, MMEs and electron microprobe analyses of K-feldspar in monzonitic rocks are available at http://www.geolsoc.org.uk/SUP18434.

The Central Anatolian Crystalline Complex (CACC) (Göncüoğlu *et al.* 1991) is a segment of the Alpine-Himalayan Belt (e.g. Çemen *et al.* 1999; Yalınız *et al.* 2000; Whitney *et al.* 2001; İlbeyli *et al.* 2004) situated in central Turkey, east of Ankara (Fig. 1). This crustal segment was formed from the amalgamation of several small terrains (Şengör & Yılmaz 1981). The CACC consists of mostly coeval plutonic and volcanic rocks intruded into ophiolitic, sedimentary and metamorphic sequences during different stages of the tectono-magmatic evolution in the area (e.g. Erler *et al.* 1991; Göncüoğlu *et al.* 1991, 1992; Akıman *et al.* 1993; Erler & Bayhan 1995; Erler & Göncüoğlu 1996; İlbeyli 2005).

Since Early Cretaceous times this region is thought to have undergone a complex magmatic evolution commencing with a subduction stage and followed by a possible docking of small colliding continents (Şengör & Yilmaz 1981; Göncüoğlu *et al.* 1991, 1992; Türeli 1991; Türeli *et al.* 1993; Erler & Göncüoğlu 1996; Yalınız *et al.* 1996; Boztuğ 1998, 2000; Boztuğ & Jonckheere 2007; Boztuğ *et al.* 2007). At the end of the orogenic cycle, crustal relaxation and underplating of mantle material contributed to the gradual collapse of the orogeny (Doglioni *et al.* 2002; İlbeyli 2005).

Different mineralization types have been reported in CACC rocks. Granitoid hosted iron mineralization (Ünlü & Stendal 1986, 1989; Stendal & Ünlü 1991; Kuşçu 2001; Kuşçu *et al.* 2002), as well as copper–molybdenum and lead–zinc, is relatively common in Central Anatolia (Kuşcu & Genç 1999; Çolakoğlu & Genç 2001). The Karacaali, Baliseyh and Basnayayla deposits, located between the cities of Kırıkkale and Yozgat (Fig. 1), are the best known examples of granitoid hosted Cu–Mo-mineralization from Central Anatolia (Karabalık *et al.* 1998; Kuşcu & Genç 1999; Kuşcu 2002; Sözeri 2003; Delibaş & Genç 2004; Delibaş 2009).

In recent years, with the development of detailed geological, petrological and geochemical information on the area, magma mixing has been recognized as a major process in the genesis of CACC

From: Sial, A. N., Bettencourt, J. S., De Campos, C. P. & Ferreira, V. P. (eds) *Granite-Related Ore Deposits*. Geological Society, London, Special Publications, **350**, 149–173.
DOI: 10.1144/SP350.9 0305-8719/11/$15.00

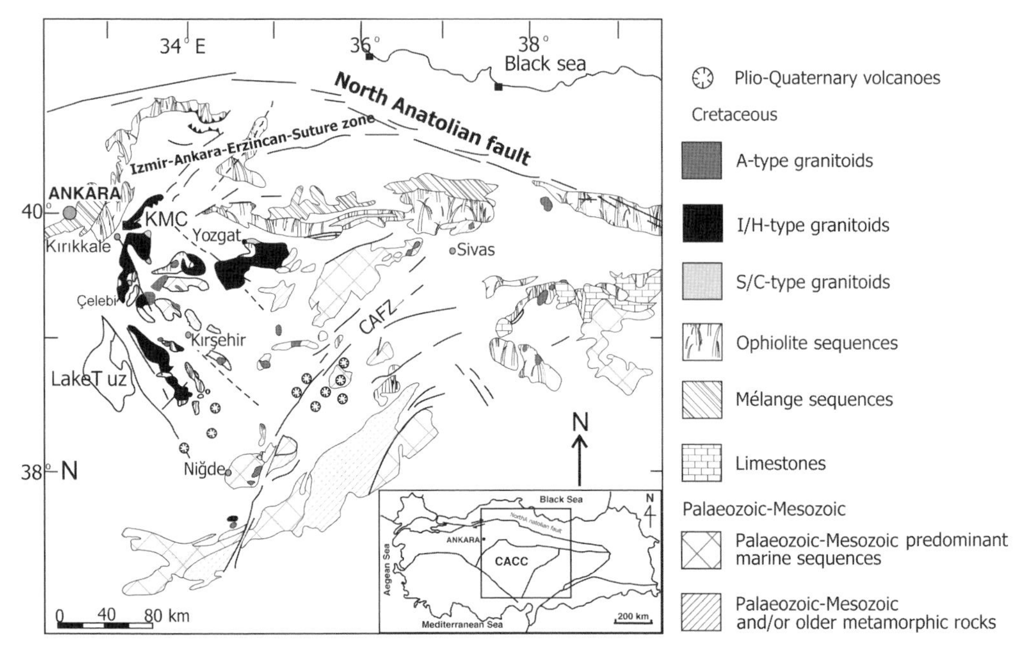

Fig. 1. Simplified regional geological map of Central Anatolia – a microcontinent in the Alpine-Himalayan belt and location map of Karacaali Magmatic Complex in Central Anatolia. CACC, Central Anatolian Crystalline Complex; KMC, Karacaali Magmatic Complex; CAFZ: Central Anatolian Fault Zone. [Modified after Ketin (1961), Bingöl (1989); granitoids classification modified after Boztuğ (1998)].

plutonic rocks (e.g. Tatar & Boztuğ 1998; Yılmaz & Boztuğ 1998; Kadioğlu & Güleç 1999; Delibaş & Genç 2004).

South of Kirikkale, forming the I-type Celebi and Behrekdag granitoid complexes (Köksal *et al.* 2004 – Fig. 1), a 100 + km-long body of plutonic rocks outcrops in a nearly continuous manner. These igneous complexes intruded the metamorphic basement and overlying ophiolitic units. Late Cretaceous to Paleocene sedimentary sequences partially cover the igneous sequences (Köksal *et al.* 2008). The Karacaali Magmatic Complex (KMC), the object of this study, is located northern of Kirikkale, about 70 km east of Ankara (Figs 1 & 2). Together with the Çelebi and Behrekdag granitoid complexes, the KMC frames the northwesternmost margin of the Central Anatolian Crystalline Complex, next to the Izmir-Ankara-Erzincan Suture zone (Fig. 1).

Despite numerous studies on the KMC granitoids (e.g. Norman 1972, 1973; Bayhan 1991; Kuşcu 2002; İşbaşarır *et al.* 2002) the metal source for different mineralization types is still under debate. One open question is the relationship between granitoids, basaltic/rhyolitic volcanism and origin of the mineralization. Delibaş & Genç (2004) and Delibaş (2009) suggested a magma mixing model for the metal enrichments found in basic and felsic rocks from this area.

In this work we will focus on new data from the Karacaali Magmatic Complex (KMC). After a review of the regional geology, we present new field, geochronological, geochemical and mineralogical data in order to better characterize KMC rocks and the associated mineralization. Based on the evidence of magma mixing and unmixing, as well as on metal partitioning between the contrasting magmas involved in this system, we discuss mineralogical–geochemical factors that may control the metal enrichment in this area. We will also propose a genetic model for the mineralization.

Regional geological setting: characterization and age of regional granitoids

The crustal segment which forms the CACC, depicts a roughly triangular block limited by three main suture zones: the Izmir-Ankara-Erzincan Suture Zone in the northern sector, the Tuz Gölü Fault to the west and the Ecemiş Fault to the east (Göncüoğlu *et al.* 1991, 1992; Erler & Bayhan 1995; Yaliniz *et al.* 1999; Fig. 1). At present, due to the complex

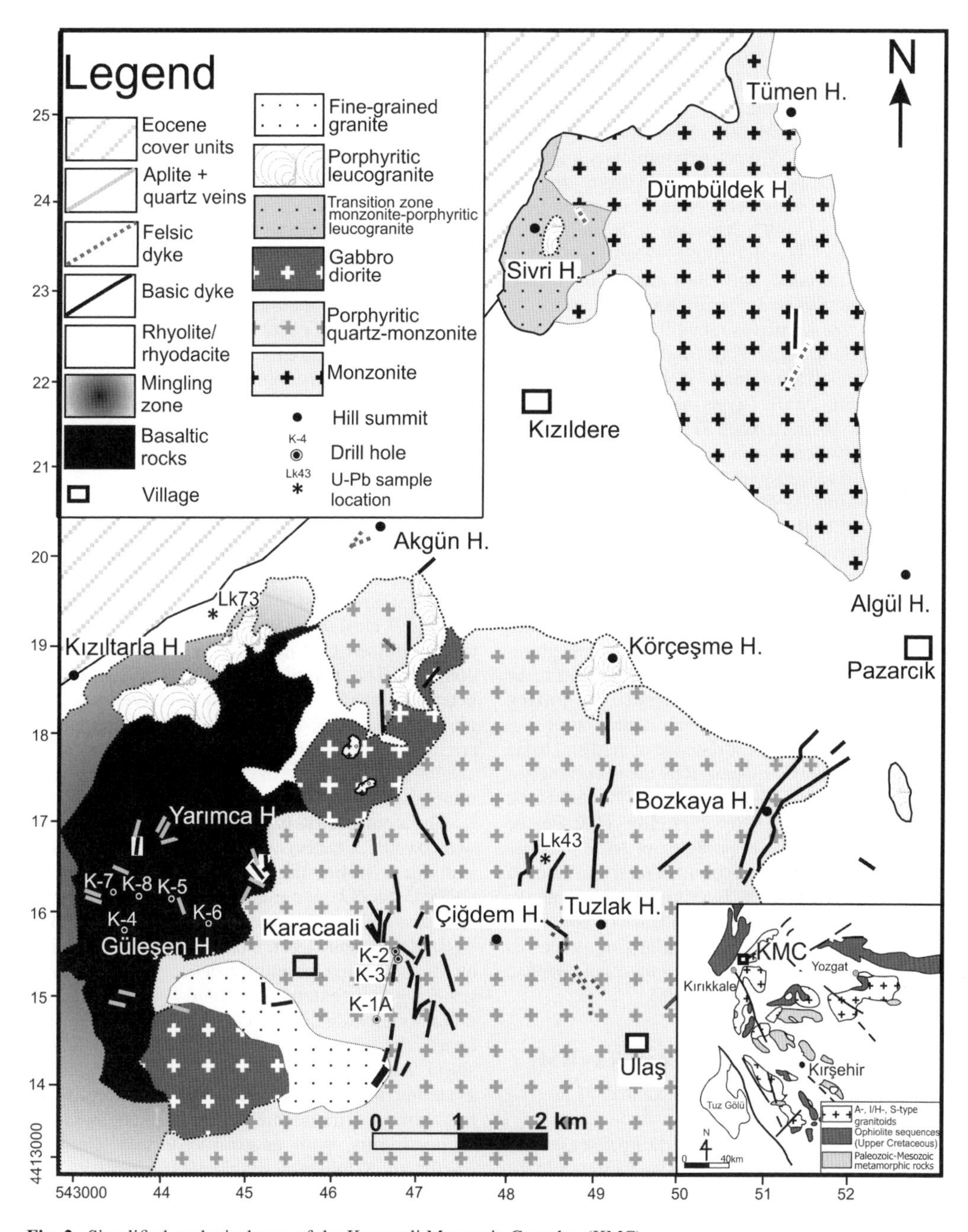

Fig. 2. Simplified geological map of the Karacaali Magmatic Complex (KMC).

tectonic environment, different crustal levels may be locally juxtaposed. The CACC consists of an assemblage of sedimentary, ophiolitic, magmatic and metamorphic rocks. The most extensive rock units are magmatic and intrude the ophiolitic and metamorphic rock sequences of the complex.

Magmatic rocks of the CACC consist essentially of granitoids and associated minor amounts of mafic rocks. These display a wide range of fabrics, mineralogies and chemical compositions, which allow their subdivision into distinct groups. Four genetically related granite types have been proposed: (1)

S- (sedimentary); (2) I- (igneous); (3) H- (hybrid); and (4) A- (alkaline) (e.g. Akıman *et al.* 1993; Aydın *et al.* 1998; Boztuğ 1998; Otlu & Boztuğ 1998; Düzgören-Aydın *et al.* 2001; İlbeyli 2005). This subdivision follows the classification proposed by Pitcher (1993). S-type, peraluminous and two mica granitoids associations have been derived from partial melting of continental crust (e.g. Göncüoğlu *et al.* 1997; Alpaslan & Boztuğ 1997; Boztuğ 2000; İlbeyli 2005). They are syn-collisional, intruded only into the metamorphic rocks and are thrusted over by ophiolitic rocks. In contrast, I-type intrusives range from monzodiorite to granite (İlbeyli *et al.* 2004) and depict metaluminous and calc-alkaline trends of subalkaline composition (Boztuğ 2000). I-type intrusions in this area are thought to have been formed from mixing/mingling between coeval underplating mafic and crustally derived felsic magmas (e.g. Boztuğ 1998; Tatar & Boztuğ 1998; Yalınız *et al.* 1999; İlbeyli *et al.* 2004; İlbeyli & Pearce 2005). I-type magmatites are considered to be post-collisional. A-type magmas intruded in a similar tectonic environment to that of I-type granitoids, as pointed out by İlbeyli *et al.* (2004) and Boztuğ (2000). Both I-type (high-K alkaline) and felsic A-type granitoids, found in Central Anatolia, are thought to be derived from hybrid melts generated by mixing between felsic crustal and mantle mafic sources.

In recent years, the coexistence of acid and basic magmas is generally supported by the occurrence of enclaves in the CACC (e.g. Yılmaz & Boztuğ 1998; Kadıoğlu & Güleç 1996, 1999; İlbeyli & Pearce 2005) reinforcing the importance of basic magmatism for the evolution of granites (e.g. Barbarin & Didier 1991; Didier & Barbarin 1991; Bateman 1995; Sha 1995). Additional evidence of both fractional crystallization and magma mixing processes in the CACC has been highlighted by Bayhan (1993) and Tatar & Boztuğ (1998).

The Karacaali Magmatic Complex (KMC)

Field relations and petrography

The KMC comprises NE–SW striking volcanic and plutonic rock units that form erosive windows in the Eocene cover units. The volcanic units consist of basalt, andesite, rhyodacite to rhyolite, whilst the plutonic units are made up of gabbro, monzonite, porphyritic quartz-monzonite, fine-grained granite (l $<$ 1 mm) and porphyritic leucogranite (Fig. 2). Towards the west–SW end of the complex, a gradational mingling zone between rhyolite and basalt is observed. In this region a profusion of late magmatic veins and dykes crosscut the plutonic rocks. The major components of veins and dykes are quartz, quartz-tourmaline and calcite, in various proportions. Porphyritic leucogranite, aplite and basalt are also present in this region as dykes.

The porphyritic quartz monzonite unit forms an over 35 km^2 pluton in the southern part of the studied area (Fig. 2). Volumetrically it is the most important rock type in the KMC. This unit is characterized by porphyritic textures, which become more pronounced towards the contact with the porphyritic leucogranite unit. Up to 5 cm long phenocrysts of alkali-feldspar are mostly subhedral, pink and generally perthitic. Plagioclase, quartz, amphibole, and biotite are the main matrix components. Pyroxene and rutile may occur in trace amounts. Apatite, zircon and titanite are the most important accessory minerals. Secondary minerals are calcite, ankerite, epidote, gypsum and anhydrite. North–south striking, nearly vertical aplite dykes crosscut this unit.

Nearly all lithological units in the KMC make contact with the porphyritic quartz monzonite unit. The contact with the gabbro–diorite unit is gradational, originating a wide range of hybrid compositions. The contact zone varies in extension from 10 to 100 m and has been mapped as the mingling unit (Fig. 2). Grain size varies from medium-grained (average 3 mm) in the porphyritic quartz monzonite to fine-grained ($<$1 mm) in the gabbro to dioritic domains. The mingling zone is therefore highly heterogeneous both in mineralogy and texture. Plagioclase, quartz, K-feldspar, amphibole and biotite are the predominant minerals, while magnetite, apatite and zircon are the most common accessory minerals.

The contacts of porphyritic quartz monzonite with basalts are also gradational, highly irregular and form pillow-like structures. Close to the contact, blocks of monzonite, generally highly altered to clay, are usually observed within the basaltic rock unit. The contacts between porphyritic quartz monzonite, porphyritic leucogranite and fine-grained granite extend from 50 to 100 m wide, with mutually intrusive relations. Increasing amounts of quartz crystals commonly occur within this transition zone. Fine-grained granite ($<$1 mm) may grade into medium-grained (0.1–5 mm) porphyritic leucogranite. The texture ranges from equigranular to porphyritic, with quartz megacrysts. Granites consist of K-feldspar, plagioclase, quartz, biotite and amphibole. Quartz megacrysts are generally rimmed by albite. Zircon is the commonest accessory mineral. In the southern and northern regions of the studied area, porphyritic leucogranite is observed as small plugs and/or as independent areas in the gabbro. In the southern region a fine-grained granitic intrusion has also been mapped as a separate intrusion.

The main gabbro–diorite unit crops out in the southwestern zone of the area as small scattered bodies along the line of contact with the porphyritic unit. Grain size ranges from fine- to medium-grained (0.1–5 mm), in contrast with the medium- to coarse-grained textures described for the porphyritic units. Plagioclase, magnesio-hornblende, clinopyroxene and magnetite are the essential minerals. In the external parts of the gabbro masses, grain size decreases and ophitic to sub-ophitic textures are observed. This has been interpreted as evidence of rapid cooling. The planar orientation of tabular plagioclase crystals defines typical flow textures in these regions. Chloritization, sericitization and epidotization are the main alteration processes in the gabbro.

Mafic micro-granular enclaves (MMEs) are abundant in the monzonitic and granitic rocks. The enclaves are oval-shaped, fine-grained, medium to dark grey in colour and have sharp contacts with their hosts. They range in size from microscopic up to over 30 cm. The largest ones are found in the monzonitic rocks. Texture of these enclaves varies from fine-grained equigranular to porphyritic. The mineralogy includes plagioclase, biotite, amphibole, ±quartz, ±K-feldspar and ±pyroxene, and accessory sphene and apatite. Therefore the mineralogy of the enclaves is identical to their host rocks, but they contain higher amounts of biotite, amphibole and magnetite. Their whole rock composition ranges from dioritic to quartz dioritic.

The rhyolites and rhyodacites are coarse-grained porphyritic rocks containing quartz (*c.* 0.5 mm) and plagioclase phenocrysts (*c.* 0.3 mm). The groundmass is made up of K-feldspar, plagioclase and quartz, with apatite and rutile as accessory minerals. Due to high magnetite and hematite contents in the groundmass, rhyolite and rhyodacite outcrops are dark grey in colour. Calcite and epidote are usually found as secondary minerals. Towards the contact with granite and monzonite units, plagioclase and K-feldspars contents gradually increase. Embayed quartz phenocrysts are common and some of those crystals are mantled with acicular K-feldspar crystals in a complex spherulitic texture.

Basaltic rocks crop out in the western part of the area. Different types of basalt, ranging from microgranular, porphyritic and brecciate have been distinguished, based on particular colour, fabric and texture. In general, basalts are fine-grained (0.1–0.3 mm) and mainly consist of euhedral plagioclase, anhedral amphibole and pyroxene. Plagioclase micro-phenocrysts (<1 mm) predominate in a groundmass of plagioclase microcrysts, vesicles and glass. Vesicles are filled with chlorite and opaque minerals. Locally magnetite and plagioclase may dominate the microgranular groundmass. In this case embayed and subhedral plagioclase may exhibit a clear flow texture. Because of the high magnetite content in the groundmass phase these rocks have been called magnetite–basalt. Apatite is an important accessory mineral, while quartz is found in trace amounts. Chlorite, calcite, epidote and actinolite are frequent secondary minerals. Magnetite-bearing basalts are often present as layers, at different levels in the basalt stratigraphy. They are also observed as micro dykes and veins, crosscutting both porphyritic and brecciated basalts.

New geochronological data: U–Pb in zircons

In this study, new conventional U–Pb geochronological information has been obtained from zircons separated from the KMC porphyritic quartz-monzonite (sample label = p.q.m, Lk43) and rhyolite/rhyodacite (sample label = r.h.y.d, Lk73), (see Fig. 2, for sampling location). U–Pb determinations (Tables 1 & 2) have been performed at the Geochronological Laboratory of the Geosciences Institute of the University of São Paulo (USP, Brazil). Analytical U–Pb data plotted in Figure 3a, b points towards following ages: 1) 73.1 ± 2.2 Ma (95% confidence) for the porphyritic quartz monzonite and 2) 67 ± 13 Ma (95% confidence) for the rhyolite/rhyodacite. According to these zircon ages, the porphyritic quartz monzonite is Late Cretaceous, broadly similar to other published ages for the Central Anatolian granitoids. Even though several geochronological studies on granitoids of CACC have been carried out, the ages of and genetic relations between the granitoids and associated volcanic rocks in CACC are still under debate. Therefore, precise age determinations for volcanic rocks in KMC are still lacking and considered to be a key to understanding the genetic relations among the granitoids.

Whole rock geochemistry of plutonic rocks

Plutonic rocks have SiO_2 contents depicting the felsic–mafic interactions described in the previous section. Major, trace and elements data for plutonic rock types are shown in Table 3. Compositions range from 75.30 to 77.08 wt% SiO_2 in granitic rocks, 55.27 to 67.89 wt% in monzonitic rocks, 49.10 to 54.67 wt% in gabbro-dioritic rocks and 55.15 to 53.00 wt% in the so-called mafic micro-granular enclaves (MMEs). In granitic rocks total Fe, as Fe_2O_3, varies from 0.50 up to 2.18 wt%, in the monzonitic rocks from 2.22 to 8.18% and in the gabbros from 6.66 up to 12.16 wt%.

In the AFM (($Na_2O + K_2O$)–FeO_t–MgO) ternary diagram (Irvine & Baragar 1971), these rocks plot mainly in the calc-alkaline field. However, a clear transition from calc-alkaline to tholeiite compositions has been observed (Delibaş 2009). The molecular A/CNK (Shand Index: Al_2O_3/($CaO + Na_2O + K_2O$)) ratios of the monzonitic rocks, gabbros and mafic micro-granular enclaves

Table 1. *U–Pb isotopic data of zircons from KMC porphyritic quartz monzonite*

	Magnetic Fraction	Weight (mg)	Pb (ppm)	U (ppm)	207/235#	Error (%)	206/238#	Error (%)	COEF.	238/206	Error (%)	207/206#	Error (%)	206/204*	206/238 Age (Ma)	207/235 Age (Ma)	207/206 Age (Ma)
ANAT-1																	
p.q.m.																	
2962	M(0) A(8zr)	0.035	33.48	2695.2	0.092438	0.67	0.115143	0.62	0.930	8.684853	0.62	0.058225	0.25	975.56	74	90	538
2963	M(0) B(15zr)	0.019	10.37	612.1	0.176462	2.98	0.012351	2.96	0.991	80.963137	2.96	0.103619	0.40	182.60	79	165	1690
2965	M (0) D(7zr)	0.023	22.83	1398.3	0.149322	1.18	0.012969	1.14	0.964	77.109920	1.14	0.083509	0.31	311.88	83	141	1281
3085	NM(−1)K(12zr)	0.032	21.24	1757.1	0.074082	0.71	0.011384	0.54	0.761	527.064776	0.54	0.047198	0.46	1787.42	73	73	59
3088	NM(−1)N(12zr)	0.014	28.69	2399.3	0.075230	0.66	0.011440	0.63	0.946	524.484694	0.63	0.047695	0.22	1718.76	73	74	84

p.q.m, porphyritic quartz monzonite; Magnetic fractions, numbers in parentheses indicated the tilt used on Frantz separator at 1.5 amp. current; # Radiogenic Pb corrected for blank and initial Pb; U corrected for blank, * Not corrected for blank or non-radiogenic Pb, Total U and Pb concentrations corrected for analytical blank; Ages, given in Ma using Ludwig Isoplot/Ex program (1998), decay constants recommended by Steiger & Jäger (1977).

Table 2. *U–Pb isotopic data of zircons from KMC rhyolite/rhyodacite*

	Magnetic Fraction	Weight (mg)	Pb (ppm)	U (ppm)	207/235#	Error (%)	206/238#	Error (%)	COEF.	238/206	Error (%)	207/206#	Error (%)	206/204*	206/238 Age (Ma)	207/235 Age (Ma)	207/206 Age (Ma)
ANAT																	
Rhyd																	
2966	NM(0) A (14ZR)	0.015	21.2	868.8	0.226358	1.61	0.022567	1.59	0.988193	44.312492	1.59	0.072748	0.25	590.0	144	207	1007
2967	NM(0) B(11ZR)	0.008	43.6	3523.7	0.092233	0.20	0.011357	0.81	0.970203	88.052973	0.81	0.058902	0.20	803.7	73	90	563
2873	NM(0) C(13ZR)	0.008	83.6	8301.1	0.054309	4.05	0.008172	1.84	0.4678	122.367568	1.84	0.044193	3.58	283.0	52	53	70
2874	NM(0) D(6ZR)	0.018	83.3	4867.6	0.065562	4.87	0.010717	0.87	0.3802	93.309695	0.87	0.044369	4.61	110.0	69	64	−90
2875	NM(0) E(6ZR)	0.025	18.6	1630.8	0.074215	3.15	0.009415	2.66	0.84864	106.209879	2.66	0.057169	1.67	289.7	60	73	498

rhyd, rhyolite/rhyodacite; Magnetic fractions, numbers in parentheses indicated the tilt used on Frantz separator at 1.5 amp. Current; # Radiogenic Pb corrected for blank and initial Pb; U corrected for blank; * Not corrected for blank or non-radiogenic Pb, Total U and Pb concentrations corrected for analytical blank; Ages, given in Ma using Ludwig Isoplot/Ex program (1998), decay constants recommended by Steiger & Jäger (1977).

Table 3. *(a) Major element (wt%)*

Sample number	Rock type*	SiO_2	TiO_2	Al_2O_3	Fe_2O_3T†	MnO	MgO	CaO	Na_2O	K_2O	Cr_2O_3	P_2O_5	LOI‡	*Total*	A/CNK§
1	**Lcg**	77.08	0.10	13.09	0.50	0.01	0.06	0.62	3.52	4.37	0.03	0.02	0.50	*99.90*	1.12
2	**Lcg**	76.72	0.24	12.61	0.95	0.02	0.61	1.26	5.53	0.62	0.02	0.04	1.40	*100.02*	1.04
3	**Fgg**	75.30	0.17	13.03	2.18	0.03	0.59	2.11	3.86	1.05	0.04	0.03	1.60	*99.99*	1.15
4	**Fgg**	76.71	0.12	13.04	0.83	0.01	0.10	1.12	3.62	3.77	0.04	0.03	0.60	*99.99*	1.08
5	**Pmnz**	65.36	0.33	15.36	4.55	0.09	1.79	4.44	2.94	4.26	0.02	0.13	0.60	*99.87*	0.87
6	**Pmnz**	65.01	0.36	15.33	4.57	0.08	1.73	4.28	3.10	4.26	0.03	0.12	0.90	*99.77*	0.87
7	**Pmnz**	63.77	0.36	16.03	4.93	0.11	1.88	4.65	2.98	4.35	0.02	0.15	0.50	*99.73*	0.88
8	**Pmnz**	65.31	0.33	15.53	4.82	0.08	1.74	4.40	2.96	4.18	0.02	0.13	0.40	*99.90*	0.89
9	**Pmnz**	67.89	0.33	16.44	2.22	0.04	0.43	3.37	3.89	3.53	0.03	0.10	1.40	*99.67*	1.01
10	**Pmnz**	66.29	0.33	15.13	3.73	0.06	1.42	4.11	2.86	4.43	0.02	0.13	1.40	*99.91*	0.89
11	**Pmnz**	63.15	0.37	16.00	4.48	0.11	1.38	4.98	3.03	4.12	0.01	0.14	2.10	*99.88*	0.86
12	**Pmnz**	64.86	0.36	16.30	4.20	0.09	1.45	4.31	3.18	4.46	0.01	0.13	0.50	*99.85*	0.91
13	**Pmnz**	63.90	0.39	16.06	4.96	0.09	1.85	4.57	2.98	4.21	0.01	0.14	0.70	*99.86*	0.90
14	**Pmnz**	66.03	0.33	15.43	3.69	0.06	1.40	4.00	2.97	4.53	0.02	0.11	1.30	*99.87*	0.90
15	**Mnz**	60.74	0.62	17.58	4.06	0.11	1.58	5.70	3.79	3.74	0.01	0.21	1.60	*99.74*	0.85
16	**Mnz**	64.27	0.46	15.88	4.80	0.09	1.47	3.58	3.46	4.58	0.01	0.16	1.00	*99.76*	0.92
17	**Mnz**	64.12	0.47	16.21	4.63	0.09	1.50	3.67	3.49	4.54	0.00	0.15	0.90	*99.77*	0.93
18	**Mnz**	55.27	0.76	17.21	8.18	0.16	3.33	6.51	2.99	2.91	0.01	0.25	2.20	*99.78*	0.86
19	**Gbr**	49.10	1.47	15.97	10.41	0.23	7.14	10.47	3.35	0.35	0.04	0.16	1.30	*100.00*	0.64
20	**Gbr**	51.36	1.33	16.67	12.16	0.19	4.11	7.18	5.24	0.25	0.01	0.12	1.40	*100.02*	0.75
21	**Gbr**	49.93	1.28	17.15	11.94	0.20	4.98	8.79	4.24	0.17	0.04	0.16	1.10	*100.00*	0.74
22	**Gbr**	49.41	1.55	16.19	7.49	0.21	8.20	13.33	1.97	0.08	0.05	0.13	1.40	*100.02*	0.58
23	**Gbr**	50.24	0.94	18.61	6.66	0.17	5.86	11.91	2.92	0.11	0.01	0.08	2.50	*100.01*	0.70
24	**Gbr**	54.67	1.25	15.59	10.38	0.19	4.87	6.68	4.62	0.36	0.01	0.18	1.20	*100.00*	0.77
25	**Mme**	55.15	0.80	17.14	8.28	0.23	3.45	6.02	3.72	3.52	0.01	0.29	1.20	*99.81*	0.82
26	**Mme**	53.00	0.70	17.90	10.15	0.21	3.89	6.60	3.56	2.40	0.01	0.18	1.30	*99.90*	0.87

*lcg, porphyritic leucogranite; fgg, fine grained granite; mnz, monzonite; pmnz, porphyritic monzonite; gbr, gabbro; MME, mafic microgranular enclaves.
†Total iron as ferric oxide.
§$Al_2O_3/(CaO + Na_2O + K_2O)$ molar ratio of $Al_2O_3/(CaO + Na_2O + K_2O)$, Shands Index.
‡Loss on ignition.

Table 3. *Continued (b) trace element (ppm) data of representative samples from plutonic rocks from the Karacaali Magmatic Complex (KMC)*

Sample number	Ba	Rb	Sr	Zr	Nb	Ni	Co	Zn	Y	Cs	Ta	Hf	Sc	Th	Ga	W	Mo	Cu	Pb	As	V
1	1152.2	120	93.1	127.5	10.8	13	0.9	12	31	1.5	0.9	4.5	3	21.1	13.2	4	3.3	4.8	10.1	6.3	4.9
2	69.2	30.4	123.7	86.9	2.5	8.3	3.3	14	31.1	1.2	0.1	3.1	10	0.6	10.8	8.6	2.3	43.3	4.5	2.3	12
3	205.5	20.1	99.5	97.4	3.2	7	4.4	20	33.6	5.4	0.1	4.6	7	2.1	11.9	4.6	11.7	43.1	7.4	5.4	20
4	445.2	56.6	46.8	93.8	3	13	2.8	12	31.3	1.2	0.2	4.2	4	2.3	11.8	3.6	5.7	19	9.1	16.9	9
5	914.3	158.4	350.1	109.4	10.6	5	11.1	16	18.8	5.8	0.9	3.7	8	34.1	13.8	2.5	1.5	47.3	9	2.7	97
6	955.1	151.8	333.3	115	10.8	8	7.9	14	20.7	3.8	0.9	3.4	7	33	13.5	2	4.3	20.2	9.8	4.2	82
7	1149.7	155.9	361.7	134.2	8.2	13	5.9	16	17.5	5.1	0.7	3.9	8	28.6	13.3	2.3	1.6	3	11	2.6	96
8	790	172.3	352	131.2	8.6	18	8.8	19	16.5	4.1	0.6	4.2	8	32.5	14	2.3	2.3	25.9	12.4	2.8	93
9	2473.3	79.1	383.6	146.8	11.3	10	2.1	14	19.8	1.9	0.9	4.3	5	34.4	13.6	4.2	2.8	9.9	20.4	6.1	38
10	666.8	192	338.7	119.5	10.2	7	9.1	13	16.3	5.7	0.8	4	8	29	14.8	1.9	2.2	10.6	9.4	4.2	83
11	869.4	173.3	380.1	123.8	10.5	6.6	10.3	46	18.7	5.8	0.7	3.9	8	30.9	14.9	4.6	1.7	34.4	35.3	5.6	94
12	1129.2	165.1	373.9	140.3	11.1	7.3	9.4	37	17.5	5.9	0.9	4.4	8	29.1	14.8	3	1.7	27.9	32.7	3.1	84
13	1200.3	166.4	393.7	133.5	9.9	4.7	9.9	28	18.7	10.2	0.8	3.9	9	30.7	15.2	2.2	1.1	5.7	14.7	3.3	102
14	990.6	160.4	353.5	123.3	10.5	4.7	4.7	20	18.6	7.7	0.8	3.7	7	38.1	13.6	1.4	1.8	11	15.1	5.9	79
15	1853.9	69.9	527.2	178.6	13.1	7	5.2	37	23.7	3.1	1	5.1	11	32.1	14.8	1.8	1.3	11	21.8	6.2	123
16	1861.9	103.8	373.3	159	11.7	12	7.6	30	22.5	4.9	0.8	5	7	30.4	14	3.4	1.9	6.6	21.3	6.1	91
17	1821.1	103.6	384.6	165.9	11.5	5	24.5	33	22.9	5.2	1.3	4.9	7	30.8	13.4	378.9	1.1	5.6	22.3	6.3	90
18	1454.2	73.1	538.5	124.3	12.3	15	19.8	56	23.9	3.4	0.9	4	16	24	16.1	2.6	1.9	30.3	17.9	10.1	189
19	70.4	7.8	199.1	96.7	4.3	32.7	36.9	19	31.9	1.2	0.3	2.7	37	0.4	18.4	1.3	0.5	27.2	8	4.3	262
20	33.9	3.4	174.5	44.5	1.3	8.7	20.1	31	21.9	0.4	<0.1	1.4	41	0.1	18.8	2	1.2	40.3	16.9	6.4	425
21	21.6	1.6	199.7	78.5	4.8	52.5	19.4	26	17.2	1.5	0.3	2.3	35	0.5	18.2	1.8	0.7	117.3	5.6	31.2	241
22	24.9	0.5	179.1	89.1	5.5	9.1	15.3	28	26.7	0.7	0.3	2.6	31	0.5	17.7	0.7	0.8	1.6	12.7	3.9	206
23	22.2	1.2	205.8	34.5	1.5	3.3	12.8	37	22	1.4	<0.1	1.2	40	0.4	18.7	3.2	0.5	0.4	19.7	4.2	277
24	68.7	7.6	173.1	58.1	3.8	7.9	12.2	34	27.1	1.7	0.2	1.8	36	0.6	17.6	2	1.3	42.2	8.2	7.5	325
25	1320.2	88.5	439.2	143.6	16	5	10.6	38	32.1	5.6	0.9	4	14	15.1	16.1	3.5	1.4	66.8	27	8.2	171
26	735.6	124.5	319.9	95.6	7.2	5	22.4	56	16.6	5.9	0.4	3.1	15	7.6	19.5	3.4	1.8	10.8	19.8	2.8	191

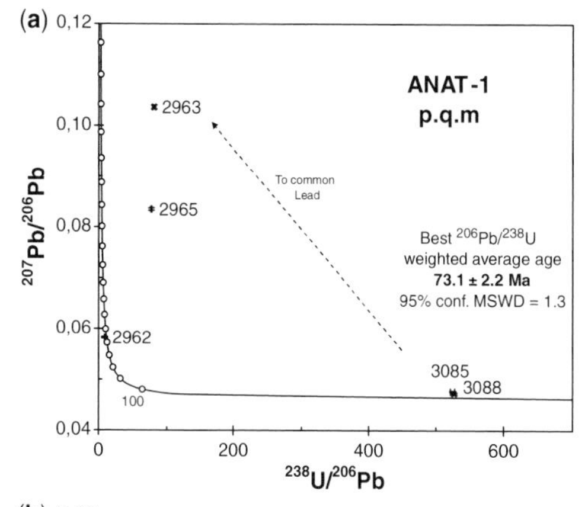

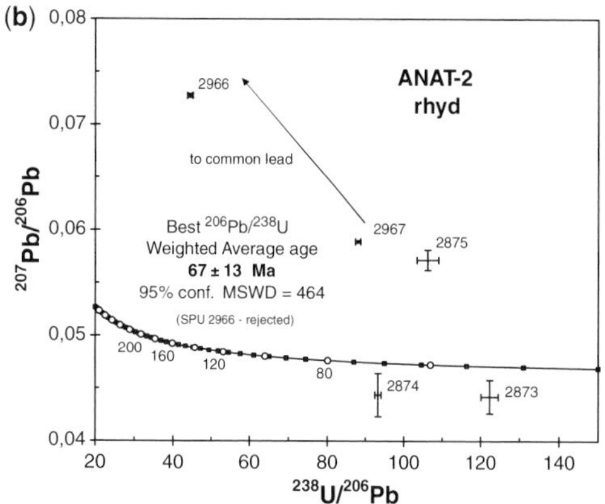

Fig. 3. (**a**) $^{238}U/^{206}Pb-^{207}Pb/^{206}Pb$ correlation diagram for the porphyritic quartz monzonite; (**b**) $^{238}U/^{206}Pb-^{207}Pb/^{206}Pb$ correlation diagram for the rhyolite.

range from 0.64–1.01; whereas the average A/CNK ratio of the granitic rocks is 1.01. These values point towards a metaluminous character for monzonites, gabbros and microgranular enclaves. In contrast, granitic rocks exhibit a transition trend between the peraluminous and the metaluminous fields (Delibaş 2009).

Variation diagrams for major and minor elements against SiO_2 show clear gaps around 57 and 73% SiO_2 (Fig. 4). Linear trends are present only for restricted values and may vary from element to element. From gabbros to granites slopes may change abruptly, as for P_2O_5 v. SiO_2. Some lateral scattering, that is, TiO_2 remaining around 0.5% while SiO_2 spreads from 62% up to 69% SiO_2, is present in all variation diagrams. Monzonitic rocks fill the compositional gap between the gabbros and granitic rocks, as expected. Linear trends in the variation diagrams for trace elements are only observed for those elements with similar diffusion coefficients compared to Si. Binary plots for most trace elements against Si do not show a good correlation. From gabbros to granites slopes can also change abruptly, as for the case of P_2O_5 v. SiO_2 (Fig. 4).

Using Zr/Y, Nb/Y and Ce/Y ratios (Taylor & McLennan 1985; Sun & McDonough 1989) to discriminate monzonitic and granitic KMC rocks, we obtained plots in the accepted field for the Average Continental Crust (ACC). On the other hand, gabbro samples have similar trace element contents as those established for the primitive mantle.

Rb/Sr ratios for monzonitic rocks are in the range between 0.13 and 0.57. Near the gabbro contacts (0.20) measured Rb/Sr ratios in granitic rocks range from 0.20 to 1.29. Rb/Sr ratios for the gabbros are very low, varying from 0.003 to 0.04, indicating a low degree of magmatic fractionation (Ishihara & Tani 2004). In granites the degree, type of fractionation and oxidation state of the magmas involved are thought to be important in determining both the potential for and the type of associated mineralization (Blevin 2003). Compatible/incompatible element ratios (Rb/Sr and K/Rb ratios) are also useful tools to determine the degree and type of fractionation of granitic magmas (Blevin & Chappell 1992; Blevin 2003; Ishihara & Tani 2004). Preliminary data from granitic rock compositions plotted in the Rb/Sr v. SiO_2 diagram depict a distinct correlation in comparison to gabbros and monzonites. This suggests an evolution for the granitic compositions by fractional crystallization processes. Granitic rocks may, therefore, be cogenetic. In contrast, a distinct differentiation process might have affected gabbroic and monzonitic rocks (Fig. 5a), confirming previous observations. In the K/Rb v. SiO_2 diagram two separate groups can be distinguished as well (Fig. 5b). While gabbros plot in the unevolved field, as expected, granitic and monzonitic rocks plot in the moderately evolved field, but follow distinct trends.

Evidence of magma mixing/mingling in the KMC

In this section we highlight and discuss the most important petrographical and geochemical evidence for magma mingling and mixing in the studied area. Evidence for mixing and/or mingling includes: (1) the existence of a mingling zone between the rhyolite/rhyodacite and basaltic units; (2) microgranular enclaves; (3) the presence of disequilibrium textures scattered throughout all units; (4) the wide range of compositional spectrum in both matrix rocks and mafic micro-granular enclaves; (5) whole rock geochemical data and (6) the mineral chemistry and Ba distribution in feldspars.

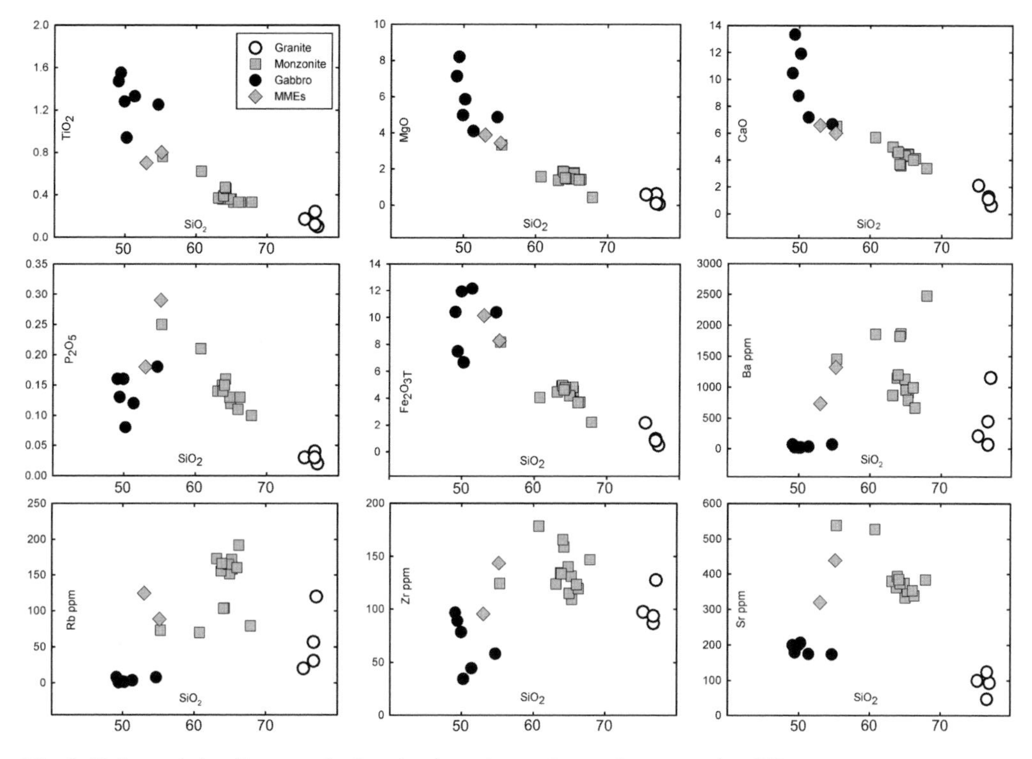

Fig. 4. Harker variation diagrams of selected major, minor and trace elements against SiO_2.

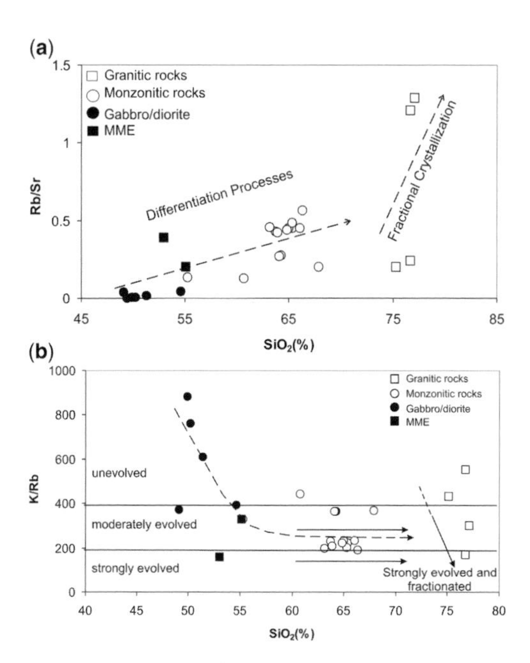

Fig. 5. (**a**) SiO_2 v. Rb/Sr diagram for plutonic rocks of the KMC; (**b**) SiO_2 v. K/Rb diagram for plutonic rocks of the KMC (modified after Blevin 2003).

All these features are indicative of two contrasting magma types, felsic and mafic, interacting to produce intermediary compositions through mixing processes.

Field evidence: the mingling zone between rhyolite/rhyodacite and basalt. Due to the intimate interfingering between rhyolite and basalt, these rock units have been mapped together as a mingling zone on a 1/25 000 map scale (Fig. 2). Contacts between rhyolite and basalt are sinuous and sharp on outcrop scale. However, reaction zones over a few centimetres wide between the rhyolite and basalt, are locally well observed (west of Güleşen Hill, Fig. 2). Tightly packed, pillow-like structures are usually noted. These structures are 1–2 m high and 2–3 m in diameter. Basaltic pillow-like structures in a granitic matrix are interpreted as flow fronts of a denser and less viscous basaltic magma ponding on the floor of a silicic magma chamber during replenishment (Wiebe *et al.* 2001). Pillow-like structures can also be highly brecciated and are displayed along flow structures. In this case it is thought that rapid decompression of viscous magmas, with contrasting gas proportions (Alıdıbırov & Dingwell 1996) should have caused the observed fragmentation.

Stretched basaltic and granitic fingers may also be observed in these regions. They may represent viscous fingering dynamics becoming dominant during replenishment, when low viscosity mafic magma intrudes high-viscosity felsic magma (Perugini & Poli 2005). The change between rounded or fingered interfaces is probably a function of the viscosity ratio between the contrasting magmas. This contrast may change with time, depending on the degree of hybridization of the system.

Brecciation of pillow-like structures, flow textures and the presence of different textures in basalt demonstrate the complexity of the process. These may represent multiple mafic magma injections into the cooler, crystal-rich felsic magma chamber and/or abrupt changes in the physicochemical parameters of the magma chamber during hybridization. Due to viscosity, heat capacity and density differences, mafic magma cannot be injected into the top of the felsic magma chamber, but tend to spread laterally, at the interface of crystal-rich and crystal-poorer rhyolitic/granitic magmas. Sinuous to sharp contacts between the rhyolite–basalt together with diffusive to transitional contacts (hybrid zones) between the granite, monzonite and gabbro support this idea. Persistent inputs of relatively dense and low viscosity mafic magma into a high viscosity chamber of felsic magma stimulates convection, diffusion and redistribution of different melts throughout the chamber (Blake 1966; Reid *et al.* 1983; Wiebe & Collins 1998; Wiebe *et al.* 2001). This process is known to be highly chaotic, therefore non-linear and fractal (e.g. Perugini & Poli 2000; Perugini *et al.* 2003; De Campos *et al.* 2008).

Micro-granular enclaves: textural and chemical evidence. Micro-granular enclaves (MMEs) occur throughout the granitic and monzonitic rocks. They show quenched textures and the presence of complexly zoned alkali feldspar indicates rapid cooling of the basic magma (Vernon 1984). Rapid cooling in the enclaves is also supported by the presence of acicular apatite crystals and disequilibrium textures. Alkali feldspar from MMEs are texturally and compositionally identical to those in the host rocks. This suggests that large alkali feldspar crystals within MMEs are xenocrysts that have been captured from the host magma during the mixing process. This may indicate that the enclaves could have been liquid blobs of basic magma, which were incorporated into the more acid magma during chaotic mixing.

Representative textures compatible with magma mixing have been extensively discussed in the literature. The most peculiar feature observed in granitic rocks is the widespread presence of basic inclusions. These inclusions are similar to the cm-sized MMEs. However, these are so small that most of them are only recognizable under the microscope, thus representing another length scale. They are usually found in the porphyritic quartz-monzonite, in the porphyritic leucogranite, in the rhyolite and in the transition zones between porphyritic quartz monzonite and rhyolite. Inclusion shapes range from rounded to ellipsoidal, elongated to lenticular and their size varies between 0.1 to 5 mm. Some of them are even filled with phenocrysts (Fig. 6a, b). Most of them show sharp contacts to the felsic host, but diffusive contacts may also be observed. They consist of biotite, skeletal magnetite and amphibole. Most mafic inclusions are thought to be samples of mafic, high temperature magma blobs chilled within a cooler, more silicic host.

Plagioclase is the most common mineral phenocryst in the monzonitic and granitic rocks. It occurs in a subhedral to euhedral fashion, and up to 1 mm long. Its composition is andesine, ranging from $An\text{-}_{31}$ to $An\text{-}_{50}$ (Supplementary Data SUP18434). They are normally zoned (core: labradorite, rim: andesine). In contrast, those from the groundmass are slightly reversely zoned. In the reaction zones between rhyolite and basalt (mingling zone) plagioclase phenocrysts commonly show dissolution textures with cavities filled with hematite and magnetite (Fig. 6a, b). In the MMEs the plagioclase composition ranges from andesine (An_{43}) to labradorite ($An_{52.5}$) (SUP18434). They are normally zoned with a labradorite (An_{63}) core and an andesine (An_{40}) rim. Some of the large plagioclase phenocrysts may contain irregular calcic zones. These zones occur as droplets, irregularly interrupted layers along the crystal margins and cores (Fig. 6d). Moreover, reverse element concentrations in feldspar phenocrysts from the monzonitic rocks contrast with normal zoning plagioclases in their coeval mafic microgranular enclaves. Plagioclase phenocrysts in both monzonitic rocks and MMEs commonly include acicular apatite and euhedral biotite inclusions. They generally follow the crystallographic direction of the plagioclase, but some of them are scattered in amphibole mantled cores of the plagioclase (Fig. 6f, g). Similarly, rounded quartz crystals are also mantled by amphibole (Fig. 6e), in this case not necessarily following main crystallographic directions of the quartz substrate.

Another additional piece of evidence for magma mixing is the presence of poikilitic quartz (Fig. 6h) in monzonitic rocks. This texture is thought to result from the late-stage crystallization of the felsic and hydrous melt. Since the more felsic system is superheated, there may be only a few quartz and K-feldspar nuclei available for further growth. Consequently, few large quartz crystals in the system are favoured as a substrate within the

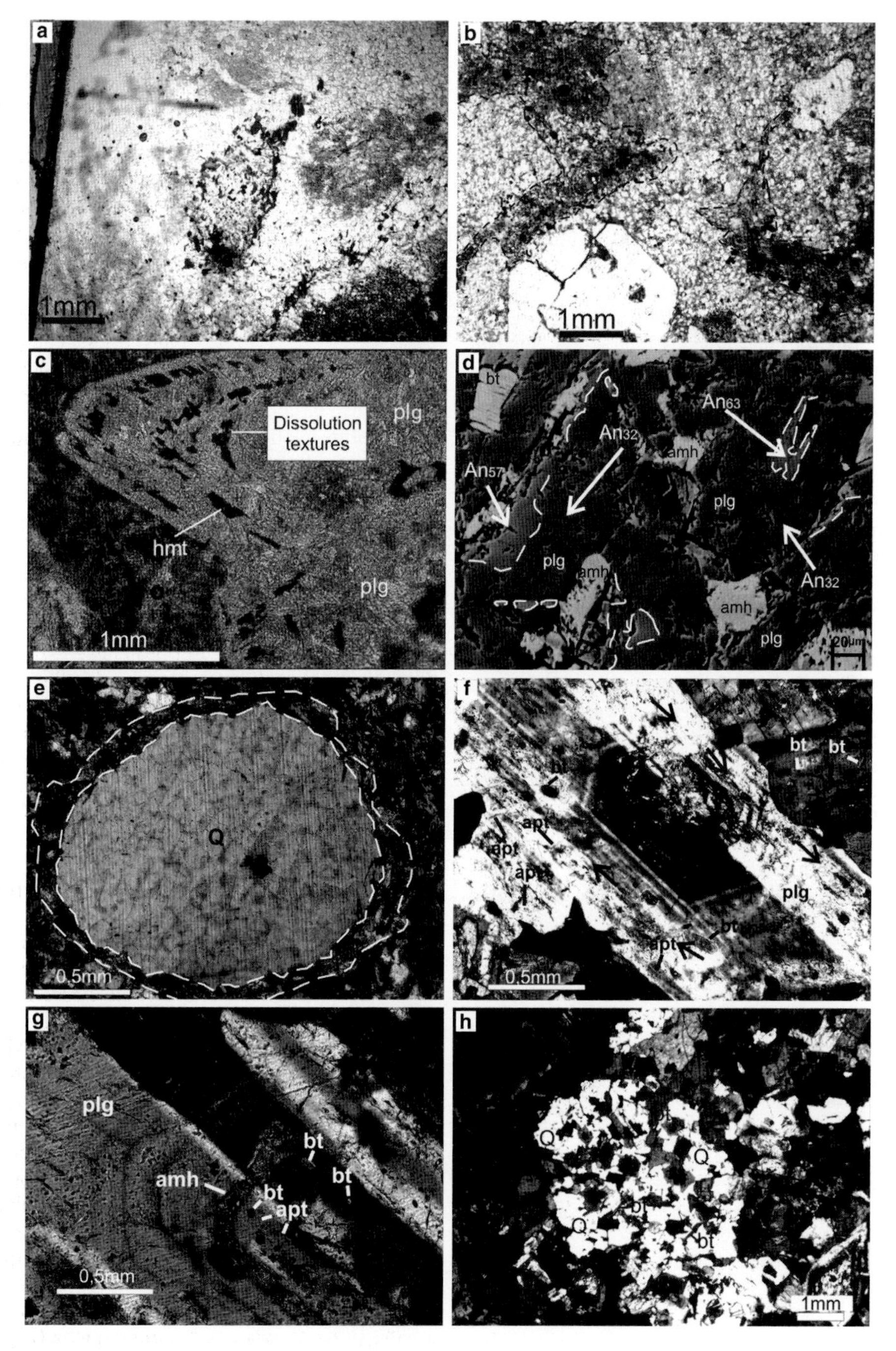

Fig. 6. Basic inclusions in granitic rocks: (**a**) Rounded, ellipsoidal mafic inclusion in the porphyritic quartz monzonite–rhyolite transition zone; (**b**) Basic inclusion from a crack in a quartz phenocryst from the transition zone porphyritic quartz monzonite–rhyolite; (**c**) Hematite-filled dissolution texture in plagioclase (hybrid zone between rhyolite and basalt); (**d**) Plagioclase containing irregular calcic zones in a MME (Electron microscope image); (**e**) Amphibole mantled rounded quartz in monzonite; (**f**) Apatite and biotite inclusions following the crystallographic directions of the plagioclase (in monzonite); (**g**) Amphibole mantled core of a plagioclase with scattered apatite and biotite inclusions (monzonite); (**h**) Poikilitic quartz in monzonite containing MMEs (Q, quartz; bt, biotite; apt, apatite; amh, amphibole; plg, plagioclase; hmt, hematite).

earlier assemblage of relatively smaller K-feldspar crystals. Biotite enrichment is restricted to the crystal borders and suggests limited K_2O and H_2O exchange between felsic and mafic magmas. This process is thought to postpone the further growth of abundant quench-generated plagioclase, hornblende and apatite crystals (e.g. Hibbard 1995).

The whole rock chemistry discussed previously points towards separate chemical evolution for mafic and felsic magmas and heterogeneous mixing between chemically contrasting melts. This partial conclusion supports field and microscopic observations for clear igneous interaction between these melts before solidification. Major evidence towards mixing is the already mentioned alkali feldspars from MMEs. The feldspar major element contents are chemically identical to those present in the host magma. They are also texturally identical. However, contrasting BaO-contents have been measured in K-feldspar phenocrysts from the monzonite. The studied K-feldspars are orthoclase with the general formula $(Ca_{0.004}Na_{0.17}K_{0.80})(Si_{3.00}Al_{0.95}Fe_{0.003})O_8$. BaO-contents range from detection limit (<0.06 wt%) up to 1.57 wt% (SUP18434). This is clear evidence for trace element zoning. Low BaO-contents in the centre contrast with higher contents towards the rims (Fig. 7a, b). In contrast, K_2O and Na_2O contents, in the same K-feldspar megacrysts, do not show a parallel significant variation. This indicates that the BaO-low cores may represent inherited xenocrysts trapped in a more mafic magma. Due to convection and diffusion in the magma chamber a subsequent BaO richer overgrowth may be related to further crystallization in a high-BaO monzonitic magma during the mixing process.

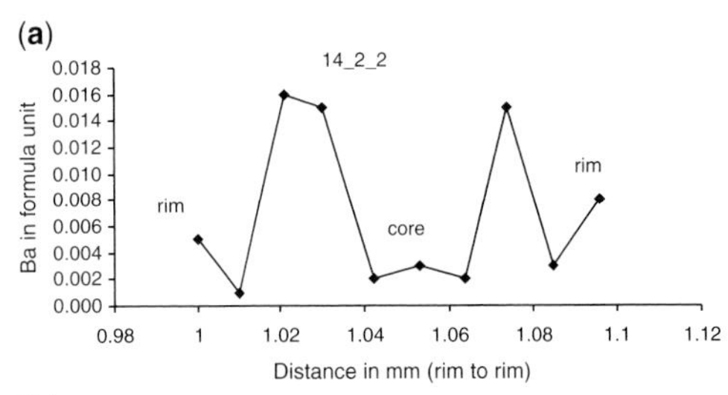

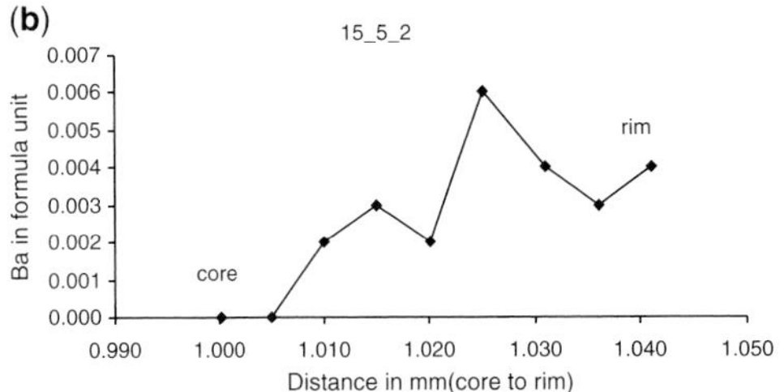

Fig. 7. Ba-content variation in K-feldspar from monzonitic rocks (**a**) rim to rim; (**b**) core to rim.

Disequilibrium textures described in this section, together with the whole rock and mineral chemistry point towards magma mixing/mingling (e.g. Hibbard 1995; Baxter & Feely 2002). From this section we conclude that magma mixing and mingling, both at the outcrop as well as at the crystal scale, is widespread documented in the KMC.

Mineralization types in the KMC

The KMC hosts both iron and copper–molybdenum mineralization. Iron mineralization is basically hosted by basaltic-andesitic rocks and consists mainly of magnetite. For this reason, this mineralization has been labelled magnetite mineralization. Based on structural and textural criteria it is possible to subdivide the magnetite mineralization into four formation settings: (a) matrix type, (b) vein type, (c) breccia-matrix type and (d) vesicle-filling type. In addition to these mineralization types, monzonite-hosted actinolite + magnetite veins are also observed. In contrast, the copper–molybdenum mineralization is related to vertical north–south trending quartz, quartz-carbonate and quartz-tourmaline veins crosscutting the monzonitic and granitic rocks. Eight different sites have been drilled in monzonitic and basaltic rocks by the General Directorate of Mineral Research and Exploration of Turkey (M.T.A.) in order to determine the ore potential of the KMC (see Fig. 2 for drill hole location). Chemical data obtained from these drill holes reveal that iron (as FeO) contents range from 15 to 60 wt% (K-4, K-5, K-6, K-7 and K-8), whereas copper, molybdenum, lead and zinc contents (K-1A, K-2, K-3) remain lower then 1.4 (Cu), 0.4 (Mo), 0.1 (Pb) and 0.2 wt% (Zn) according to İşbaşarır *et al.* (2002) and Kuşcu (2002).

Matrix- and vein-type magnetite mineralization

The matrix-type magnetite mineralization has been sampled from drill cores at different levels within the microgranular and porphyritic basalts (Fig. 8). This mineralization type is also found as micro dykes and veins, crosscutting both porphyritic and brecciated basalts. Therefore, both mineralization types will be described together in this session.

Oriented plagioclase microlites floating within the vesicle-rich magnetite groundmass are typical for this magnetite enrichment. Plagioclase microlites contain numerous magnetite inclusions. The matrix mineralogy is dominated by magnetite but also frequently contains actinolite together with apatite, chlorite and minor calcite. Magnetite

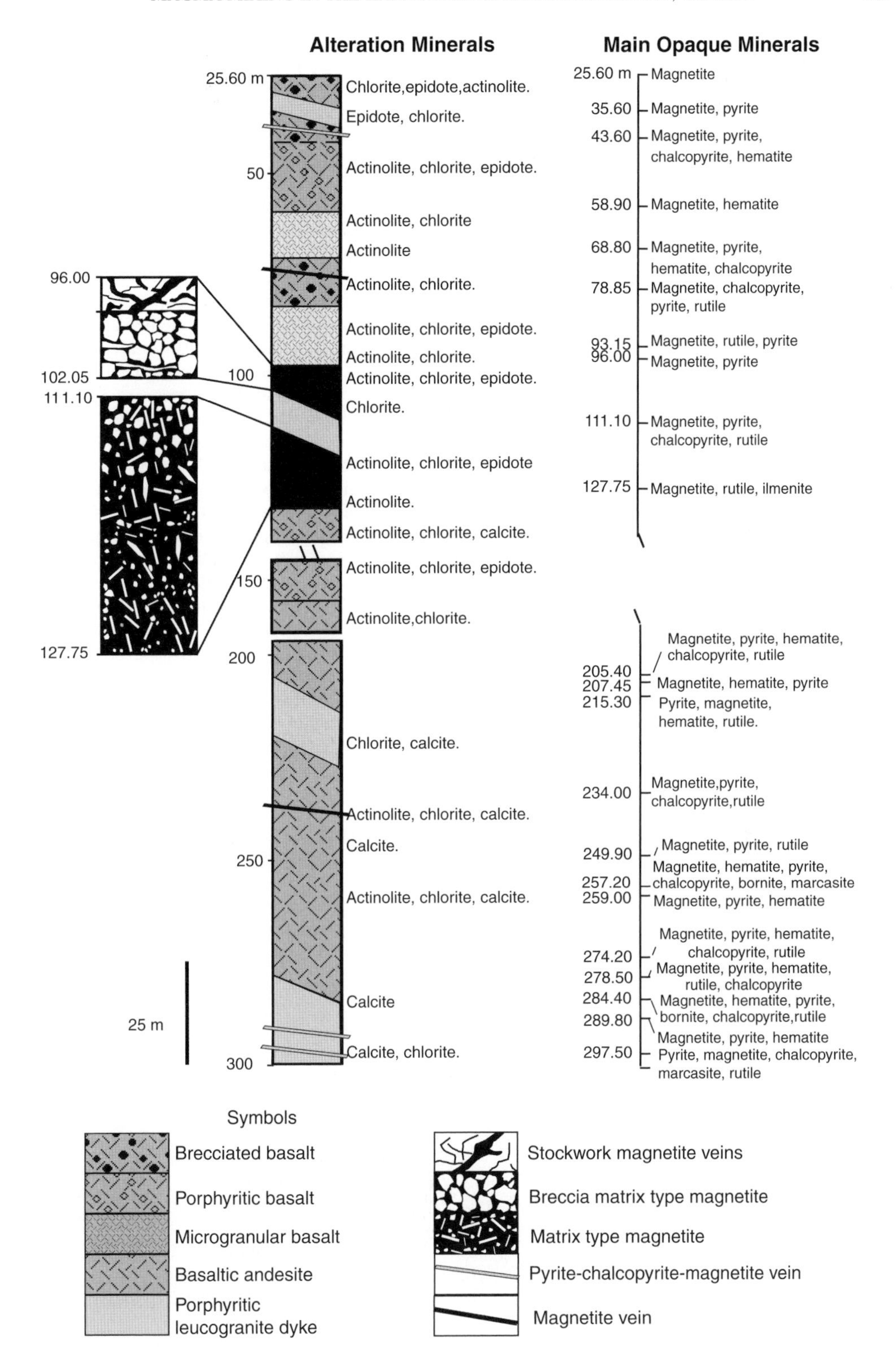

Fig. 8. Simplified geological log of drill hole K-4.

crystals are usually fine grained (0.1–0.3 mm), anhedral and fractured. Hematite and rutile may also be present in the matrix as accessory minerals. Fine-grained pyrite and chalcopyrite inclusions have been frequently recognized in the magnetite matrix. Stockwork-type pyrite and chalcopyrite veins crosscut the magnetite matrix. Actinolite is the main gangue mineral of these stockworks and fills the fractures and vesicles of the wall rock.

Fine magnetite veins may crosscut equally basaltic-andesitic rocks and range from <1 mm to several centimetres thick. They also occur as stockwork and isolated veins within the fault zones (Fig. 9a, b). Vein infill contains magnetite, actinolite, hematite, ilmenite, rutile, pyrite, chalcopyrite, calcite, ankerite and dolomite. Galena and sphalerite are additionally found in carbonate-rich magnetite veins. Hematite is a secondary replacement product of magnetite. Rutile and ilmenite may also be observed as vein infillings and as small (10–250 μm) disseminated anhedral grains in the basaltic-andesitic host rocks.

Isolated massive magnetite veins observed in fault zones range from 0.5 up to 1 m thickness and are intensely mylonitized. They have sharp contacts with their basaltic-andesitic host rocks and are crosscut by actinolite–calcite, pyrite–chalcopyrite and limonite stockwork veins. On both sides of the vein walls, host rocks are commonly enriched in actinolite and calcite. Alteration intensity progressively decreases outwards from the veins through the host rocks. The main ore mineral is euhedral/subhedral magnetite. Pyrite and chalcopyrite have also been detected in minor amounts.

In general, ore samples from magnetite veins display both primary and secondary textures and structures. Magnetite shows typical foam texture. In actinolite-rich veins, pyrite fills open spaces and fractures in magnetites. In actinolite-poor samples magnetite envelops and/or forms pyrite and chalcopyrite intergrowths. In carbonate-rich magnetite veins (with calcite, dolomite and ankerite) chalcopyrite droplets are commonly found as inclusions in magnetite. Complementary minerals in carbonate-rich magnetite veins are sphalerite, galena and pyrite. In this case sphalerite usually contains widespread chalcopyrite exsolution droplets, pointing towards high temperature processes.

According to microprobe data, matrix-type magnetite contains low SiO_2 (0.07–0.98%), but is highly enriched in V_2O_3 (0.06–0.62%) and TiO_2 (0.09–0.65%). However, anhedral magnetite inclusions in plagioclase have high SiO_2 (1.08 and 2.69%) content. Moreover, they contain low TiO_2 (0.03 and 0.36%) and V_2O_3 is below the detection limit (*c.* 4 ppm in Table 4). Although element zoning between cores and rims is rare, it has been detected for magnetite inclusions in feldspars. TiO_2 contents may reach up to 0.59% in core and may be as low as 10 ppm at the crystal rim. Magnetite in actinolite-rich and in massive veins is equally V_2O_3 (0.04%) and TiO_2 (0.01–0.27%) poor (Table 4).

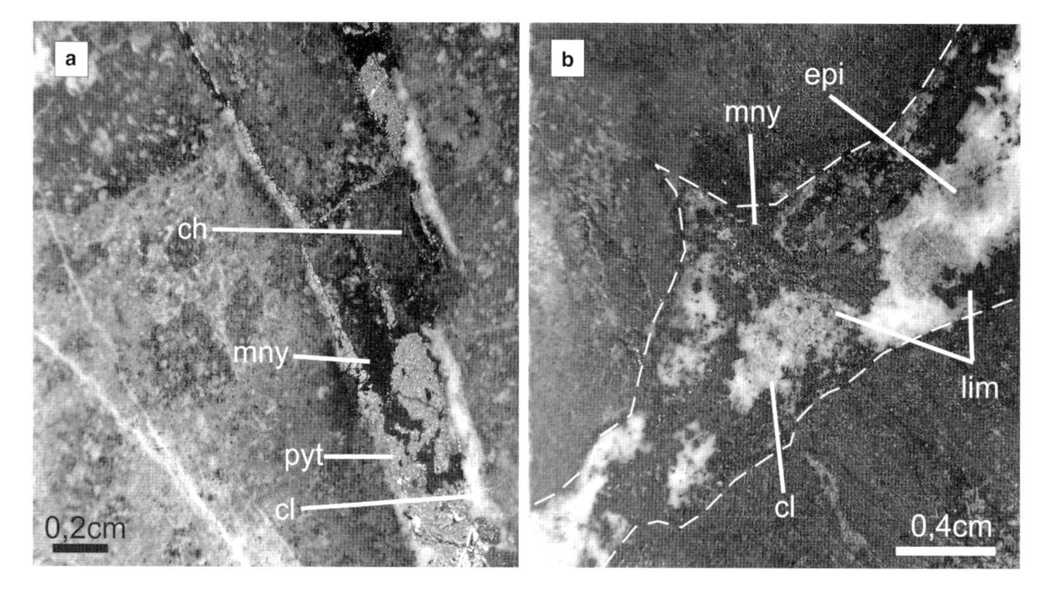

Fig. 9. Vein-type mineralization (**a**) Magnetite-pyrite-calcite veins and chloritic alteration zones; (**b**) Magnetite-calcite-epidote vein; (mny, magnetite; pyt, pyrite; act, actinolite; lim, limonite; ch, chlorite; cl, calcite; epi, epidote).

Table 4. *Chemical data for magnetites from the matrix- and vein-type mineralization*

Sample	SiO_2	MgO	TiO_2	V_2O_3	Al_2O_3	MnO	$Fe_2O_3^{T§}$	Total
M-120c*	0.34	0.16	0.18	bd‡	0.45	0.45	96.37	97.95
M-120r*	0.31	0.18	0.12	bd	0.19	0.02	96.82	97.64
M-122c*	0.50	0.13	0.34	bd	0.17	0.12	96.29	97.54
M-122r*	0.35	0.10	0.16	bd	0.17	0.33	96.58	97.69
M-129c*	0.19	0.14	0.34	0.28	0.23	0.54	96.32	98.03
M-129r*	0.26	0.01	bd	bd	0.01	bd	97.61	97.89
M-131c*	0.27	0.06	bd	bd	0.27	bd	97.16	97.76
M-131r*	0.23	0.14	0.14	bd	0.14	0.09	96.72	97.46
M-133c*	0.07	0.09	bd	bd	0.19	bd	97.48	97.83
M-133r*	0.26	0.06	0.09	0.10	0.12	0.33	96.90	97.85
M-116	0.19	bd	0.19	0.21	0.05	0.69	96.88	98.21
M-140i†	1.08	0.25	0.36	bd	0.61	0.19	94.69	97.17
M-141i†	2.69	0.63	0.03	bd	2.06	0.09	93.63	99.14
V-368	0.23	0.24	0.16	0.04	0.32	0.58	97.821	99.34
V-369	0.22	0.21	bd	bd	0.31	0.21	98.834	99.79
V-370	0.29	0.18	bd	bd	0.16	0.00	97.782	98.41
V-371	0.07	0.16	0.10	bd	0.19	0.06	97.687	98.28
V-372	0.17	0.05	bd	bd	0.14	0.41	97.842	98.62
V-373	0.28	0.10	0.27	0.20	0.23	0.45	97.349	98.69
V-374	0.23	0.29	bd	bd	0.25	0.26	97.3	98.33
V-376	0.28	0.22	0.12	bd	0.24	0.32	97.326	98.50
V-378	0.26	0.21	0.04	bd	0.32	0.23	97.461	98.53
V-379	0.28	0.33	bd	bd	0.34	0.28	97.662	98.90
V-380	0.02	0.05	bd	bd	0.05	0.00	98.243	98.36
V-381	0.03	0.10	bd	bd	0.15	0.41	97.729	98.42
V-382	0.34	0.14	0.01	bd	0.09	0.39	98.159	99.12

M samples, Matrix-type magnetite; *c, core; r, rim. †magnetite inclusions in plagioclase.
V samples, Vein-type magnetite; ‡bd, below detection limit. §total iron as ferric oxide.

Breccia-matrix type magnetite mineralization

Basaltic-andesitic rocks of the KMC contain many different types of breccias. Breccias are generally observed within the mingling zone of andesitic and rhyodacitic rocks. They are both monolithic and/or heterolithic in nature. Their fragment composition depends in part on their location within the mingling zone. Near the contact between rhyolite and basaltic rocks, breccias are heterolithic, with rounded to angular fragments of both rock types. The fragments are cemented by an andesitic/rhyodacitic magnetite-rich matrix. Both rhyolitic and basaltic fragments contain disseminated magnetite. On the other hand, breccias related to pillow-like structures are monolithic and contain highly rounded, nearly spherical basaltic blobs, which suggest that they formed *in-situ* by mechanical fragmentation. In breccias related to pillow-like structures, magnetite and actinolite predominate in the dark-green matrix. Small amounts of chlorite and apatite associated with magnetite, and subordinate amounts of quartz are also present in the matrix. In breccias related to pillow-like structures, fragments are sericitized and chloritized, either completely or only along their borders, resulting in a characteristic pale white-greenish colour. Breccias are commonly crosscut by stockwork-type actinolite–magnetite and calcite veins. This mineralization type is generally located on the upper part of the matrix-type magnetite mineralization (Fig. 8).

Magnetite found in breccia-type mineralization is anhedral and intensely fractured. Fractures are filled with pyrite, rutile and chalcopyrite. Hematite is an oxidation product of magnetite along fractures and mineral margins. Magnetite is generally V_2O_3 and TiO_2 poor and relatively MgO rich in comparison to vein-type magnetites (Table 5).

Vesicle-filling type magnetite mineralization

Vesicles are common in nearly all basaltic–andesitic rocks in KMC. Vesicle dimensions range from microscopic (300–500 μm) to macroscopic scales (1–10 cm). They are filled with magnetite, chlorite, calcite, actinolite and quartz. Vesicle infillings are characterized by a thin calcite-rich margin that varies inwards though a chloritic zone and finally to a magnetite–actinolite-rich core. Magnetite crystals are anhedral to subhedral and are martitized along mineral margins. Ilmenite and rutile occur together with magnetite and some of them are also

Table 5. *Chemical data for magnetites from the breccia-type mineralization*

Sample No	SiO_2	MgO	TiO_2	V_2O_3	Al_2O_3	MnO	Fe_2O_3 T‡	Total
K5-312	0.32	0.29	0.06	bd†	0.28	0.50	98.722	100.16
K5-313	0.24	0.25	bd	bd	0.24	0.28	99.322	100.33
K5-338	0.29	0.27	bd	bd	0.42	bd	99.304	100.29
K5-314c*	0.39	0.22	bd	bd	0.23	0.45	98.944	100.24
K5-314r*	0.26	0.40	0.53	bd	0.16	0.57	98.293	100.21
K5-316c*	0.28	0.29	0.26	bd	0.37	0.31	98.861	100.36
K5-316r*	0.17	0.16	0.36	bd	0.24	0.26	98.814	100.00
K5-318c*	0.20	0.20	bd	bd	0.36	0.33	99.359	100.45
K5-320c*	0.21	0.23	bd	bd	0.28	0.26	99.11	100.09
K5-320r*	0.34	0.30	0.01	bd	0.35	0.26	99.279	100.54
K5-322c*	0.29	0.36	0.02	bd	0.31	0.28	99.076	100.32
K5-322r*	0.29	0.26	bd	bd	0.20	0.14	99.535	100.42
K5-324c*	0.34	0.29	bd	bd	0.44	0.30	99.014	100.39
K5-327r*	0.21	0.27	bd	bd	0.17	0.26	99.11	100.02
K5-335c*	0.53	0.49	bd	bd	0.42	0.16	98.863	100.46
K5-335r*	0.22	0.20	bd	0.03	0.18	0.41	99.054	100.06
K5-340c*	0.18	0.39	0.03	bd	0.31	0.27	98.934	100.12
K5-340r*	0.37	0.32	0.26	0.31	0.20	0.61	98.146	99.92
K5-343c*	0.07	0.18	0.11	bd	0.14	0.39	98.81	99.71
K5-343r*	0.08	0.08	0.46	0.38	0.11	0.69	98.107	99.53

*c, core; r, rim.
†bd, below detection limit.
‡total iron as ferric oxide.

replaced by limonite. Magnetite found in vesicles has very high SiO_2 and MgO, and low TiO_2 contents. On the other hand, in comparison with other types of mineralization described in this work, they are relatively enriched in V_2O_3 (Table 6).

Considerations on the genesis of the iron mineralization. Basaltic rocks from the KMC are generally H_2O-rich. Volatile-rich minerals such as actinolite, calcite and apatite are often widely present. These rocks are not only H_2O but also CO_2, P, Cl^- and F^- rich. Rocks containing actinolite, calcite, and apatite are known to be crystallized from volatile-rich magmas (Philpotts 1967; Kolker 1982; Naslund 1983; Nyström & Henríquez 1994; Rhodes *et al.* 1999; Naslund *et al.* 2000).

Table 6. *Chemical data for magnetites from the vesicles-filling type mineralization*

Sample No	SiO_2	MgO	TiO_2	V_2O_3	Al_2O_3	MnO	Fe_2O_3 T‡	Total
K5–145	0.48	0.36	0.16	bd†	0.41	0.27	98.23	99.91
K5-146	0.50	0.38	0.30	bd	0.34	0.37	98.14	100.03
K5-147	0.28	0.37	0.07	bd	0.27	0.48	98.44	99.91
K5-150	0.15	0.34	bd	bd	0.31	0.12	98.53	99.45
K5-151	0.48	0.39	0.29	bd	0.47	0.24	98.78	100.66
K5-152	0.52	0.36	0.09	0.24	0.40	0.43	98.34	100.14
K5-154	0.84	0.60	bd	0.08	0.60	0.22	98.07	100.33
K5-156	0.45	0.48	0.07	bd	0.40	0.21	98.55	100.15
K5-157	0.88	0.59	bd	0.14	0.41	0.43	97.85	100.16
K5-159	0.41	0.53	0.25	0.14	0.46	0.22	98.31	100.17
K5-160	0.36	0.40	0.03	bd	0.43	bd	98.14	99.37
K5-161	0.45	0.41	0.35	bd	0.37	0.35	97.53	99.45
K5-143c*	0.37	0.46	bd	bd	0.56	0.15	98.83	100.37
K5-143r*	0.29	0.53	0.38	bd	0.41	0.32	98.58	100.51

*c, core; r, rim.
†bd, below detection limit.
‡total iron as ferric oxide.

Hydrothermal and magmatic origins for magnetite can be discriminated according to amounts of trace element contents. Hydrothermal magnetites have high Mg, Mn, Si and Cr, whereas magmatic magnetites have low Mg, Mn, Si and high V, Cu, Ti, P, Ni, U contents (Naslund 1983; Nyström & Henríquez 1994; Naslund *et al.* 2000). The high V_2O_3 and TiO_2 and relatively low MgO, SiO_2 and MnO contents of magnetite from the matrix type mineralization in the KMC suggests that they are primary (magmatic) iron oxide mineralizations.

Silicate liquid immiscibility is a known differentiation process in Fe-rich basalts (Roedder 1979). Recent experimental data (Veksler & Thomas 2002; Veksler *et al.* 2002) shows that magmatic differentiation may also lead to liquid immiscibility between high-silica and hydrous melts. These results are of major importance for the genesis of metal-rich mineralization. Our data supports the hypothesis that the iron-rich mineralization in KMC is the result of liquid immiscibility.

Monzonite-hosted Cu–Mo mineralization

The monzonite-hosted Cu–Mo mineralization is widespread in both granitic and monzonitic rocks and in north–south striking vertical/sub-vertical quartz, carbonate and tourmaline veins. Primary vein-type Cu–Mo mineralization is generally recognized along drill cores, but has seldom been observed in outcrops.

The main ore minerals are: chalcopyrite, molybdenite, galena, sphalerite, pyrite, magnetite, hematite, rutile, covelline and bornite. Limonite, malachite and azurite are also observed in fractures of the monzonitic rocks. These last minerals indicate the development of a later, secondary oxidation zone. Chalcopyrite and molybdenite are usually enriched in quartz veins while sphalerite and galena are commonly observed in carbonate-rich veins. Around veins, pyrite, chalcopyrite, sphalerite and magnetite are also found impregnating monzonitic and granitic rocks. In quartz-calcite veins crosscutting the monzonitic and granitic rocks chalcopyrite predominates over pyrite. Chalcopyrite inclusions in sphalerite and magnetite are frequent in both carbonate- and quartz-rich veins. In magnetite-rich veins, chalcopyrite may be replaced by covellite along fractures. In deeper parts of the sulphide-rich vein system, magnetite–actinolite veins crosscut the monzonite. Calcite, scapolite and epidote are common alteration products around veins. The main ore mineral in the veins is magnetite, although chalcopyrite may also occur in minor amounts. Along fractures hematite is a usual oxidation product of magnetite. These magnetites contain low TiO_2 (0.01–0.55%) and V_2O_3 (0.08–0.4%), but are highly enriched in SiO_2 (0.08–0.69%) and MgO (0.07–0.34%), (Table 7).

The source of Cu and Mo. Despite the very close spatial relationship (Blevin & Chappell 1992 1995), the correlation between granite composition and ore elements may be highly complex. This is mainly due to different physico-chemical characteristics of different metals (Fe, Cu and Mo). Another explanation for the non-correlation is the chaotic non-linear nature of the mixing process (Perugini & Poli 2000; De Campos *et al.* 2008; Perugini *et al.* 2008) already discussed previously, notably in the section title '*Field evidence: the mingling zone between rhyolite/rhyodacite and basalt*'. The mixing process is the interplay between thermal and/or compositional convection and chemical diffusion (Mezic *et al.* 1996), is largely non-linear and dependent on the viscosity and density of the end members involved.

The calc-alkaline monzonite units and most of the granitic rocks of the KMC show transitional characteristics. Petrographical and geochemical

Table 7. *Chemical data for magnetites from the magnetite-actinolite veins crosscutting monzonitic rocks*

Sample No	SiO_2	MgO	TiO_2	V_2O_3	Al_2O_3	MnO	Fe_2O_3 ^{T‡}	Total
B11-540	0.36	0.19	0.01	bd†	0.20	0.26	98.91	99.93
B11-549	0.08	0.07	0.02	bd	0.09	0.09	98.88	99.23
B11-550	0.33	0.21	0.18	0.08	0.22	0.39	97.73	99.15
B11-552	0.20	bd	0.09	0.40	0.08	0.46	98.24	99.47
B11-556	0.49	0.28	bd	bd	0.23	0.05	98.52	99.58
B11-558	0.62	0.33	0.05	bd	0.23	0.14	97.64	99.01
B11-560	0.69	0.29	0.55	bd	0.37	0.31	97.14	99.35
B11-561	0.45	0.34	bd	bd	0.21	0.24	97.84	99.08
B11-562	0.33	0.33	bd	bd	0.19	0.39	97.94	99.18

*c, core; r, rim.
†bd, below detection limit.
‡total iron as ferric oxide.

(major and trace element) data indicate an origin from highly evolved magma batches. Considering that this type of granite magma may contain Cu–Mo and Au, granite magmas may be a potential source for Cu and Mo (Ishihara 1981; Blevin & Chappell 1992; Blevin *et al.* 1996; Sillitoe 1996). However, when Cu and Mo contents of gabbroic, monzonitic, granitic and hybrid rocks in the area are plot against SiO_2, different trends are observed (Fig. 10a, b, Table 3).

Cu contents in gabbroic rocks increase with increasing SiO_2. This means that, with increasing fractional crystallization, from gabbro to diorite rocks, Cu is enriched in the melt and remains for incorporation in later-forming mineral phases. The occurrence of chalcopyrite droplet inclusions in magnetite and high Cu values (117 ppm) yielded by gabbroic rocks point towards a primary Cu-rich mafic melt. In monzonitic and granitic rocks, Cu-contents, although variable, tend to decrease with increasing SiO_2-values. This is further evidence for the importance of fractional crystallization in the differentiation processes. During crystallization Cu partitioning is higher for intermediary minerals like amphibole, biotite and magnetite, so that the Cu-contents of residual SiO_2-richer phases decrease. Thus the Cu source is likely to be the mafic and not the granitic magma.

Molybdenum contents show a positive correlation with SiO_2 from gabbros to granites (Fig. 10a, b, Table 3). Since Goldschmidt (1958), Mo enrichment in later phases due to igneous fractionation is a well known process. Highly silicic granites are thus the main host for molybdenite mineralization (Ishihara & Tani 2004). Solution/mineral partition coefficient for Mo ranges from 50 to 80 [DMo (solution/mineral) = 50–80], while its solution/melt partition coefficient is 2.5 [DMo (solution/melt) = 2.5] (Candela & Holland 1983; McDonough 2003; Palme & O'Neil 2003). As a result, during crystallization, Mo cannot be accommodated into the structure of crystallizing minerals. Thus, it becomes enriched in residual solutions and the source of Mo is the granitic magma. In this case, the molybdenite veins are formed at later stages of fractional crystallization of a highly silicic magma.

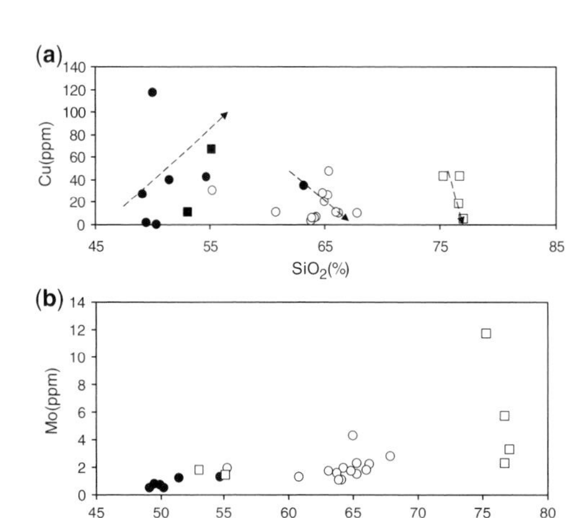

Fig. 10. (**a**) SiO_2–Cu variation diagrams for gabbroic, granitic and monzonitic rocks; (**b**) SiO_2–Mo variation plots for gabbroic, granitic and monzonitic rocks of KMC (symbols are the same as in Fig. 5).

The source of sulphur. Sulphur-solubility depends on the iron content of the silicate melt (Hattori & Keith 2001). Therefore, in comparison to mafic magmas, sulphur solubility in felsic magmas is lower. Sulphate and sulphide compound formation depends on the oxidation level of the silicate magma. The crystallization of anhydrite in a felsic system requires both oxidation and additional S from an external source (Hattori 1993, 1996; Pallister *et al.* 1996; Kress 1997; Hattori & Keith 2001). This external source most likely arises due to the mafic magma. Reduced magmas are S^{2-} rich and contain sulphide minerals. Oxidized magmas, on the other hand, are generally $(SO_4)^{2-}$ rich and commonly contain anhydrite (Whitney & Stormer 1983).

In the KMC the average Fe_2O_3-content for granitic rocks is 3 wt%, for monzonitic rocks 7.5 wt%, and 9.2 wt% for gabbroic rocks. According to the Fe_2O_3 contents, the highest S-solubility is expected to be found in more basic magma batches. Also, the usual presence of gypsum and anhydrite within the quartz monzonite–gabbro transition zone in the KMC can be explained by the crystallization of two different magmatic sources: an S-rich basic and an O-rich felsic system. SO_2 released from the basic magma intrudes into the partly crystallized acid magma and is converted into H_2S. The conversion of SO_2 to H_2S in the felsic magma system may cause oxidation of the felsic magma and crystallization of anhydrite (Hattori 1993, 1996). According to this assumption, sulphur for the monzonite related Cu–Mo mineralization should originate from the basic magma.

A proposal for a genetic model of the mineralization

As a summary, for the genesis of the KMC deposits, our data show that a H_2O-, CO_2-, Cl-, F-, Cu-, Fe- and S-rich basic magma intruded into a semi-crystallized and highly evolved, oxidized acid magma. Abrupt pressure, temperature and compositional changes in the basic layer of the magma caused fragmentation and the separation of an iron oxide-rich phase from the silicate-rich phase. Matrix-type Fe-mineralization, hosted in magnetite basalts, must have been developed as a result of this sudden separation. This mineralization shows clear flow textures, orbicular-vesicles, micro-scaled dykes and high apatite-contents. Abnormal metal enrichment together with igneous features, actinolite alteration and actinolite vein formation in host rocks could indicate sudden unmixing of a fluid- and iron-oxide rich melt from a silica-rich melt (Fig. 11). Magnetite compositional variations also support this hypothesis as well as recent experimental data showing that liquid immiscibility between hydrous and high-silica melts is likely to occur during differentiation processes (Veksler & Thomas 2002; Veksler *et al.* 2002).

In reduced environments inside granitic and monzonitic rocks SO^{2-}, Ca^{2+}, Mg^{2+}, Cl^-, CO_2, Cu, Pb, Zn and H_2O rich solutions coming from basic magma may have formed sulphide, carbonate and quartz veins. These solutions are considered to play an essential role in the re-mobilization of existing Mo-enrichments in highly differentiated felsic phases. It also seems to have an important contribution to the formation of younger molybdenite-quartz veins. These same solutions are also thought to have caused, at low temperatures and pressures, the formation of argillic alteration zones in rhyolite/rhyodacite and monzonite units.

Final discussion and conclusions

An argument against the hypothesis of magma mixing could be the non-linear behaviour of interelemental plots. Linear correlations in bivariate element–element diagrams from volcanic and plutonic rock suites are generally believed to be an excellent indicator of mixing relationships (e.g. Langmuir *et al.* 1978; Fourcade & Allègre 1981). However, new insights from experimental research refute the identification of magma mixing based solely on the presence of straight lines on interelemental plots (De Campos *et al.* 2008; Perugini *et al.* 2008).

Field and petrographical observations, together with chemical data from the KMC suggest continuous inputs of mafic magma into a semi-evolved acid magma chamber. Field relations and geochronological studies on single zircons (U–Pb) point towards a coeval relationship between plutonic and

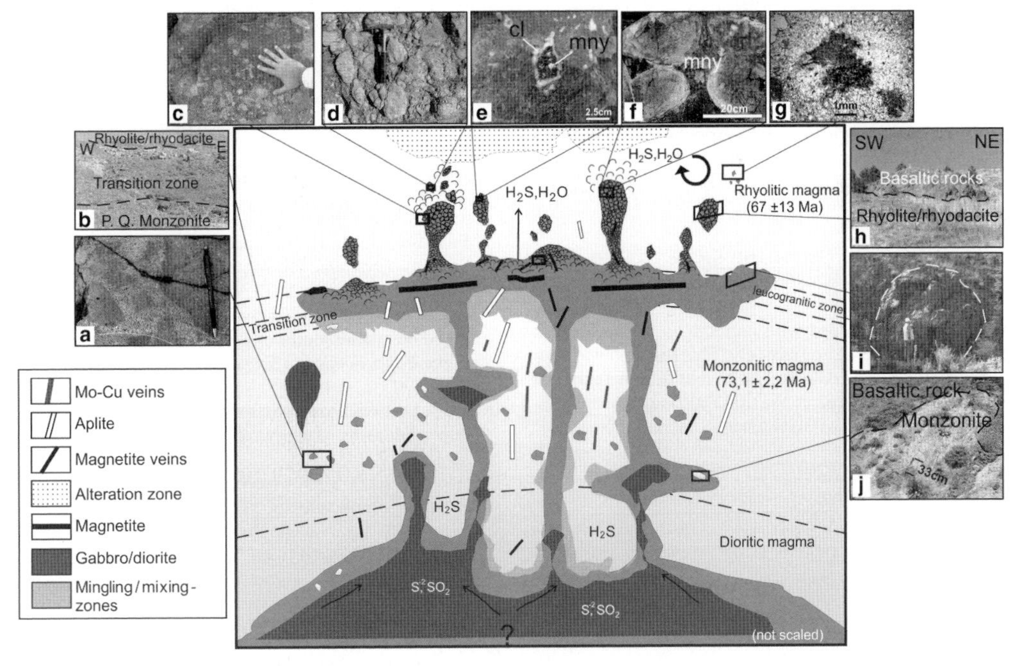

Fig. 11. Model for metal enrichment processes in the KMC (**a**) MMEs in monzonite; (**b**) Transition zone between porphyritic quartz monzonite and rhyolite/rhyodacite; (**c**) Heterolithic breccia; (**d**) Monolithic, highly rounded, nearly spheroidal basaltic breccia; (**e**) Vesicle-filling type magnetite mineralization; (**f**) Breccia-matrix type magnetite mineralization; (**g**) Basic inclusions in rhyolite/rhyodacite; (**h**) Brecciated pillow-like structure in rhyolite/rhyodacite; (**i**) Pillow-like structure in mingling zone; (**j**) Monzonitic block within the basaltic rock (mny, magnetite; cl, calcite). Discussion and explanation in the text.

volcanic rocks. The relatively overlapping ages between monzonite (73.1 ± 2.2 Ma) and rhyolitic rocks (67 ± 13 Ma) reflect a long lasting magma production and crystallization in one or several zoned magma chambers. Progressive transitional contacts from the plutonics into the volcanic rocks indicate that the granitic magma crystallized in a sub-volcanic environment. Sinuous to sharp contacts between rhyolite and basalt, together with diffusive to transitional contacts (hybrid zone) between the granite, monzonite and gabbro, support this hypothesis. Brecciation producing pillow-like structures, flow textures and the presence of different type of basalts are signs of the complexity of this process.

The successive replenishment of granitic magma batches with basic magma may have caused the sudden separation or unmixing of iron-oxide-rich melts, which can have resulted in the matrix type iron mineralizations at high temperatures, while vein type iron mineralization occurred at lower temperatures. Gas vesicles and fragmentation triggered the occurrence of fractures and brecciated zones in the basalt and thus provided a suitable environment for further iron enrichments. SO_2 from the basic magma could have supplied enough sulphur for the sulphide mineralization in quartz and carbonate veins.

We propose that the intrusion of an oxidized, Fe- and Cu-rich basic magma into a partially crystallized acid magma caused partial mixing and may have triggered the abrupt separation of a Fe-oxide-rich melt.

Our results highlight the importance of magma mixing and metal unmixing, possibly associated with stress relaxation during post-collisional evolution.

We thank M. Basei (USP/Brazil) for the U–Pb determinations. We are very grateful to two anonymous reviewers for the improvements to an earlier version. J. Hanson and M. Tekin Yürür helped with corrections to the English and useful comments, respectively. This research was supported by the TUBITAK (Turkish Scientific and Technical Research Council, Project No: 104Y019) and by the University of Munich (LMU), Germany.

References

Akiman, O., Erler, A. *et al.* 1993. Geochemical characteristics of the granitoids along the western margin of the Central Anatolian crystalline complex and their tectonic implications. *Geological Journal*, **28**, 371–382.

Alidibirov, M. & Dingwell, D. B. 1996. Magma fragmentation by rapid decompression. *Nature*, **380**, 146–148.

Alpaslan, N. & Boztuğ, D. 1997. The co-existence of syn-COLG and post-COLG plutons in the Yildızeli area (W-Sivas). *Turkish Journal of Earth Sciences*, **6**, 1–12.

Aydin, N. S., Göncüoğlu, M. C. & Erler, A. 1998. Latest Cretaceous magmatism in the Central Anatolian crystalline complex: brief review of field, petrographic and geochemical features. *Turkish Journal of Earth Science*, **7**, 258–268.

Barbarin, B. & Didier, J. 1991. Conclusions: enclaves and granite petrology. *In*: Didier, J. & Barbarin, B. (eds) *Enclaves and Granite Petrology*. Elsevier Science Publication, New York, 545–549.

Bateman, R. 1995. The interplay between crystallization, replenishment and hybridization in large felsic magma chambers. *Earth Science Reviews*, **39**, 91–106.

Bayhan, H. 1991. Petrographical and chemical–mineralogical characteristics of Karacaali pluton (Kırıkkale). *Isparta Mühendislik Fakültesi Dergisi*, **5**, 121–131 [in Turkish with English abstract].

Bayhan, H. 1993. Ortaköy granitoyidinin (Tuzgölü doğusu) petrografik ve kimyasal-mineralojik özellikleri. *Doğa-Türk Yerbilimleri Dergisi*, **2**, 147–160 [in Turkish].

Baxter, S. & Feely, M. 2002. Magma mixing and mingling textures in granitoids: examples from the Galway Granite, Connemara, Ireland. *Mineralogy and Petrology*, **76**, 63–74.

Bingöl, E. 1989. *1:2.000 000 ölçekli Türkiye Jeoloji Haritası*, General Directorate of Mineral Research and Exploration (MTA) publication Ankara [in Turkish].

Blake, D. H. 1966. The net-veined complex of the Austurnhorn Intrusion, South-eastern Iceland. *Journal of Geology*, **74**, 891–907.

Blevin, P. 2003. Metallogeny of granitic rocks, The Ishihara Symposium. *Granites and Associated Metallogenesis*, **14**, 5–8.

Blevin, P. L. & Chappell, B. W. 1992. The role of magma sources, oxidation states and fractionation in determining the granite metallogeny of eastern Australia. *Transaction of the Royal Society of Edinburgh: Earth Sciences*, **83**, 305–316.

Blevin, P. L. & Chappell, B. W. 1995. Chemistry origin and evolution of mineralized granites in the Lachlan Fold Belt, Australia: the metallogeny of I- and S-type granites. *Economic Geology*, **90**, 1604–1619.

Blevin, P. L., Chappell, B. W. & Allen, C. M. 1996. Intrusive metallogenic provinces in eastern Australia based on granite source and composition. *Transactions of the Royal Society of Edinburgh: Earth Sciences*, **87**, 281–290.

Boztuğ, D. 1998. Post collisional central Anatolian alkaline plutonism, Turkey. *Turkish Journal of Earth Sciences*, **7**, 145–165.

Boztuğ, D. 2000. S-I-A-type intrusive associations: geodynamic significance of synchronism between metamorphism and magmatism in central Anatolia, Turkey. *In*: Bozkurt, E., Winchester, J. & Piper, J. A. (eds) *Tectonics and Magmatism in Turkey and the Surrounding Area*. Geological Society, London, Special Publications, **173**, 407–424.

Boztuğ, D. & Jonckheere, R. C. 2007. Apatite fission track data from central Anatolia granitoids (Turkey): Constraints on Neo-Tethyan closure. *Tectonics*, **26**, 1–18.

Boztuğ, D., Tichomirowa, M. & Bombach, K. 2007. $^{207}Pb-^{206}Pb$ single-zircon evaporation ages of some granitoid rocks reveal continent-oceanic island arc collision during the Cretaceous geodynamic evolution of the central Anatolian crust, Turkey. *Journal of Asian Earth Sciences*, **31**, 71–86.

Candela, P. A. & Holland, H. D. 1983. The partitioning of copper and molybdenum between silicate melts and aqueous fluids. *Geochimica et Cosmochimica Acta*, **48**, 373–380.

Çemen, İ., Göncüoğlu, M. C. & Dirik, K. 1999. Structural evolution of the Tuzgölü basin in central Anatolia, Turkey. *Journal of Geology*, **107**, 693–706.

Çolakoğlu, A. R. & Genç, Y. 2001. Macro-micro textures and genetic evoluation of lead-zinc deposits of Akdağmadeni (Yozgat) region. *Geological Bulletin of Turkey*, **44**, 45–56 [in Turkish with English abstract].

De Campos, C. P., Dingwell, D. B., Perugini, D., Civetta, L. & Fehr, T. K. 2008. Heterogeneities in magma chambers: insights from the behavior of major and minor elements during mixing experiments with natural alkaline melts. *Chemical Geology*, **256**, 131–145.

Delibaş, O. 2009. *The role of magma mixing processes in the formation of iron, copper, molybdenum and lead mineralizations of Kırıkkale-Yozgat region*. PhD Thesis, Hacettepe University, Turkey.

Delibaş, O. & Genç, Y. 2004. Origin and formation processes of iron, copper–molybdenum and lead mineralizations of Karacaali (Kırıkkale) Magmatic Complex. *Geological Bulletin of Turkey*, **47**, 47–60 [in Turkish with English abstract].

Didier, J. & Barbarin, B. 1991. The different types of enclaves in granites-nomenclature. *In*: Didier, J. & Barbarin, B. (eds) *Enclaves and Granite Petrology*. Elsevier Science Publication, Amsterdam, Developments in Petrology, **13**, 19–23.

Doglioni, C., Agostini, S. *et al.* 2002. On the extension in western Anatolia and the Aegean Sea. *Journal of Virtual Explorer*, **7**, 117–131.

Düzgören-Aydin, N. W., Malpas, W., Göncüoğlu, M. C. & Erler, A. 2001. Post collisional magmatism in the central Anatolia, Turkey: Field petrographic and geochemical constraints. *International Geology Reviews*, **43**, 695–710.

Erler, A. & Bayhan, H. 1995. General evaluation and problems of the Central Anatolian granitoids. *Yerbilimleri*, **17**, 49–67 [in Turkish with English abstract].

Erler, A. & Göncüoğlu, M. C. 1996. Geologic and tectonic setting of the Yozgat Batholith. Northern Central Anatolian Crystalline Complex, Turkey. *International Geology Reviews*, **38**, 714–726.

Erler, A., Akiman, O. *et al.* 1991. Petrology and geochemistry of the magmatic rocks of the Kırşehir Massif at Kaman (Kırşehir) and Yozgat. *Doğa, Turkish Journal of Engineering and Environmental Sciences*, **15**, 76–100 [in Turkish with English abstract].

Fourcade, S. & Allègre, C. F. 1981. Trace elements behavior in granite genesis: A case study the calc-alkaline plutonic association from the Querigut complex (Pyrenees, France). *Contributions to Mineralogy and Petrology*, **76**, 177–195.

Goldschmidt, V. M. 1958. *Geochemistry*. Oxford University, Clarendon Press.

Göncüoğlu, M. C., Toprak, V., Kuşçu, İ., Erler, A. & Olgun, E. 1991. *Orta Anadolu Masifi'nin batı bölümünün jeolojisi, Bölüm 1: Güney Kesim*. Turkish Petroleum Corporation (TPAO) Report. No. 2909 [in Turkish].

Göncüoğlu, M. C., Erler, A., Toprak, V., Yaliniz, K., Olgun, E. & Rojay, B. 1992. *Orta Anadolu Masifi'nin batı bölümünün jeolojisi, Bölüm 2: Orta Kesim*. Turkish Petroleum Corporation (TPAO) Report. No. 3155 [in Turkish].

Göncüoğlu, M. C., Köksal, S. & Floyd, P. A. 1997. Post-collisional A-type magmatism in the Central Anatolian Crystalline Complex; petrology of the Idişdağı Intrusives (Avanos, Turkey). *Turkish Journal of Earth Sciences*, **6**, 65–76.

Hattori, K. 1993. High-sulfur magma, a product of fluid discharge from underlying mafic magmas: evidence from Mount Pinatubo, Philippines. *Geology*, **21**, 1083–1086.

Hattori, H. K. 1996. Occurrence of sulfide and sulfate in the 1991 Pinatubo eruption products and their origin. *In*: Newhall, C. G. & Punongbayan, R. S. (eds) *Fire and Mud: Eruptions and Lahars of Mount Pinatubo, Philippines*. University of Washington Press, Seattle, 807–824.

Hattori, H. K. & Keith, D. J. 2001. Contribution of mafic melt to porphyry copper mineralization: evidence from Mount Pinatubo, Philippines, and Bingham Canyon, Utah, USA. *Mineralium Deposita*, **36**, 799–806.

Hibbard, M. J. 1995. *Petrography to Petrogenesis*. Prentice-Hall, Englewood Cliff, NewJersey.

Irvine, T. N. & Baragar, W. R. A. 1971. A guide to the chemical classification of the common volcanic rocks. *Canadian Journal of Earth Sciences*, **8**, 523–548.

Ishihara, S. 1981. The granitoid series and mineralization. *In*: Skinner, B. S. (ed.) *Economic Geology 75th Anniversary Volume*. Society of Economic Geologists, Littleton, CO, 458–484.

Ishihara, S. & Tani, K. 2004. Magma mingling/mixing v. magmatic fractionation: Geneses of the Shirakawa Mo-mineralized granitoids, Central Japan. *Resource Geology*, **54**, 373–382.

İlbeyli, N. 2005. Mineralogical–geochemical constraints on intrusives in central Anatolia, Turkey: Tectono-magmatic evolution and characteristics of mantle source. *Geological Magazine*, **142**, 187–207.

İlbeyli, N. & Pearce, J. A. 2005. Petrogenesis of igneous enclaves in plutonic rocks of the Central Anatolian Massif, Turkey. *International Geology Reviews*, **47**, 1011–1034.

İlbeyli, N., Pearce, J. A., Thirlwall, M. F. & Mitchell, J. G. 2004. Petrogenesis of collision-related plutonics in Central Anatolia, Turkey. *Lithos*, **72**, 163–182.

İşbaşarir, O., Arda, N. & Tosun, S. 2002. *Kırıkkale-Karacaali demir cevherleşmesi ve Orta Anadolu manyetik anamoli sahaları jeoloji ve jeofizik raporu*, General Directorate of Mineral Research and Exploration (MTA), Turkey. Report No. 10534 [in Turkish].

Kadioğlu, Y. K. & Güleç, N. 1996. Ağaçören Granitoidinde yeralan gabro kütlelerinin yapısal konumu: Jeolojik ve Jeofizik (Özdirenç) verilerinin yorumu. *Doğa Türk yerbilimleri dergisi*, **5**, 153–159 [in Turkish].

KADIOĞLU, Y. K. & GÜLEÇ, N. 1999. Types and genesis of the enclaves in central Anatolian granitoids. *Geological Journal*, **34**, 243–256.

KARABALIK, N., YÜCE, N. & ŞARDAN, S. 1998. *Kırşehir, Kırıkkale yöresi genel jeokimya, Karaahmetli ile Dağevi sahaları maden jeolojisi raporu*. General Directorate of Mineral Research and Exploration (MTA), Turkey. Report No. 10142 [in Turkish].

KETIN, İ. 1961. *1:500 000 ölçekli Türkiye Jeoloji Haritası, Sinop paftası*. General Directorate of Mineral Research and Exploration (M.T.A) publication Ankara [in Turkish].

KÖKSAL, S., ROMER, L. R., GÖNCÜOĞLU, M. C. & KÖKSAL, T. F. 2004. Timing of post-collisional H-type to A-type granitic magmatism: U–Pb titanite ages from the Alpine Central Anatolian Granitoids (Turkey). *International Journal of Earth Sciences (Geologisches Rundschau)*, **93**, 974–989.

KÖKSAL, S., GÖNCÜOGLU, M. C., TOKSOY-KÖKSAL, F., MÖLLER, A. & KEMNITZ, H. 2008. Zircon typologies and internal structures as petrogenetic indicators in contrasting granitoid types from central Anatolia. *Mineralogy and Petrology*, **93**, 185–211.

KOLKER, A. 1982. Mineralogy and geochemistry of Fe–Ti oxide and apatite (nelsonite) deposits and evaluation of the liquid immiscibility hypothesis. *Economic Geology*, **77**, 1146–1158.

KRESS, V. 1997. Magma mixing as a source for Pinatubo sulphur. *Nature*, **389**, 591–593.

KUŞÇU, İ. 2001. Geochemistry and mineralogy of the skarns in the Çelebi District, Kırıkkale, Turkey. *Turkish Journal of Earth Sciences*, **10**, 121–132.

KUŞCU, E. 2002. *Karacaali (Kırıkkale) Cu–Mo cevherleşmesi jeokimya ve maden jeolojisi raporu*. General Directorate of Mineral Research and Exploration (MTA), Turkey. Report No. 10635 [in Turkish].

KUŞCU, E. & GENÇ, Y. 1999. Başnayayla (Yozgat) molybdenum-copper mineralization. *Geological Bulletin of Turkey*, **42**, 115–134 [in Turkish with English abstract].

KUŞÇU, İ., YILMAZER, E. & DEMIRELA, G. 2002. Oxide–Cu–Au (Olympic Dam Type) perspective to skarn type iron oxide mineralization in Sivas-Divriği region. *Geological Bulletin of Turkey*, **45**, 33–47 [in Turkish with English abstract].

LANGMUIR, C., VOCKE, R., HANSON, G. & HART, S. 1978. A general mixing equation with applications to Icelandic basalts. *Earth and Planetary Sciences Letters*, **37**, 380–392.

MEZIC, I., BRADY, J. F. & WIGGINS, S. 1996. Maximal effective diffusivity for time periodic incompressible fluid flows. *SIAM, Journal of Applied Mathematics*, **56**, 40–57.

MCDONOUGH, W. F. 2003. Compositional model for the Earth's core. *In*: CARLSON, R. W. (ed.) Elsevier-Pergamon, Oxford, *Treatise on Geochemistry*, **2**, 547–568.

NASLUND, H. R. 1983. The effect of oxygen fugacity on liquid immiscibility in iron-bearing silicate melts. *American Journal of Science*, **283**, 1034–1059.

NASLUND, H. R., AGUIRRE, R., DOBBS, F. H., HENRIQUEZ, F. & NYSTRÖM, J. O. 2000. The origin, emplacement and eruption of ore magmas. *IX Conreso Geologico Chileno Actas*, **2**, 135–139.

NORMAN, T. 1972. Ankara Yahşıhan bölgesinde Üst Kretase-Alt Tersiyer İstifinin Stratigrafisi. *Geological Bulletin of Turkey*, **15**, 180–277 [in Turkish with English abstract].

NORMAN, T. 1973. Late Cretaceous-Early Tertiary sedimentation in Ankara Yahşihan Region. *Geological Bulletin of Turkey*, **16**, 27–41 [in Turkish with English abstract].

NYSTRÖM, J. O. & HENRÍQUEZ, F. 1994. Magmatic features of iron ores of the Kiruna type in Chile and Sweden: ore textures and magnetite geochemistry. *Economic Geology*, **89**, 820–839.

OTLU, N. & BOZTUĞ, D. 1998. The coexistence of the silica oversaturated (ALKOS) and undersaturated alkaline (ALKUS) rocks in the Kortundağ and Baranadağ plutons from the Central Anatolian alkaline plutonism, E. Kaman/NW Kırşehir, Turkey. *Turkish Journal of Earth Science*, **7**, 241–257.

PALLISTER, J. S., HOBLITT, R. P., MEEKER, G. P., KNIGHT, R. J. & SIERNS, D. F. 1996. Magma mixing at Mount Pinatubo: Petrogenetic and chemical evidence from the 1991 deposits. *In*: NEWHALL, C. G. & PUNONGBAYAN, R. S. (eds) *Fire and Mud: Eruptions and Lahars of Mount Pinatubo, Philippines*. University of Washington Press, Seattle, 687–731.

PALME, H. & O'NEIL, H. C. 2003. Cosmochemical estimates of mantle composition. *Treatise on Geochemistry, Mantle and Core*, **2**, 1–35.

PERUGINI, D. & POLI, G. 2000. Chaotic dynamics and fractals in magmatic interaction processes: a different approach to the interpretation of mafic microgranular enclaves. *Earth and Planetary Science Letters*, **175**, 93–103.

PERUGINI, D. & POLI, G. 2005. Viscous fingering during replenishment of felsic magma chambers by continuous inputs of mafic magmas: Field evidence and fluid-mechanics experiments. *Geology*, **33**, 5–8.

PERUGINI, D., BUSÀ, T., POLI, G. & NAZZARENI, S. 2003. The role of chaotic dynamics and flow fields in the development of disequilibrium textures in volcanic rocks. *Journal of Petrology*, **44**, 733–756.

PERUGINI, D., DE CAMPOS, C., DINGWELL, D. B., PETRELLI, M. & POLI, G. 2008. Trace element mobility during magma mixing: Preliminary experimental results. *Chemical Geology*, **256**, 146–157.

PHILPOTTS, A. R. 1967. Origin of certain iron–titanium oxide and apatite rocks. *Economic Geology*, **62**, 303–315.

PITCHER, W. S. 1993. *The Origin and Nature of Granite*. Chapman & Hall, London.

REID, B., JR., EVANS, C. O. & FATES, G. D. 1983. Magma mixing in granitic rocks of the Central Sierra Nevada, California. *Earth and Planetary Science Letters*, **66**, 2543–2561.

RHODES, A. L., ORESKES, N. & SHEETS, S. 1999. Geology and rare earth element geochemistry of magnetite deposits at El Laco, Chile. *In*: SKINNER, B. (ed.) *Geology and Ore Deposits of the Central Andes*. Society of Economic Geologists, Littleton, CO, Special Publication, **7**, 299–332.

ROEDDER, E. 1979. Silicate liquid immiscibility in magmas. *In*: YODER, H. S., JR (ed.) *The Evolution of Igneous Rocks*. Princeton University Press, Princeton, New Jersey.

ŞENGÖR, A. M. C. & YILMAZ, Y. 1981. Tethyan evolution of Turkey: a plate tectonic approach. *Tectonophysics*, **75**, 181–241.

SHA, L. K. 1995. Genesis of zoned hydrous ultramafic/mafic silicic intrusive complexes: an MHFC hypothesis. *Earth Science Reviews*, **39**, 59–90.

SILLITOE, R. H. 1996. Granites and metal deposits. *Episodes*, **19**, 126–133.

SÖZERI, K. 2003. *Geology, geochemistry and petrology of Balışeyh (Kırıkkale) molybdenum deposit and around region*. PhD Thesis, Ankara University.

STEIGER, R. H. & JÄGER, E. 1997. Subcommission on Geochronology: convention on use of decay constants in geo- and cosmochronology. *Earth and Planetary Science Letters*, **36**, 359–362.

STENDAL, H. & ÜNLÜ, T. 1991. Rock geochemistry of an iron ore field in the Divriği region, Central Anatolia, Turkey. A new exploration model for iron ores in Turkey. *Journal of Geochemical Exploration*, **40**, 281–289.

SUN, S. S. & MCDONOUGH, W. F. 1989. Chemical and isotopic systematics of ocean basalts; implications for mantle composition and processes. *In*: SAUNDERS, A. D. & NORREY, M. J. (eds) *Magmatism in Ocean Basins*. Geology Society, London, Special Publications, **42**, 313–345.

TATAR, S. & BOZTUĞ, D. 1998. Fractional crystallization and magma mingling/mixing processes in the monzonitic association in the SW part of the composite Yozgat batholith (Şefaatli-Yerköy, SW Yozgat). *Turkish Journal of Earth Sciences*, **7**, 215–230.

TAYLOR, S. R. & MCLENNAN, S. M. 1985. *The Continental Crust: Its Composition and Evolution*. Blackwell, Oxford, 27–72.

TÜRELI, T. K. 1991. *Geology, petrography and geochemistry of Ekecikdag Plutonic Rocks (Aksaray Region-Central Anatolia)*. PhD Thesis, METU, Ankara.

TÜRELI, T. K., GÖNCÜOGLU, M. C. & AKIMAN, O. 1993. Petrology and genesis of the Ekecikdag Granitoid (western part of the Central Anatolian Crystalline Complex). *General Directorate of Mineral Research and Exploration (MTA) Bulletin*, **115**, 15–28 [in Turkish with English abstract].

ÜNLÜ, T. & STENDAL, H. 1986. Geochemistry and element correlation of iron deposits in the Divriği Region, Central Anatolia, Turkey. *Jeoloji Mühendisliği*, **28**, 5–19 [in Turkish with English abstract].

ÜNLÜ, T. & STENDAL, H. 1989. Rare earth element (REE) geochemistry from the iron ores of the Divriği region, Central Anatolia, Turkey. *Geological Bulletin of Turkey*, **32**, 21–38 [in Turkish with English abstract].

VEKSLER, I. V. & THOMAS, R. 2002. An experimental study of B-, P- and F-rich synthetic granite pegmatite at 0.1 and 0.2 gpa. *Contributions to Mineralogy and Petrology*, **143**, 673–683.

VEKSLER, I. V., THOMAS, R. & SCHMIDT, C. 2002. Experimental evidence of three coexisting immiscible fluids in synthetic granite pegmatite. *American Mineralogist*, **87**, 775–779.

VERNON, R. H. 1984. Microgranitoid enclaves in granites-globules of hybrid magma quenched in a plutonic environment. *Nature*, **309**, 438–439.

WHITNEY, D. L., TEYSSIER, C., DILEK, Y. & FAYON, A. K. 2001. Metamorphism of the Central Anatolian crystalline complex, Turkey: Influence of orogen-normal collision v. Wrench dominated tectonics on P-T-t paths. *Journal of Metamorphic Geology*, **19**, 411–432.

WHITNEY, J. A. & STORMER, J. C., JR. 1983. Igneous sulfides in the fish canyon tuff and the role of sulfur in calc-alkaline magmas. *Geology*, **11**, 99–102.

WIEBE, R. A. & COLLINS, W. J. 1998. Depositional features and stratigraphic sections in granitic plutons: Implications for the emplacement and crystallization of granitic magma. *Journal of Structural Geology*, **20**, 1273–1289.

WIEBE, R. A., FREY, H. & HAWKINS, D. P. 2001. Basaltic pillow mounds in the Vinalhaven Intrusion, Maine. *Journal of Volcanology and Geothermal Research*, **107**, 171–184.

YALINIZ, M. K., AYDIN, N. S., GÖNCÜOGLU, M. C. & PARLAK, O. 1999. Terlemez quartz monzonite of central Anatolia (Aksaray–Sarıkaraman): age, petrogenesis and geotectonic implications for ophiolite emplacement. *Geological Journal*, **34**, 233–242.

YALINIZ, M. K., FLOYD, P. A. & GÖNCÜOĞLU, M. C. 1996. Supra-subduction zone ophiolites of Central Anatolia:- Geochemical evidence from the Sarıkaraman ophiolite, Aksaray, Turkey. *Mineralogical Magazine*, **60**, 697–710.

YALINIZ, M. K., FLOYD, P. & GÖNCÜOĞLU, 2000. Geochemistry of volcanic rocks from the Çiçekdağ ophiolite, Central Anatolia, Turkey, and their inferred tectonic setting within the northern branch of the Neotethyan ocean. *In*: BOZKURT, E., WINCHESTER, J. & PIPER, J. A. (eds) *Tectonics and Magmatism in Turkey and the Surrounding Area*. Geological Society, London, Special Publications, **173**, 203–218.

YILMAZ, S. & BOZTUĞ, D. 1998. Petrogenesis of the Çiçekdağ Igneous Complex, North of Kırşehir, Central Anatolia, Turkey, Turkish. *Journal of Earth Sciences, Special Issue on Alkali Magmatism*, **7**, 185–199.

Geochemical indicators of metalliferous fertility in the Carboniferous San Blas pluton, Sierra de Velasco, Argentina

J. N. ROSSI[1]*, A. J. TOSELLI[1], M. A. BASEI[2], A. N. SIAL[3] & M. BAEZ[1]

[1]*Facultad de Ciencias Naturales, UNT. Miguel Lillo 205, CP. 4000, Tucumán, Argentina*

[2]*Instituto de Geociencias, Universidade de São Paulo, Rua do Lago 562, São Paulo, Brazil*

[3]*NEG-LABISE, Department de Geologia, Universidade Federal de Pernambuco, C.P. 7852, Recife, PE, 50670-000, Brazil*

**Corresponding author (e-mail: juanitarossi@gmail.com)*

Abstract: In the Sierra de Velasco, northwestern Argentina, undeformed Lower Carboniferous granitoids (350–334 Ma) intrude deformed Lower Ordovician granites and have been emplaced by passive mechanisms, typical of tensional environments. The semi-elliptic, about 300 km^2 shallow-emplaced San Blas pluton is 340–330 Ma old, with εNd_t between −1.3 and −1.8 which indicates that, different from the nearby Famatinian–Ordovician granitoids, the San Blas pluton had a relatively brief crustal residence, with an interaction between asthenospheric material and greywackes. The cupola of the pluton was almost totally eroded down during the Upper Carboniferous.

The San Blas pluton is a porphyritic granite composed mainly of monzogranite to syenogranite and shows graphic intergrowth and miarolitic cavities up to 5 cm in diameter, filled with quartz. Two different textures are recognized: perthitic microcline megacrysts (30–45 vol%) set in a medium- to coarse-grained groundmass of quartz, microcline and oligoclase, with sericitic alteration. Biotite, muscovite, apatite, zircon, fluorite and opaque minerals are the accessory phases. The other textural variation consists in microcline megacrysts (10%–15 vol%) and a fine-grained groundmass, of quartz, microcline and oligoclase, biotite, apatite, muscovite, zircon and magnetite.

The average SiO_2 content in this pluton is 74.94%, the ASI = 1.1, CaO and MgO are less than 1%, total Fe_2O_3 and P_2O_5 contents are low, and $K_2O > Na_2O$. Low Ba, Sr and high Rb contents, coupled with Sn contents (*c.* 15 ppm), W (*c.* 380 ppm), Nb, Y, Ta, Th and U confirm this is a special granite. The K/Rb ratio (*c.* 75) indicates that Rb has been fractionated to the residual melt whereas the Zr/Hf (*c.* 25) demonstrates that hydrothermal alteration occurred. The Sr/Eu ratio of *c.* 75 along other geochemical features characterize this pluton as a fertile evolved granite.

The chondrite-normalized rare earth element (REE) diagram shows the tetrad effect that allows the subdivision of the lanthanides into four groups.

In general, the tetrad effect is recognized in evolved granites and products of hydrothermal alteration such as greisens. The above-mentioned features show that the San Blas granite is fertile, and the absence of ore deposits has been probably caused by erosion of a mineralized cupola during Carboniferous and Cenozoic exhumation. The finding of alluvial cassiterite and wolframite in drainage systems is the first evidence of the fertile character of this granite.

At the Sierra de Velasco, extensive exposures of Lower Ordovician, I- and S-type granitoids are observed. They are mainly syn- to late- and post-kinematic, and are now orthogneisses of the Famatinian orogenic cycle, a wide magmatic arc in northwestern Argentina (Pankhurst *et al.* 2000; Toselli *et al.* 2002, 2007). Ordovician to Devonian deformation bands and NW–SE trending lineaments have affected the whole geological setting (Toselli *et al.* 2007).

In the northern and central parts of the Sierra de Velasco, these granitoids are intruded by three bodies of mainly Lower Carboniferous age, the Sanagasta, Huaco and San Blas plutons. The deformation structures are crosscut by some of these younger plutons.

The Huaco and Sanagasta granitoids are two adjacent sub-ellipsoidal, undeformed plutons (Grosse *et al.* 2009), and the San Blas pluton is a notable semi-elliptic shaped granite (Báez 2006).

The three granites share common features, being porphyritic monzo- to syeno-granites which are geochemically evolved with SiO_2 contents between 73 and 75%.

From: Sial, A. N., Bettencourt, J. S., De Campos, C. P. & Ferreira, V. P. (eds) *Granite-Related Ore Deposits.* Geological Society, London, Special Publications, **350**, 175–186.
DOI: 10.1144/SP350.10 0305-8719/11/$15.00

The main objective of this contribution is to discuss in detail the field relationships, petrography, mineralogy, geochemical evolution and fractionation of the San Blas pluton (Fig. 1). The potential metalliferous fertility of this pluton will be also considered. Lannefors (1929) and some unpublished reports of the Mining and Geological Survey of Argentina indicate the finding of alluvial cassiterite and wolframite in drainage systems developed within this pluton. It was the first evidence of a fertile character in this granitic pluton.

Rb–Sr and Sm–Nd isotopic data are used to determine crustal residence age, the origin and possible contribution from juvenile material to allow a comparison with nearby Ordovician granitoids of the Pampean Ranges.

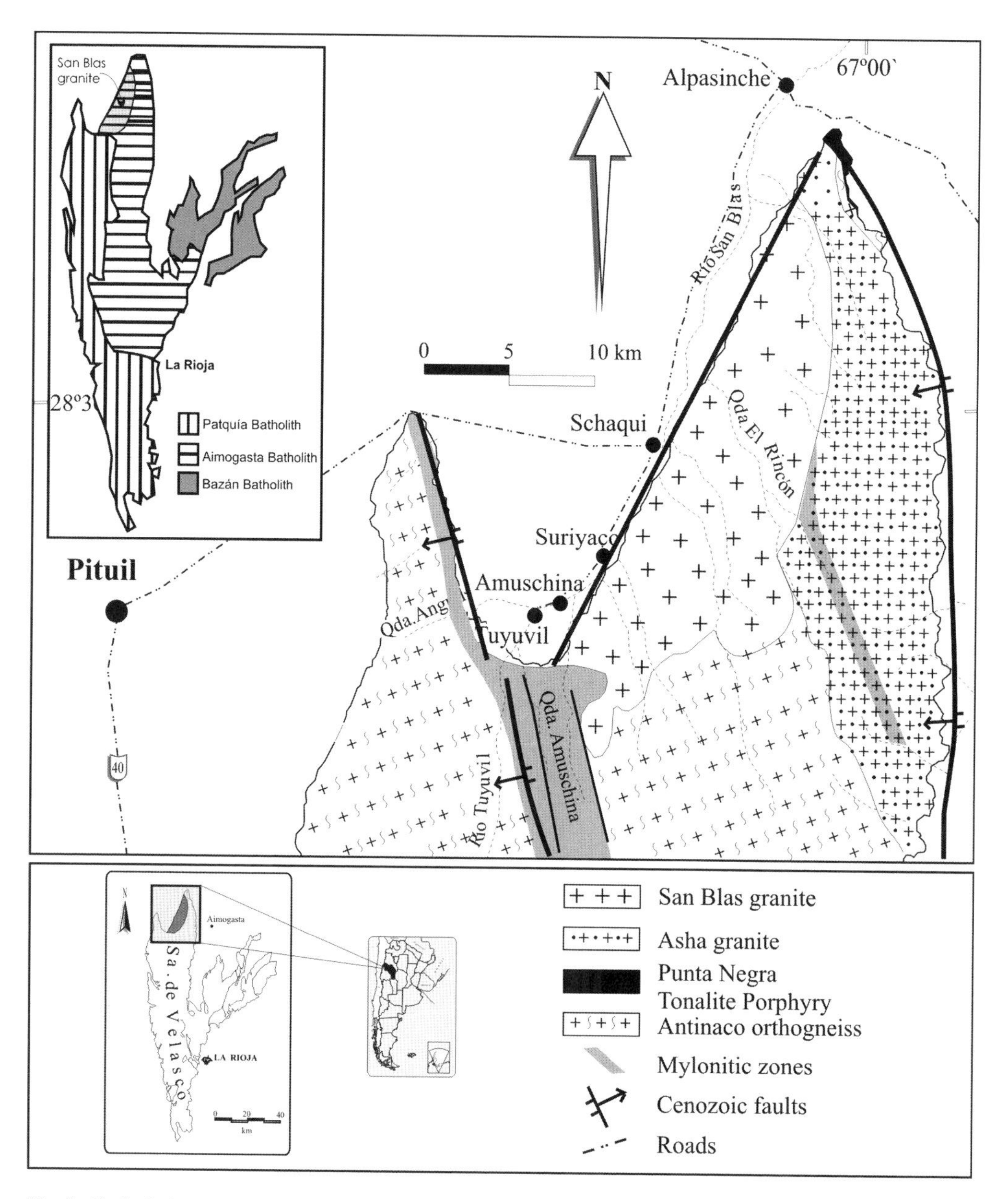

Fig. 1. Geological sketch of the Sierra de Velasco and San Blas Granite, northwestern Argentina.

Analytical techniques

Major elements were determined by inductively coupled plasma - atomic emission spectroscopy (ICP-AES), and trace elements by inductively coupled plasma mass spectrometry (ICP-MS) at the Activation Laboratories (Canada), using lithium metaborate/tetraborate fusion.

Isotopic analyses were performed at the Zentrallabor für Geochronologie des Departments für Geo- und Umweltwissenschaften der Ludwig Maximilians Universität, Munich. Powdered samples for Rb–Sr and Sm–Nd analyses were dissolved in a mix of HF, HNO_3 and $HClO_4$. Before dilution, samples for Rb and Sr analyses were treated separately with a spike. Chemical separation of Rb and Sr used cation exchange technique on DOWEX AG 50Wx8 resin. Isotopic ratio measurements were made with Finnigan THQ mass spectrometer. The measured isotopic ratios were normalized to $^{88}Sr/^{86}Sr = 8.3752094$.

For Sm and Nd, samples were treated with a combined Sm–Nd–Sr spike. The separation of the REEs was performed with cationic exchange techniques using DOWEX AG 50W×8 resin. The separation of Sm from Nd was done using with the H_3PO_4 HDEHP ester on teflon powder as the cation exchange resin. The isotopic ratios were normalized to $^{146}Nd/^{144}Nd = 0.7219$. Isotopic ratio measurements were performed in a MAT 261/262 mass spectrometer, with a confidence interval of 95% (2σ).

Geological and petrographic characteristics

The San Blas body is a semi-elliptic pluton about 28 km long and 8 km wide, located to the north of the Sierra de Velasco, with its major axis trending NE–SW (Figs 1 & 2a).

The surface morphology shows a smooth westward tilted erosion surface exposed since the Upper Carboniferous (Jordan & Allmendinger 1986). At the southwestern boundary, this pluton intrudes the Ordovician Antinaco orthogneiss, and contains incorporated xenoliths of the country rock; northwards it intrudes the Ordovician Punta Negra porphyry tonalite (Fig. 1), and to the east it shows sharp contacts against the supposedly Ordovician La Costa pluton (Toselli *et al.* 2006). The absence of deformation at its contacts zones and the incorporation of xenoliths by stopping suggest for the San Blas granite a post-tectonic passive intrusion mechanism in a brittle shallow setting.

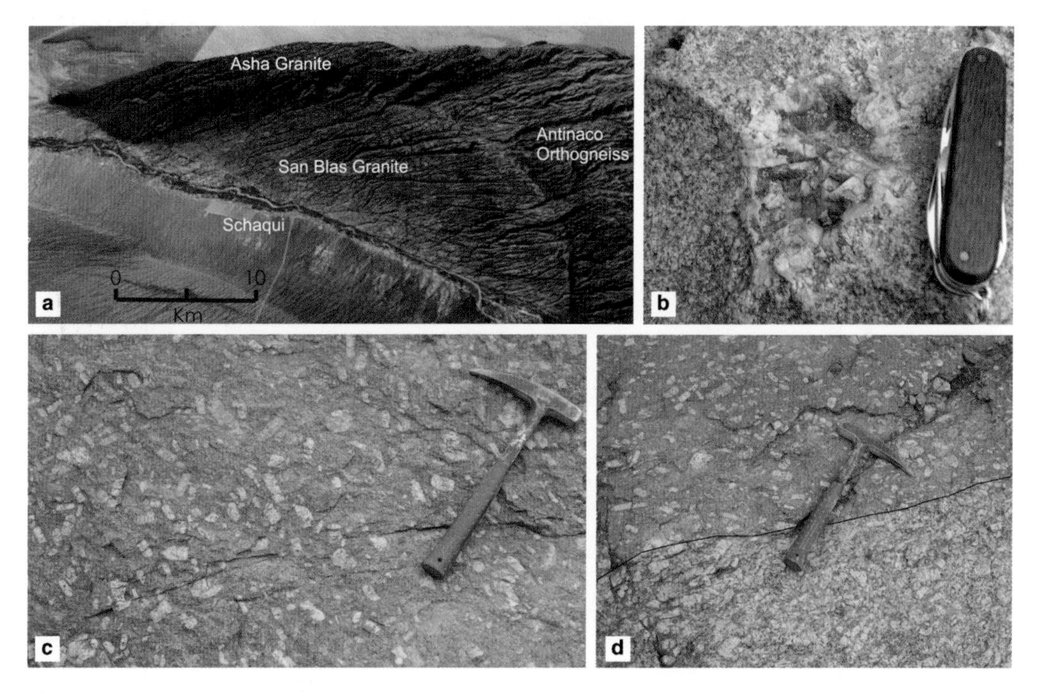

Fig. 2. (**a**) Panoramic overview of the outcrop of the San Blas pluton and its Carboniferous erosion surface; (**b**) miarolitic cavities; (**c**) microcline megacrysts in a fine-grained groundmass; (**d**) contact zone between medium- to coarse-grained granite and fine grained matrix with megacrysts.

Table 1. *Whole rock chemical analysis. Major elements in %. Traces in ppm*

Sample	5795	6336	6340	6445	6447	6452	6511	6513	6514	6515	6523	6749	6754
SiO_2	75.20	70.78	73.93	73.73	75.71	75.23	75.47	76.59	74.45	76.86	75.26	74.93	76.11
TiO_2	0.10	0.34	0.22	0.12	0.12	0.21	0.17	0.10	0.07	0.02	0.19	0.03	0.12
Al_2O_3	12.39	14.31	13.24	14.74	13.15	12.63	12.65	12.71	12.29	13.08	13.07	14.64	12.40
Fe_2O_3t	1.44	2.80	1.99	1.23	1.46	2.02	1.88	1.27	1.60	0.40	1.88	0.60	1.76
MnO	0.04	0.05	0.03	0.13	0.04	0.03	0.04	0.05	0.04	0.07	0.09	0.12	0.03
MgO	0.07	0.32	0.12	0.18	0.24	0.14	0.09	0.22	0.02	0.07	0.09	0.12	0.03
CaO	0.60	1.16	0.66	0.69	0.62	0.60	0.74	0.48	0.77	0.37	0.52	0.32	0.63
Na_2O	3.16	3.48	3.11	3.80	2.92	2.95	3.15	2.88	3.69	4.53	2.93	4.12	2.89
K_2O	4.82	5.78	5.78	4.37	4.68	5.46	4.90	4.05	4.88	4.30	4.52	3.76	5.53
P_2O_3	0.02	0.19	0.10	0.28	0.13	0.09	0.06	0.15	0.02	0.06	0.10	0.30	0.03
LOI	0.99	1.05	1.06	1.14	1.31	0.81	1.04	0.95	1.00	0.52	1.06	0.67	0.86
Total	98.33	100.26	100.24	100.41	100.38	100.18	100.18	99.44	98.83	100.21	99.78	99.51	100.40
Rb	583	594	460	465	473	446	496	426	771	576	395	689	546
Sr	19	66	56	29	18	38	34	21	8	16	46	28	18
Ba	46	179	188	95	37	121	128	36	20	36	203	20	73
Zr	103	284	231	51	59	200	227	50	211	70	153	18	167
Cs	n.a.	39.2	18.4	61.2	21.4	16.7	40.6	37.7	41.2	22.2	19.6	13.3	21.7
Hf	4.90	8.4	6.7	2.00	2.1	5.5	7.5	1.9	11.9	6.4	5.3	1.2	6.5
Y	88	72	37.4	23.1	25.7	54	61.6	23	192	58	63	3	89
Nb	74.0	47.2	34.9	25.5	9.7	25.1	43	10.7	102	48.8	31.0	25.2	64.5
Ta	14.70	14.2	9.75	10.2	4.29	6.31	13.6	8.00	17.3	33	6.74	14.0	11.6
Ga	26	27	23	22	12	22	24	15	35	35	16	20	29
Th	50.80	49.5	40.6	11.20	7.18	87.1	60.8	6.43	97.6	18.3	46.0	4.15	57.6
U	17.0	5.11	7.69	4.62	2.10	8.60	6.63	3.01	10.4	2.53	10.3	5.22	18.8
Sn	16	21	14	9	6	10	14	13	20	8	17	2	6
W	320	568	609	490	254	305	579	284	595	325	242	177	178

Table 2. *Lanthanides, K, Rb, Sr, Zr and Hf of San Blas granite. C_1 Chondrite of Sun & McDonough (1989). Data in ppm*

Sample	5795	6445	6447	6452	6511	6513	6514	6515	6523	6749	6754	C_1
La	35.60	13.2	8.93	89.9	44.3	6.92	44.2	23.9	38.2	1.37	53.2	0.237
Ce	82.90	28.0	20.5	1.97	94.6	17.2	108	63.9	86.8	2.87	120	0.612
Pr	9.24	3.47	2.35	20.6	11.7	2.05	13.4	7.82	9.70	0.30	14.4	0.095
Nd	39.10	12.2	8.74	69.4	40.3	7.46	56.4	28.3	34.9	1.06	56.1	0.467
Sm	10.60	3.09	2.43	12.8	9.33	2.29	16.0	8.52	8.22	0.23	12.6	0.153
Eu	0.25	0.41	0.205	0.58	0.66	0.172	0.17	0.12	0.61	0.01	0.49	0.058
Gd	10.70	2.65	2.54	9.26	8.0	2.21	17.2	7.84	7.36	0.19	12.2	0.2055
Tb	2.39	0.62	0.63	1.66	1.73	0.58	3.88	1.96	1.64	0.04	2.32	0.0374
Dy	15.0	3.89	4.15	9.85	10.8	3.73	25.5	13.3	10.2	0.33	14.8	0.2540
Ho	2.97	0.76	0.82	1.82	2.08	0.74	5.78	2.68	2.03	0.07	3.07	0.0566
Er	10.0	2.22	2.74	5.34	5.85	2.47	17.6	9.85	6.39	0.27	9.71	0.1655
Tm	1.67	0.42	0.479	0.783	0.98	0.464	2.89	1.95	0.97	0.06	1.55	0.0255
Yb	10.20	2.77	3.20	4.68	5.74	3.27	17.6	15.4	6.21	0.55	9.96	0.170
Lu	1.41	0.40	0.491	0.631	0.79	0.518	2.48	2.51	0.91	0.01	1.41	0.0254
K	40012	36277	38850	45325	40675	33619	40510	35696	37525	31216	45910	545
Rb	583	465	473	446	496	426	771	576	395	689	546	2.32
Sr	19	29	18	38	34	21	8	16	46	28	18	7.26
Zr	103	51	59	200	227	50	211	70	153	17	167	3.87
Hf	4.90	2.0	2.1	5.5	7.5	1.9	11.9	6.4	5.3	1.2	6.5	0.1066

The western boundary, covered by modern sediments, is sharply defined by a north–southward trending fault (Figs 1 & 2a).

The San Blas pluton is composed essentially of porphyritic monzogranites and, less often, of syenogranites. Within the borders and in the central area, semi spherical miarolitic cavities have been recognized, up to 5 cm sized, some of them filled with quartz crystals and others contain graphic intergrowth of quartz and microcline (Fig. 2b).

Two textural phases have been recognized: (*a*) a fine grained groundmass with scarse phenocrysts content, at the northeastern and southwestern borders, and (*b*) a medium- to coarse-grained groundmass with abundant phenocrysts in the central zone of San Blas pluton. There is no sharp textural transition between the two phases which appear to be intermingled sometimes (Fig. 2c, d).

In the phase *b*, perthitic microcline megacrysts constitute 30–45% of the total volume of the rock. They are tabular shaped, euhedral to subhedral, from 3 to 15 cm long, and often contain inclusions of biotite and quartz.

In thin section, the groundmass of the porphyritic granite is inequigranular, medium- to coarse-grained. It is composed of quartz, microcline and oligoclase with some alteration to sericite. Biotite, in higher modal content than muscovite, is flake-shaped (1–3 mm long), shows corroded borders and appears in clusters associated with opaque minerals. Accessory minerals are apatite in round-shaped grains and subhedral zircon. Fluorite is interstitial and relatively scarce.

In some places, the abundance of microcline megacrysts decrease to 10% and the groundmass changes to a more fine-grained nearly aplitic texture and the fluorite being more abundant and secondary muscovite increases. In these weakly altered zones, neither topaz nor tourmaline has been found. Zircon grains are prismatic shaped (0.01–0.04 mm long).

The phase *a* contains less microcline megacrysts (10%–15 vol%) set in a fine-grained inequigranular groundmass (0.2–1 mm). Megacrysts are euhedral to subhedral, 1 to 5 cm long, often exhibiting Carlsbad twins. In thin sections, microcline displays albite–pericline twins, and often inclusions of biotite and quartz. The groundmass is composed of quartz, microcline, sodic oligoclase as essential minerals and apatite, muscovite, zircon and magnetite as accessory minerals.

Geochemistry

Whole-rock major, minor and trace chemistry data are presented in Tables 1 and 2. The content of SiO_2: (71–76%) has a mean value of 74.9%;

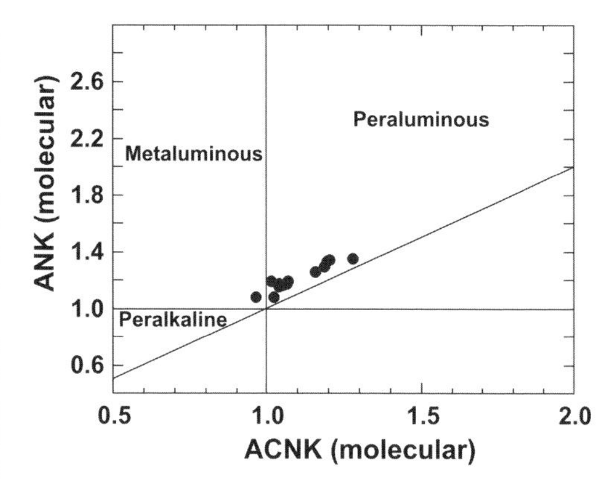

Fig. 3. ($Al_2O_3/CaO + Na_2O$) v. ACNK ($Al_2O_3/CaO + Na_2O + K_2O$), from Maniar & Piccoli (1989). Granite samples plot mainly in the peraluminous field.

TiO_2, CaO and MgO are lower than 1%; Fe_2O_{3total}: (0.4–2.8%) has a mean value of 1.56%; $K_2O > Na_2O$, with a mean value $K_2O < 5\%$. P_2O_5 contents are low (average value around 0.12%). All of these data suggest significantly evolved granite magma. The aluminum average of saturation index (ASI = molar $[Al_2O_3/(CaO + Na_2O + K_2O)] = 1.10$) indicates moderate peraluminosity (Fig. 3).

The trace element chemistry points to high evolution of the San Blas pluton better than major elements. A low mean content for Ba (77 ppm), Sr (31 ppm), high average value content of Rb (532 ppm) and of Cs (29.4 ppm) are observed but even so the HFSE (Nb, Ta and Y) have high average values. Th and U contents are also high and indicate their fractionation in REE, Y, Th, U-rich accessory minerals as monazite, zircon and apatite (Bea 1996). High content of Sn (mean value 12 ppm) and W (mean value 379 ppm) confirms the specialized character of the San Blas granite (Table 1).

The K/Rb ratio has been used to characterize the evolution state of granitic melts. $K/Rb < 100$ ratios are regarded as indication of highly evolved granitic melts. The mean value of $K/Rb = 76$ indicates that Rb tends to fractionate in residual melt or the fractionation between a silicate melt and an aqueous fluid phase (Clarke 1992); the mean value of $Zr/Hf = 24$ is lower than the chondritic ratio (36.4). $Zr/Hf < 20$ ratios are characteristic for strong magmatic – hydrothermal alteration (Irber 1999). Recently, Bea *et al.* (2006) have demonstrated that Zr/Hf significantly lower than chondrite results from zircon fractionation.

Sr/Eu ratio ranges from 37 to 138, with an average value of 77, being most values lower than the chondritic ratio (125). The observed range suggests fractionation of both, Sr and Eu^{2+} and so

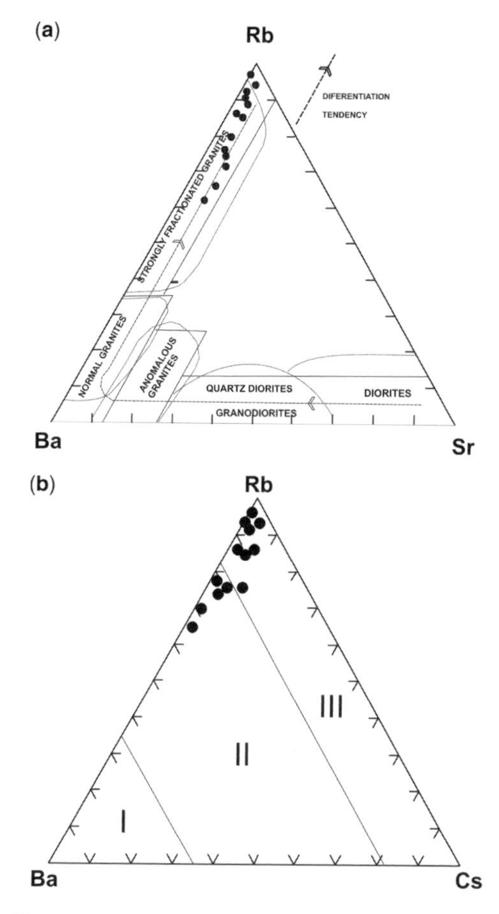

Fig. 4. (**a**) Rb–Ba–Sr plot (El Bouseily & El Sokkary 1975). Plotted points shift to strongly differentiation granites; (**b**) Rb–Ba–Cs plot (El Bousely & El Sokkary 1975), granites plot in fields II and III that indicate different potentials of mineralization of Sn and W.

the Sr/Eu ratio increases sympathetically with the evolution of the granitic magma.

The character of evolved granite is well illustrated in the Rb–Ba–Sr triangular diagram (El Bouseily & El Sokkary 1975). The plotted data points shift towards the Rb apex (Fig. 4a). A Rb–Ba–Cs complementary diagram (El Bouseily & El Sokkary 1975) distinguishes three fields: Ba decreases and Rb increases from field I to field III (Fig. 4b). Most samples in field III are Cs-enriched (mean 31.2 ppm) as those of field II (mean 26.9 ppm). Neiva (1984) had observed that samples within field III have the highest content of Sn (Sn > 18 ppm) while samples within field II have Sn > 15 ppm. In the present case, the situation is variable since the mean content of Sn within field III is 9 ppm, and in samples of field II, mean Sn is 15 ppm. Samples within field I belong to normal granites with average contents of Sn of 3 ppm (Fig. 4b) with regard to W, the samples within field II have a mean of 461 ppm and those of the field III have 328 ppm.

The Rb–K_2O diagram of Tuach *et al.* (1986) shows separate samples with the highest Sn and W contents belonging to the cupola zone (specialized granite) from those granites deeper, developed (not specialized granite), which indicates a geochemically layered pattern for the magmatic chamber (Fig. 5).

Main geochemical characteristics of the San Blas pluton are those of a fertile granite (with mean values of Sn = 13 ppm and $W = 379$ ppm), but the lack of ore veins and greisen in actual outcrops can be explained by a deep erosion of the granite cupola during the Carboniferous exhumation.

The diagram $CaO/(FeO + MgO + TiO_2)$ v. $CaO + FeO + MgO + TiO_2$ of Patiño Douce

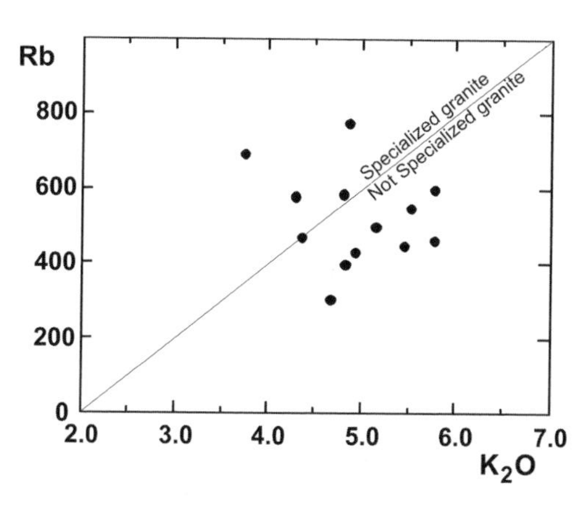

Fig. 5. Rb v. K_2O plot (Tuach *et al.* 1986) discriminating specialized and non-specialized granites.

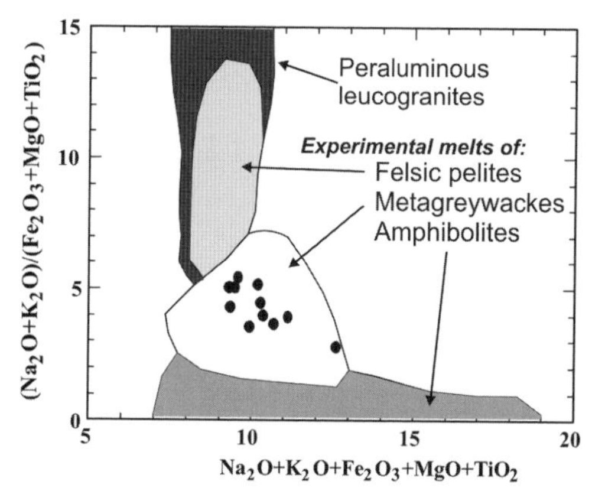

Fig. 6. Diagram $CaO/(FeO + MgO + TiO_2)$ v. $CaO + FeO + MgO + TiO_2$ (Patiño Douce 1999) which suggests a melt of crustal metagreywackes magma source.

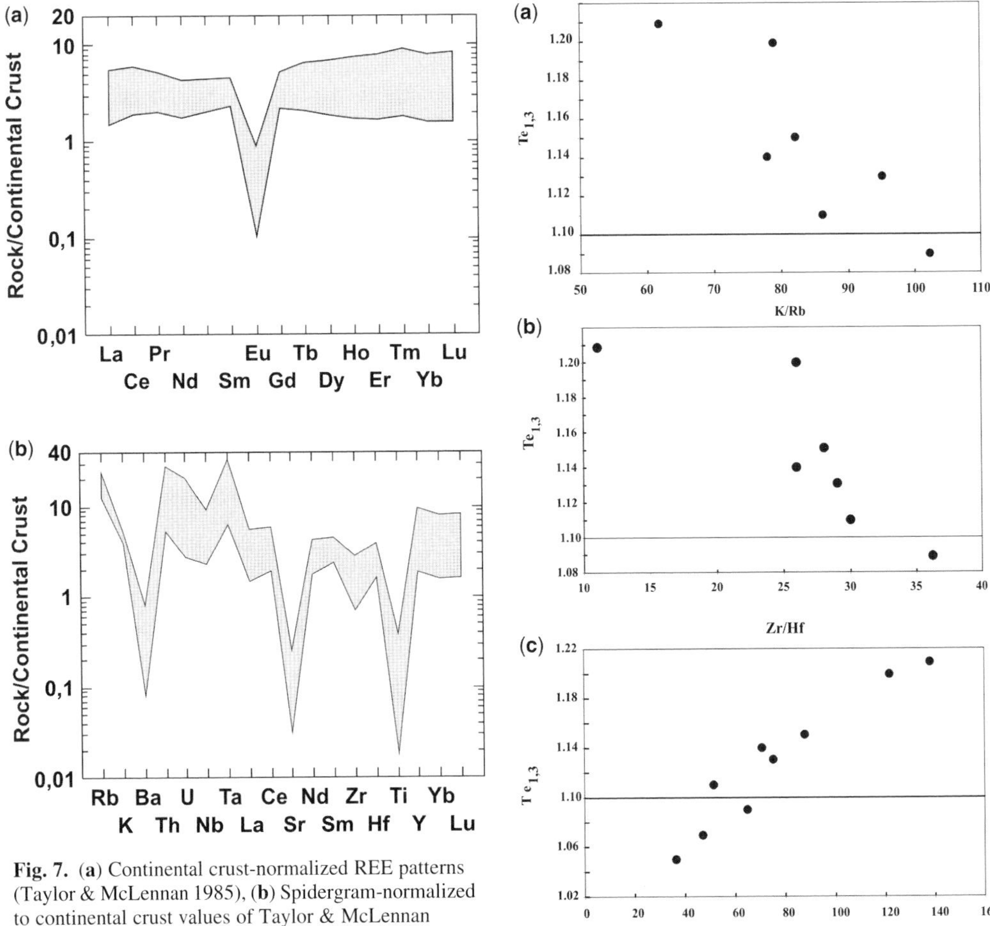

Fig. 7. (**a**) Continental crust-normalized REE patterns (Taylor & McLennan 1985), (**b**) Spidergram-normalized to continental crust values of Taylor & McLennan (1985).

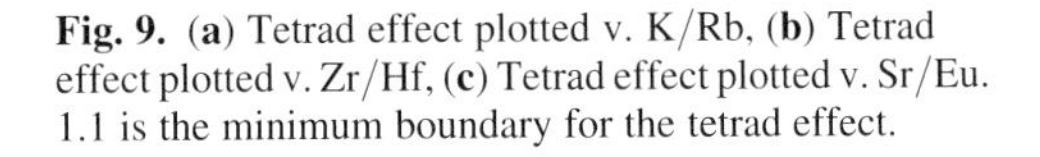

Fig. 9. (**a**) Tetrad effect plotted v. K/Rb, (**b**) Tetrad effect plotted v. Zr/Hf, (**c**) Tetrad effect plotted v. Sr/Eu. 1.1 is the minimum boundary for the tetrad effect.

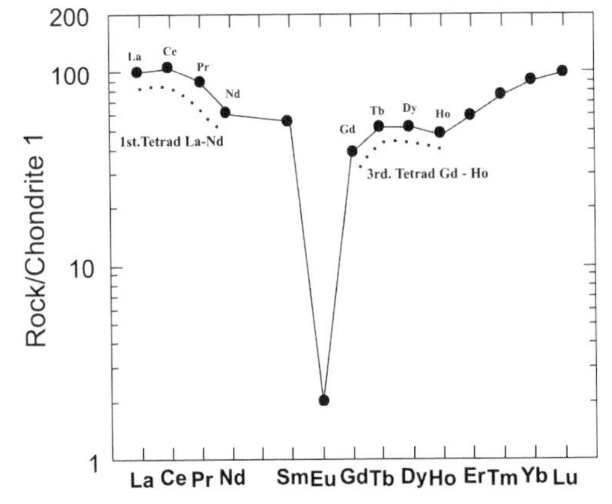

Fig. 8. Tetrad effect La–Nd (T_1), Nd–Gd (T_2), Gd–Ho (T_3), showed in sample 6515 from the San Blas pluton. Note the strong Eu negative anomaly.

(1999) shows evidence of a melt of crustal metagreywackes magma source. The experiments indicate that the metagreywackes contain biotite + plagioclase, but no aluminosilicates. The physical conditions of formation correspond to magmas formed by hybridization in continental crust of normal thickness at depths of 30 km or less (Fig. 6).

Continental crust-normalized REE patterns (according to Taylor & McLennan (1985) values) show remarkable negative Eu anomalies and flat patterns of distribution for the other REEs. These patterns suggest feldspar fractionation in the source (Fig. 7a).

Continental crust-normalized spidergrams (according to Taylor & McLennan (1985) values) show strong depletion of Ba, Sr and Ti. The depletion of Ba, Sr and Eu (in REE diagram) is

Table 3. *Values of the tetrad effect and ratios K/Rb; Sr/Eu; Zr/Hf and Eu/Eu**

Sample	$TE_{1,3}$	K/Rb	Sr/Eu	Zr/Hf	Eu/Eu*
5795	1.04	69	76	21	0.07
6445	1.14	78	70.7	26	0.44
6447	1.15	82	87.8	28	0.25
6452	1.09	102	65.5	36	0.16
6511	1.11	86	51.5	30	0.23
6513	1.20	79	122	26	0.23
6514	1.07	53	47	18	0.03
6515	1.21	62	138	11	0.04
6523	1.13	95	75	29	0.24
6749	1.08	45	–	15	0.15
6754	1.05	84	37	26	0.12
Average value	–	75	77	26	–

Table 4. *Data of Rb and Sr, in ppm*

Sample	Rb	Sr	$^{87}Rb/^{86}Sr$	Error	$^{87}Sr/^{86}Sr$	Error
6336	594	66	263 512	0.5270	0.830105	0.000032
6340	460	56	240 285	0.4806	0.820571	0.000027
6511	496	31	430 409	0.8608	0.909522	0.000049

not only originated by fractionation of feldspars but the effect is strongly enhanced by a late stage melt-fluid interaction. The spidergrams show also strong enrichment for Rb, Th, U, Nb, Ta and Y with respect to the continental crust; moreover Hf $>$ Zr, and Rb $>$ K. These enrichments can be explained by a late stage melt-fluid interaction, so high-field strength elements such as Nb, Ta, Zr, Hf, Y and REE may form complexes with a variety of bonds (F, B) whose stability is no longer constrained by the charge and ionic radius (Keppler 1993; Bau 1996) (Fig. 7b).

Some lanthanides in the San Blas granite show a 'tetrad effect', the most remarkable is shown in Figure 8. The tetrad effect was described for the first time by Fidelis & Siekierski (1966) and it is referred to a subdivision of the lanthanides in four groups, in a pattern of chondritic-normalized distribution: (1) La–Nd, (2) Pm–Gd, (3) Gd–Ho and (4) Er–Lu, and each group forms a convex pattern M-type (magmatic) or concave W-type (aqueous), (Masuda *et al.* (1987)). The four groups are separated in the boundary points located in between Nd and Pm, Gd, and in between Ho and Er, which correspond to a $\frac{1}{4}$, $\frac{1}{2}$, and $\frac{3}{4}$, filled 4f electronic shell.

The tetrad effect has been recognized in high evolved granitic rocks, hydrothermal alteration and mineralization. In the practice, only tetrads 1 and 3 are significant as shown in Figure 8.

The more evolved and those members with weak hydrothermal alteration (greisenization) in the San Blas pluton show a measurable tetrad effect. The measure of this effect is obtained by the mean deviation of the tetrad pattern with respect to the normal pattern of the REE without that effect. The first and the third tetrad have been obtained using Irber's (1999) formula. The data are shown in Table 2 and the results of the calculation in Table 3. Zr/Hf ratios in Table 2 are smaller than the chondritic ratio of 36.3 from Sun & McDonough (1989). That is a consequence of the larger solubility of $ZrSiO_4$ than $HfSiO_4$ in polymerized evolved meta- or peraluminous granitic melts, that results in a depletion of the Zr/Hf ratio (Linnen & Keppler 2002). The same conclusion was reached by Bea *et al.* (2006).

The tetrad effect ($Te_{1,3}$) must be higher than 1.1 for the reproduction of the characteristic pattern. ($Te_{1,3}$) was plotted against K/Rb, Zr/Hf and Sr/Eu (Fig. 9a–c). K/Rb and Zr/Hf ratios show negative correlation with the tetrad effect, while Sr/Eu ratio has positive correlation. The Eu/Eu* ratios are variable and show no correlation with the tetrad effect (Eu*, expected concentration from interpolating the normalized values of Sm and Gd). In spite of the weak tetrad effect, there is a correlation with the hydrothermal alteration shown by the rocks.

Isotopic geochemistry

Whole-rock Rb–Sr isotope data are presented in Table 4. The $^{87}Rb/^{86}Sr$ and $^{87}Sr/^{86}Sr$ of the three analysed samples define an isochron, obtained

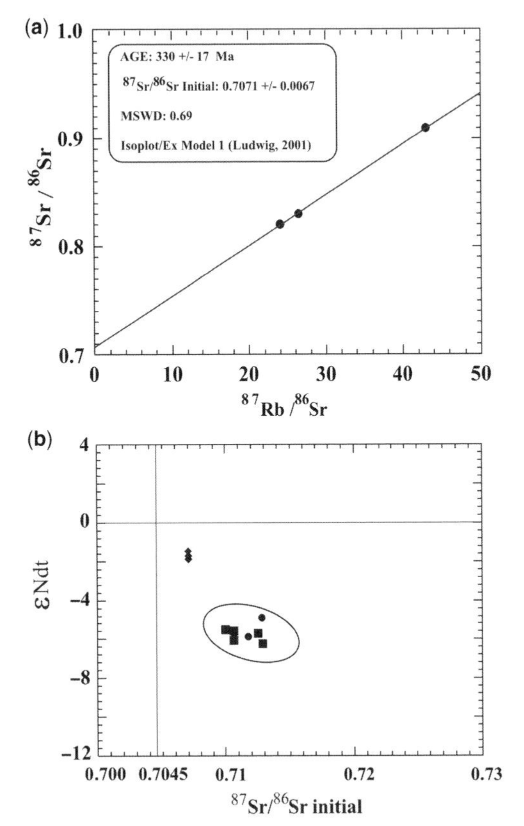

Fig. 10. (**a**) Rb–Sr Isochron for the San Blas pluton; (**b**) εNd_t v. $(^{87}Sr/^{86}Sr)_i$. Symbols: rhombs: San Blas granite; circles: Cerro Negro and Punta Negra Ordovician dacite porphyries (Toselli and Rossi, unpublished data); squares: Fiambalá and Copacabana Ordovician granitoids (Höckenreiner 2003); ellipse: Ordovician I and S granitoids (Pankhurst *et al.* 2000).

using the program Isoplot/Ex Model 1 (Ludwig 2001). The isochron yielded an age of 330 ± 17 Ma with an initial $^{87}Sr/^{86}Sr$ ratio of 0.7071 ± 0.0067 and MSWD 0.69 (Fig. 10a).

A previous age determination (U–Pb on zircon, TIMS technique) for the San Blas pluton indicated 334 ± 5 Ma (Báez & Basei 2005) and a U–Pb SHRIMP zircon age of 340 ± 3 Ma (Dahlquist *et al.* 2006).

The Sm and Nd whole-rock isotope analyses are shown in Table 5. Table 6 presents the Sm and Nd isotopic ratios for the samples, the chondritic uniform reservoir (CHUR), depleted mantle (DM) and continental crust (CC). The values of $(^{143}Nd/^{144}Nd)_{CHUR}$ are from Goldstein *et al.* (1984), $(^{147}Sm/^{144}Nd)_{CHUR}$ from Peucat *et al.* (1988), $(^{143}Nd/^{144}Nd)_{DM}$, $(^{147}Sm/^{144}Nd)_{DM}$ and $(^{147}Sm/^{144}Nd)_{CC}$ are from Liew & Hofmann (1988).

The data and isotopic two-stage model ages T_{DM} of 1,16–1,19 Ga, were obtained from Liew & Hofmann (1988) and the decay constant used: $\lambda_{Sm} = 6.54 \times 10^{-12} \times a^{-1}$. Figure 10b shows the plot of εNd_t v. $(^{87}Sr/^{86}Sr)_i$. Samples of the San Blas pluton plot on a restrict field with values of εNd_t between -1.3 to -1.8 and initial $(^{87}Sr/^{86}Sr)$ 0.707. The εNd_t and the model age T_{DM} data indicate that the petrogenesis of the San Blas granite involved a significant mantle component with an upper Mesoproterozoic crust. These data are different from those obtained for the Ordovician Famatinian granitoids whose εNd_t vary between -4 and -7, $(^{87}Sr/^{86}Sr)_i$ between 0.708 and 0.716 and their model age T_{DM}, between 1.5 and 1.7 Ga (Upper Palaeoproterozoic–Lower Mesoproterozoic) (Pankhurst *et al.* 2000; Höckenreiner 2003).

Table 5. *Data of Sm and Nd, in ppm*

Sample	Sm	Nd	$^{147}Sm/^{144}Nd$	$^{143}Nd/^{144}Nd$	Error 2σ
6336	11.40	53.70	0.1283	0.512398	0.000009
6340	8.60	45.70	0.1138	0.512364	0.000006
6511	9.60	39.52	0.1469	0.512452	0.000010

Table 6. *Isotopic ratios of Sm–Nd in the samples. Initial ratios of Nd isotope in the samples for t = 340 Ma, labelled εNd_t. CHUR, chondritic uniform reservoir; DM, depleted mantle; TDM, 2 stage model age.*

Sample	6336	6340	6511	CHUR	DM	Continental crust
$^{143}Nd/^{144}Nd$	0.512398	0.512364	0.512452	0.512638	0.513151	–
$^{147}Sm/^{144}Nd$	0.1283	0.1138	0.1469	0.1967	0.219	0.12
$(^{143}Nd/^{144}Nd)_t$	0.512120	0.512117	0.512144	0.512212	–	–
εNd_t	−1.79	−1.8	−1.3	–	–	–
T_{DM} (Ga)	1187	1190	1166	–	–	–

Discussion and conclusion

The geochronological data indicate that the San Blas granite is a Lower Carboniferous, undeformed pluton which intruded passively into Ordovician granitoids of variable grade of deformation. This fact makes this pluton equivalent in this aspect to other Carboniferous granites of the Sierra de Velasco (e.g. Huaco and Sanagasta plutons) (Toselli *et al.* 2006; Grosse *et al.* 2008, 2009; Söllner *et al.* 2007).

Plutons of this age are known in the northwestern Pampean Ranges, for example the La Quebrada pluton in Sierra de Mazan (Lazarte *et al.* 2006); Los Ratones and other minor bodies in the Sierra de Fiambalá (Arrospide 1985; Grissom *et al.* 1998; Lazarte *et al.* 2006); Papachacra granite in the Sierra de Papachacra (Lazarte *et al.* 2006) and in the Sierra de Zapata the Quimivil granite (Lazarte *et al.* 1999). With exception of Huaco and Sanagasta plutons, all the other above-mentioned plutons share common features with the San Blas granite in that they have geochemical signatures belonging to fertile granites with Sn and W mineralizations and constituting undeformed, nearly circular plutons, generally intruded into Ordovician granitoids with variable deformation grades, or as the case of the Los Ratones granite, intruded into a basement of low metamorphic grade and mylonite zones (Neugebauer 1995).

A remarkable characteristic of the Sierra de Velasco granites is a 'magmatic hiatus', namely, the absence of an intervening Devonian magmatism, with a gap of more than 100 Ma between the Ordovician and the Carboniferous magmatism. On the contrary, in the eastern and southern Pampean Ranges, Devonian granitic magmatism is dominant. It is recorded by large batholiths as Achala, Comechingones, Cerro Aspero-Alpa Corral, Córdoba province; and Las Chacras, Renca and others in the province of San Luis, while Ordovician batholiths are missing although abundant and voluminous to the north. Likewise in the south, there is a lack of Carboniferous granites, while in the north new Carboniferous granites have been discovered. The resolution of this important geological problem about the origin of the distribution timing of the plutonic magmatism during the Palaeozoic in the Pampean Ranges will be one of the main objectives of our future research.

The San Blas Carboniferous pluton is, for the moment, the only fertile granite in this area to which Sm–Nd model ages show participation of sharp different crustal–mantle sources. Future research should be extended to others Carboniferous granites, fertile or not, to find out if in their tectono-magmatic pattern, they share geological and genetic processes.

This research was supported by the Research Council of the National University of Tucumán and by the Superior Institute for Geological Correlation (INSUGEO). Special thanks go to the Institute of Geosciences of the University of São Paulo and to F. Söllner and P. Grosse, for isotope analyses at the Zentrallabor für Geochronologie, Munich, Germany. We thank two anonymous reviewers for their critical reviews and useful suggestions that have improved the manuscript.

References

ARROSPIDE, A. 1985. Las manifestaciones de greissen en la Sierra de Fiambalá, Catamarca. *Revista de la Asociación Geológica Argentina*, **40**, 97–113.

BÁEZ, M. A. 2006. *Geología, Petrología y Geoquímica del basamento igneo-metamórfico del sector norte de la Sierra de Velasco, Provincia de La Rioja. Facultad de Ciencias Exactas, Físicas y Naturales.* Universidad Nacional de Córdoba. PhD Thesis, National University of Córdoba, Argentina.

BÁEZ, M. A. & BASEI, M. A. 2005. El plutón San Blas, magmatismo postdeformacional carbonífero en la Sierra de Velasco. *Serie Correlación Geológica*, **19**, 239–246.

BAU, M. 1996. Controls on the fractionation of isovalent trace elements in magmatic and aqueous systems: evidence from Y/Ho, Zr/Hf, and lanthanide tetrad effect. *Contributions to Mineralogy and Petrology*, **123**, 323–333.

BEA, F. 1996. Residence of REE, Y, Th and U in granites and crustal protoliths; implications for the chemistry of crustal melts. *Journal of Petrology*, **37**, 521–552.

BEA, F., MONTERO, P. & ORTEGA, M. 2006. A LA-ICP-MS evaluation of Zr reservoirs in common crustal rocks: implications for Zr and Hf geochemistry, and zircon-forming processes. *The Canadian Mineralogist*, **44**, 693–714.

CLARKE, D. B. 1992. The mineralogy of peraluminous granites: a review. *The Canadian Mineralogist*, **19**, 3–17.

DAHLQUIST, J. A., PANKHURST, R. J., RAPELA, C. W., CASQUET, C., FANNING, C. M., ALASINO, P. H. & BÁEZ, M. 2006. The San Blas pluton: an example of the carboniferous plutonism in the Sierras Pampeanas, Argentina. *Journal of South American Earth Sciences*, **20**, 341–350.

EL BOUSEILY, A. M. & EL SOKKARY, A. A. 1975. The relation between Rb, Ba and Sr in granitic rocks. *Chemical Geology*, **16**, 207–219.

FIDELIS, I. & SIEKIERSKI, S. 1966. The regularities in stability constants of some Rare Earth complexes. *Journal of Inorganic Nuclear Chemistry*, **28**, 185–188.

GOLDSTEIN, S. L., O'NIONS, R. K. & HAMILTON, P. J. 1984. A Sm–Nd study of atmospheric dusts and particulates from major river systems. *Earth and Planetary Science Letters*, **70**, 221–236.

GRISSOM, G. C., DEBARI, S. M. & SNEE, L. W. 1998. Geology of the Sierra de Fiambalá, northwest Argentina: implications for Early Palaeozoic Andean Tectonics. *In*: PANKHURST, R. J. & RAPELA, C. W. (eds) *The Proto-Andean Margin of Gondwana*. Geological Society, London, Special Publications, **142**, 297–323.

GROSSE, O., BÁEZ, M. A., TOSELLI, A. J., BELLOS, L. I., ROSSI, J. N. & SARDI, F. G. 2008. Caracterización petrológica de los granitos Carboníferos (en contraposición con los granitoides Ordovícicos) de la Sierra de Velasco, Sierras Pampeanas. *Actas del XVII Congreso Geológico Argentino*, **III**, 1357–1358.

GROSSE, P., SÖLLNER, F., BÁEZ, M. A., TOSELLI, A. J., ROSSI, J. N. & DE LA ROSA, J. D. 2009. Lower Carboniferous post-orogenic granites in Central Eastern Sierra de Velasco, Sierras Pampeanas, Argentina: U–Pb monazite geochronology, geochemistry and Sr–Nd isotopes. *International Journal of Earth Sciences*, **98**, 1001–1025.

HÖCKENREINER, M. 2003. Die Tipa – Scherzone (Unterdevon, NW – Argentinien): Geochronologie, Geochemie und Strukturgeologie. *Münchner Geologische Hefte. Reihe A*, **34**, 1–92.

IRBER, W. 1999. The lanthanide tetrad effect and its correlation with K/Rb, Eu/Eu*, Sr/Eu, Y/Ho and Zr/Hf of evolving peraluminous granite suites. *Geochimica et Cosmochimica Acta*, **63**, 489–508.

JORDAN, T. E. & ALLMENDINGER, R. W. 1986. The Sierras Pampeanas of Argentina: a modern analogue of rocky mountain foreland deformation. *American Journal of Science*, **286**, 737–764.

KEPPLER, H. 1993. Influence of fluorine on the enrichment of high field strength trace elements in granitic rocks. *Contributions to Mineralogy and Petrology*, **114**, 479–488.

LANNEFORS, N. A. 1929. Informe sobre las minas de estaño de Mazán y algunos otros trabajos mineros en la sierra de Velasco, Provincia de La Rioja. *Dirección General de Minas, Geología e Hidrología*, Buenos Aires, Publicación no 54..

LAZARTE, J. E., FERNÁNDEZ TURIEL, J. L., GUIDI, F. & MEDINA, M. E. 1999. Los granitos Río Rodeo y Quimivil: dos etapas del magmatismo paleozoico de Sierras Pampeanas. *Revista de la Asociación Geológica Argentina*, **54**, 333–352.

LAZARTE, J. E., AVILA, J. C., FOGLIATA, A. S. & GIANFRANCISCO, M. 2006. Granitos evolucionados relacionados a mineralización estanno-wolframífera en Sierras Pampeanas Occidentales. *Serie Correlación Geológica*, **21**, 75–104.

LIEW, T. C. & HOFMANN, A. W. 1988. Precambrian crustal components, plutonic associations, plate environment of the Hercynian Fold Belt of Central Europe: indications from a Nd and Sr isotopic study. *Contributions to Mineralogy and Petrology*, **98**, 129–138.

LINNEN, R. L. & KEPPLER, H. 2002. Melt composition control of Zr/Hf fractionation in magmatic processes. *Geochimica et Cosmochimica Acta*, **66**, 3293–3301.

LUDWIG, K. R. 2001. *Using Isoplot/Ex Geochronological Toolkit for Microsoft Excel*. Berkeley Geochronological Center, Berkeley, Special Publication no. 1.

MANIAR, P. D. & PICCOLI, P. M. 1989. Tectonic discrimination of granitoids. *Geological Society of America Bulletin*, **101**, 635–643.

MASUDA, A., KAWAKAMI, O., DOHMOTO, Y. & TAKENAKA, T. 1987. Lanthanide tetrad effects in nature: two mutually opposites types, W and M. *Geochemical Journal*, **21**, 119–124.

NEIVA, A. M. R. 1984. Geochemistry of tin-bearing granitic rocks. *Chemical Geology*, **43**, 241–256.

NEUGEBAUER, H. 1995. *Die Mylonite von Fiambalá – Strukturgeologische und petrographische Untersuchungen and der Ostgrenze des Famatina-Systems, Sierra de Fiambalá*, NW-Argentinien. PhD Thesis, Munich University, Munich.

PANKHURST, R. J., RAPELA, C. W. & FANNING, C. M. 2000. Age and origin of coeval TTG, I- and S-type granites in the Famatinian belt of NW Argentina. *Transactions of the Royal Society of Edinburgh: Earth Sciences*, **91**, 151–168.

PATIÑO DOUCE, A. E. 1999. What do experiments tell us about the relative contributions of crust and mantle to the origin of granitic magmas. *In*: CASTRO, A., FERNÁNDEZ, C. & VIGNERESSE, J. L. (eds) *Understanding Granites: Integrating New and Classical Techniques*. Geological Society, London, Special Publications, **168**, 55–75.

PEUCAT, J. J., VIDAL, P., BERNARD-GRIFFITHS, J. & CONDIE, K. C. 1988. Sr, Nd and Pb isotopic systematics in the Archean low- to high-grade transition zone of southern India: syn-accretion vs. post-accretion granulites. *Journal of Geology*, **97**, 537–550.

SÖLLNER, F., GERDES, A., GROSSE, P. & TOSELLI, A. J. 2007. *U–Pb age determinations by LA-ICP-MS on zircons of the Huaco granite, Sierra de Velasco (NW-Argentina): A long-term history of melt activity within an igneous body*. Abstracts 20th Colloquium on Latin American Earth Sciences, Kiel.

SUN, S. S. & MCDONOUGH, W. F. 1989. Chemical and isotopic systematics of oceanic basalts: implications for mantle composition and processes. *In*: SAUNDERS, A D. & NORREY, M. J. (eds) *Magmatism in Ocean Basins*. Geological Society of London, Special Publications, **42**, 313–345.

TAYLOR, S. R. & MCLENNAN, S. M. 1985. *The Continental Crust: Its Composition and Evolution*. Blackwell, Oxford.

TOSELLI, A. J., SIAL, A. N. & ROSSI, J. N. 2002. Ordovician magmatism of the Sierras Pampeanas, Famatina System and Cordillera Oriental, NW Argentina. *In*: ACEÑOLAZA, F. G. (ed.) *Aspects of the Ordovician System in Argentina*. Serie Correlación Geológica, Instituto Superior de Correlación Geológica, Tucumán, Argentina, **16**, 7–16.

TOSELLI, A. J., ROSSI, J. N., BÁEZ, M. A., GROSSE, P. & SARDI, F. 2006. El batolito carbonífero Aimogasta, Sierra de Velasco, La Rioja, Argentina. *Serie Correlación Geológica*, **21**, 137–154.

TOSELLI, A. J., MILLER, H., ACEÑOLAZA, F. G., ROSSI, J. N. & SÖLLNER, F. 2007. The Sierra de Velasco of Northwest Argentina, Argentina. An example for polyphase magmatism at the margin of Gondwana. *Neues Jarhbuch für Geologisch und Paläontologisch Abhandlungen*, **246**, 325–345.

TUACH, J., DAVENPORT, P. H., DICKSON, W. L. & STRONG, D. F. 1986. Geochemical trends in the Ackley Granite, southeast Newfoundland: their relevance to magmatic-metallogenic processes in high-silica systems. *Canadian Journal of Earth Sciences*, **23**, 747–765.

Index

Page numbers in *italic* denote figures. Page numbers in **bold** denote tables.